AF812598

Learning and Analytics in Intelligent Systems 51

Series Editors

George A. Tsihrintzis, *University of Piraeus, Piraeus, Greece*
Maria Virvou, *University of Piraeus, Piraeus, Greece*
Lakhmi C. Jain, *KES International, Shoreham-by-Sea, Australia*

The main aim of the series is to make available a publication of books in hard copy form and soft copy form on all aspects of learning, analytics and advanced intelligent systems and related technologies. The mentioned disciplines are strongly related and complement one another significantly. Thus, the series encourages cross-fertilization highlighting research and knowledge of common interest. The series allows a unified/integrated approach to themes and topics in these scientific disciplines which will result in significant cross-fertilization and research dissemination. To maximize dissemination of research results and knowledge in these disciplines, the series publishes edited books, monographs, handbooks, textbooks and conference proceedings.

Indexed by EI Compendex.

Fred Paas · Srikanta Patnaik · Taosheng Wang
Editors

New Directions in Educational Technology and Administration

Proceedings of the 4th International Conference on Educational Technology and Administration

Springer

Editors
Fred Paas
Department of Psychology, Education
and Child Studies
Erasmus University Rotterdam
Rotterdam, The Netherlands

Srikanta Patnaik
InterScience Institute of Management
and Technology (IIMT)
Bhubaneswar, Odisha, India

Taosheng Wang
School of Business
Hunan International Economics University
Changsha, China

ISSN 2662-3447 ISSN 2662-3455 (electronic)
Learning and Analytics in Intelligent Systems
ISBN 978-3-031-95251-7 ISBN 978-3-031-95252-4 (eBook)
https://doi.org/10.1007/978-3-031-95252-4

© The Editor(s) (if applicable) and The Author(s), under exclusive license
to Springer Nature Switzerland AG 2025

This work is subject to copyright. All rights are solely and exclusively licensed by the Publisher, whether the whole or part of the material is concerned, specifically the rights of translation, reprinting, reuse of illustrations, recitation, broadcasting, reproduction on microfilms or in any other physical way, and transmission or information storage and retrieval, electronic adaptation, computer software, or by similar or dissimilar methodology now known or hereafter developed.
The use of general descriptive names, registered names, trademarks, service marks, etc. in this publication does not imply, even in the absence of a specific statement, that such names are exempt from the relevant protective laws and regulations and therefore free for general use.
The publisher, the authors and the editors are safe to assume that the advice and information in this book are believed to be true and accurate at the date of publication. Neither the publisher nor the authors or the editors give a warranty, expressed or implied, with respect to the material contained herein or for any errors or omissions that may have been made. The publisher remains neutral with regard to jurisdictional claims in published maps and institutional affiliations.

This Springer imprint is published by the registered company Springer Nature Switzerland AG
The registered company address is: Gewerbestrasse 11, 6330 Cham, Switzerland

If disposing of this product, please recycle the paper.

Preface

Educational technology and administration have emerged as pivotal domains in reshaping the global education landscape. In the present scenario, educational institutions are increasingly integrating advanced technologies such as artificial intelligence, machine learning, and data analytics to enhance teaching, learning, and administrative processes. Virtual classrooms, adaptive learning platforms, and smart assessment tools are transforming traditional education into dynamic, student-centered experiences. Simultaneously, administrative functions are being streamlined through cloud-based systems, real-time analytics, and digital resource management, ensuring efficiency and transparency. However, challenges such as equitable access, ethical considerations, and digital literacy gaps remain significant. As these fields continue to evolve, they hold immense potential to create more inclusive, efficient, and innovative education systems worldwide, empowering learners and educators alike.

It is with great enthusiasm that we present the proceedings of the International Conference on Educational Technology and Administration (EduTA2024). This esteemed gathering brings together a vibrant community of scholars, practitioners, educators, and policymakers to explore the transformative potential of technology and innovation in education and administration. The conference aims to serve as a platform for sharing insights, fostering collaborations, and inspiring solutions that will shape the future of education in a rapidly evolving world. At its core, EduTA2024 is dedicated to addressing the integration of cutting-edge educational technology and the advancement of administrative practices to create dynamic, inclusive, and effective learning environments. The conference embraces a multidisciplinary approach, focusing on themes such as digital classroom innovations, smart learning technologies, data-driven decision-making in education, and the ethical and social implications of technological integration. It also delves into areas like learning analytics, personalized learning, virtual and augmented reality, and the use of artificial intelligence in educational administration.

The primary aim of EduTA2024 is to bridge the gap between technology and pedagogy, equipping educators and administrators with tools and strategies to meet the diverse needs of learners in the twenty-first century. By fostering dialogue and collaboration, EduTA2024 aspires to contribute to a global education ecosystem that is more efficient, equitable, and forward-thinking. The benefits of this conference extend beyond academia to society at large. As technology continues to revolutionize education, it becomes a catalyst for addressing critical global challenges, such as access to quality education, lifelong learning opportunities, and the development of skills necessary for thriving in the digital age. EduTA2024 provides a unique platform for thought leaders to share best practices and explore innovative solutions that promote social mobility, economic growth, and cultural enrichment. By empowering educators and administrators, the conference aims to impact learners at every stage of their journey, from preschool to higher education and beyond.

EduTA2024 also embraces a futuristic vision, anticipating the needs of learners and institutions in an era of rapid technological change. Educational technology and administration hold the transformative potential to redefine the landscape of modern education, fostering innovative learning environments and efficient administrative frameworks. By bridging cutting-edge technologies with effective management practices, this field paves the way for equitable, inclusive, and sustainable educational systems. As we stand on the brink of new possibilities, the collective efforts of educators, researchers, and policymakers will be pivotal in shaping a future where technology and administration empower learners and institutions alike. This conference serves as a vital platform to exchange ideas, collaborate, and inspire transformative changes. Together, let us embrace this opportunity to contribute to the evolution of education in the digital age.

Srikanta Patnaik
Program Chair

Editorial

Educational technology and administration stand at the forefront of modern education, addressing the pressing needs of an evolving educational landscape. The futuristic approach in this field encompasses integrating artificial intelligence, learning analytics, immersive technologies, and collaborative tools to enhance teaching methodologies, personalize learning experiences, and streamline administrative processes. These advancements are not just reshaping how education is delivered but are also ensuring greater inclusivity, accessibility, and equity across global learning environments

The International Conference on Educational Technology and Administration (EduTA2024) hosted by Zhuhai Research Institute of Hubei University, China, from December 13–15, 2024. This conference provided an exceptional platform for researchers, educators, administrators, and industry professionals to engage in meaningful discourse on the transformative potential of educational technology and its administration in shaping the future of learning and institutional management

The impact of educational technology and administration on today's educational sector is profound. Learning analytics is enabling data-driven decision-making to optimize curriculum design and student engagement. Online and distance learning have emerged as vital components of modern education, breaking geographical barriers and providing opportunities to learners worldwide. Moreover, leadership in modern education has become pivotal in driving systemic change, fostering a culture of innovation, and ensuring the ethical and responsible deployment of educational technologies. EduTA2024 witnessed a focused exploration of these themes, with the organizing committee receiving 88 submissions. After rigorous review and editing, 38 papers were accepted for presentation and publication, of which 29 papers were officially registered and are included in these proceedings. These contributions are categorized into three major sections: (i) Learning analytics and curriculum design, (ii) online and distance learning, and (iii) leadership in modern education

Learning Analytics and Curriculum Design

The track *Learning Analytics and Curriculum Design* encompasses a diverse range of research contributions aimed at enhancing educational practices through data-driven insights and innovative curriculum strategies. Articles in this track explore how technology can redefine learning processes and improve teaching outcomes. For instance, studies like "The Study of Computer-assisted Technology on the Quality of English Writing" and "AI-Driven Adaptive Learning Systems in University English Writing Instruction" focus on leveraging intelligent systems for personalized and adaptive learning experiences. Contributions such as "The Personalized Recommendation System for TESOL Learners" and "Design and Implementation of College Oral English Teaching System" highlight the application of optimization algorithms and web technologies in tailoring learning approaches to individual needs. Moreover, works like "Construction of a Digital Platform for the Preschool Education Professional Course Cluster" and "Analysis of Physical Education Teaching Evaluation System" delve into creating digital ecosystems for streamlined course delivery and intelligent evaluation. The exploration of large AI models in designing courses and innovative teaching methods, as seen in "Designing Courses on Foundations and Applications of Large AI Models" and "Exploration of Innovation in Practical Courses," underscores the evolving role of artificial intelligence in reshaping educational paradigms. The inclusion of research on multi-interactive teaching models, such as "Exploration of English Multi-interactive Teaching Mode" and "Research on Multi-interactive Teaching Model of College English Listening and Speaking," emphasizes collaborative and engaging learning frameworks. Finally, articles like "Research on User Demands of Online Open Courses" and "Research on innovative design and evaluation system construction of international foreign language teaching" present forward-looking solutions to meet global educational demands. Collectively, these studies contribute to advancing the field of learning analytics and curriculum design, paving the way for more inclusive, efficient, and futuristic educational systems

Online and Distance Learning

The Second track *Online and Distance Learning* highlights the transformative impact of advanced technologies in shaping flexible and accessible educational frameworks. Articles like "English Speech Segmentation and Keyword Extraction Based on Deep Learning" and "Intelligent Music Content Generation and Learning Strategy Optimization Combined with Bayesian Optimization" delve into the use of cutting-edge AI techniques to enhance learning content creation and delivery. The focus on personalized learning experiences is further reinforced through works such as "Automatic Generation of Online English Course Content Based on Natural Language Processing" and "Research and Application of Online Learning Platform Resource Recommendation Based on Personalization," which emphasize the importance of tailored educational resources. The effectiveness of AI tools is showcased in "A Study on the Effectiveness of AI-Assisted Teaching Tools for English Speaking," where AI-driven evaluations foster improved language proficiency. Innovative models like the "English Virtual Teacher System Based on Neural Network Model" and the "Integrated Application of Augmented Reality and Artificial Intelligence into English Learning" illustrate how virtual environments and immersive technologies can bridge the gap between learners and instructors. Together, these studies provide a forward-looking vision for online and distance learning, focusing on accessibility, scalability, and adaptability in education

Leadership in Modern Education

The concluding track *Leadership in Modern Education* explores innovative strategies and technological advancements that empower educators and institutions to shape the future of learning. Articles such as "From Manual to Semi-Automatic—Artificial Intelligence Language Instruction in the Space Design Education of Film and Television Major Universities" showcase how AI is transforming specialized education fields. The integration of data-driven approaches is evident in works like "Fusion Strategy of Preschool Education and Music Education Combined with Data Mining Technology," which demonstrates the potential of interdisciplinary innovation. Similarly, "Research on Tiered Service Demand Forecasting Model for Vocational Education Based on Machine Learning Algorithms" emphasizes the role of predictive analytics in tailoring education to diverse learner needs. The focus on practical applications is further reflected in "Application and Effect Evaluation of AI in Practice Teaching—Taking the Course of Big Data Platform Deployment and Operation and Maintenance as an Example," which highlights the role of AI in bridging theoretical knowledge and real-world skills. The importance of collaboration among academia, industry, and education is underscored in "Collaborative Innovation of 'Science, Education, and Industry': New Paths and Strategy Optimization for the Cultivation of Applied Talents in Higher Vocational Artificial Intelligence Technology." Lastly, "Probe Into the Path of Teachers' Professional Development and Ability Improvement for Digital Education" explores the critical role of educators in adapting to digital transformation. Together, these articles present a comprehensive view of leadership in modern education, emphasizing innovation, collaboration, and adaptability.

EduTA2024 reflects the collaborative spirit and intellectual rigor of the participants. The presented research and insights highlight the dynamic interplay of technology and administration in addressing global educational challenges and paving the way for a sustainable, equitable, and future-ready educational ecosystem.

We extend our heartfelt gratitude to all authors, reviewers, and attendees for their valuable contributions and active participation. We also acknowledge the dedication and hard work of the organizing committee and our host, Zhuhai Research Institute of Hubei University, in making this event a resounding success. We hope that this collection of research serves as a valuable resource and inspiration for continued innovation, collaboration, and progress in the field of educational technology and administration. Together, let us drive the transformation of education for a brighter, more inclusive future.

Acknowledgements. We would like to express our sincere appreciation and gratitude to all those who have contributed to the success of the International Conference on Educational Technology and Administration (EduTA2024).

First of all, we would like to express our heartfelt thanks to all the contributors of EduTA2024, who have contributed their research findings in this volume. Their active involvement, insightful discussions, and valuable contributions made the conference a truly enriching experience. The papers presented in this conference explore emerging trends, such as adaptive learning systems, immersive educational experiences, and decentralized learning networks. It emphasizes the importance of preparing students for an interconnected, technology-driven world while fostering critical thinking, creativity, and ethical responsibility. Through these discussions, the conference aspires to set a roadmap for the evolution of education in the coming decades. The ideas, research, and collaborations generated during EduTA2024 will leave a lasting legacy, inspiring transformative changes in educational technology and administration. As we navigate the challenges and opportunities of the digital era, this conference serves as a beacon for innovation, inclusion, and excellence in education.

Our deepest appreciation goes out to the program and technical committee for their diligent work in reviewing and selecting the papers that were presented at the conference. Their expertise and commitment to academic excellence helped to ensure that the papers presented were of the highest quality. Furthermore, we extend our deepest appreciation to the organizing committee for their hard work and dedication in planning and executing EduTA2024. Their meticulous attention to detail, tireless coordination, and outstanding organizational skills contributed significantly to the smooth running of the conference. In particular, we wish to express our sincere appreciation to Prof. Hao Huang, Zhuhai Research Institute of Hubei University (China), who delivered an opening address for the conference, Prof. Rocky Lam, City University of Hong Kong (China), Prof. Timothy Teo, The Chinese University of Hong Kong (China), Prof. Lianne Lam, The Hong Kong Polytechnic University (China), Prof. Na Dong, Tianjin University (China), and Prof. Kuo-Kun Tseng, Harbin institute of technology in Shenzhen (China), who made very beneficial speeches for the conference audience, which are titled "AI adoption in Enterprise," "Research on AI in Education: Current Status and Future Directions," "Opportunities and Challenges in Space-Based Solar Power," "Construction Scheme for Project based Course System of "Computer Vision Principles and Applications" and "Multimodal Large Language Model for 3D Content Generation and Simulation," respectively. Thanks should also go to 2025 Projects of Hunan Provincial Social Science Achievement Review Committee (XSP25YBC470) and 2024 Research Projects on Ideological and Political Work in Colleges and Universities in Hunan Province (24C39), which sponsored the conference. We would also like to extend our appreciation to Zhuhai Research Institute of Hubei University, China, Investment Promotion Center of Xiangzhou District in Zhuhai, China, Bureau of Science, Technology and Industrial Information Technology of Xiangzhou District in Zhuhai, China, and LiPC Space Resources Alliances Education and Culture Centre in Hong Kong, China, which provided support for the conference.

Last but not least, we extend our gratitude to the Springer Series Editors of Learning and Analytics in Intelligent Systems Prof. George A. Tsihrintzis, Prof. Maria Virvou and Prof. Lakhmi C. Jain for their support and guidance in publishing the conference proceedings "**New Directions in Educational Technology and Administration**" in their esteemed book series. We also extended our sincere thanks to Dr. Thomas Ditzinger, Editorial Director of **Springer Interdisciplinary and Applied Sciences and Engineering division**.

We are grateful to all the individuals and organizations that played a part in making EduTA2024 a successful and memorable event. Without your support and dedication, this conference would not have been possible. Thank you all for your contributions and commitment through which we were able to explore the New Directions in Educational Technology and Administration.

Fred Paas
Srikanta Patnaik
Taosheng Wang

EduTA2024 Conference Committee

General Chair

Fred Paas	Erasmus University Rotterdam, Netherlands

Program Chair

Srikanta Patnaik (Director)	I.I.M.T., Bhubaneswar, Odisha, India

Organising Chairs

Taosheng Wang	Hunan International Economics University, China
Hamadullah Kakepoto	University of Sindh, Pakistan
Guomin Chen	Guilin University of Aerospace Technology, China
Hao Huang	Zhuhai Research Institute, Hubei University, China

Publication Chairs

Andrew W. H. Ip	University of Saskatchewan, Canada
Lianne Lam	Hong Kong Polytechnic University, China

Finance Chair

Dhananjaya Sarangi	I.I.M.T., Bhubaneswar, Odisha, India

International Advisory Board

Robyn Gillies	School of Education, The University of Queensland, St. Lucia, QLD 4072, Australia

Alex Bowers	Associate Professor of Educational Leadership, Teachers College, Columbia University New York, USA
Nir Friedman	The Rachel and Selim Benin School of Computer Science and Engineering The Alexander Silberman Institute of Life Sciences, The Hebrew University of Jerusalem Israel
Jerusalem Yibeltal Yizengaw	Bahir Dar University, Ethiopia, Addis Ababa University, Ethiopia
S. Srinivas Kumar (Vice-chancellor)	JNT University Anantapur, India
Navin Sheth (Vice-chancellor)	Gujarat Technological University, India

Technical Program Committee

David Ben-Arieh	Kansas State University, USA
Tony Brooks	Aalborg University, Denmark
B. S. Chowdhry	Mehran University of Engineering and Technology, Pakistan
Thierry Courbis	Leader Health, Switzerland
O. Ferrer-Roca	University of La Laguna, Spain
Hee-Cheol Inje Kim	University, South Korea
Sukanta Kumar Sabut	Siksha 'O' Anusandhan University, India
Sugam Sharma	Iowa State University, USA
Kenei, Shie	ViewSend ICT Co., Ltd., Japan
O. Paolo Antonio S. Silva	Joslin Diabetes Center, USA
M., Simonov	Politecnico di Milano, Italy
Mayank Singhal	Methodist University, USA
Sittiprapaporn, Wichian	Mahasarakham University, Thailand
Wu-Chen, Su	Kaohsiung Chang Gung Memorial Hospital, Taiwan
Yonghang Tai	Yunnan Normal University, China
B. J. F. Van Beijnum	University of Twente, Netherlands
Khin Than Win	University of Wollongong, Australia
Jialing Zhang	Yale University, USA
Liang Zhou	Shanghai Jiao Tong University, China
S. Chakraverty	National Institute of Technology Rourkela, India
Li Cheng	The Hong Kong Polytechnic University, Hong Kong
Jeng-Haur Horng	National Formosa University, Yunlin, Taiwan
Abderrafiaa Koukam	Université de Technologie de Belfort-Montbéliard, France

Pedro U. Lima	Institute for Systems and Robotics, Lisbon, Portugal
Mun-Kew Leong	National University of Singapore
Luis Martínez López	University of Jaén
Vladimir Mazalov	Karelia Research Center of Russian Academy of Sciences, Russia
Muhammad Nur	Diponegoro University, Semarang, Indonesia
Luca Oneto	University of Genoa, Italy
B. K. Panigrahi	IIT Delhi, India
Kay Chen Tan	National University of Singapore, Singapore
Sarma Yadavalli	University of Pretoria, South Africa
Yeon-Mo Yang	Kumoh National Institute of Technology, Gumi, South Korea
Liangchi Zhang	The University of New South Wales, Australia
Baojiang Zhong	Soochow University, Suzhou, China
Ahmed Zobaa	Brunel University, Uxbridge, Middlesex, UK
Hakim Bendjenna	University of Larbi Tebessi, Algeria
Sanjiv K. Bhatia	University of Missouri - St. Louis, USA
Debasis Chaudhuri	Defence Electronics Applications Laboratory, India
R.I., Damper	University of Southampton, UK
Guilherme N. DeSouza	University of Western Australia, Australia
Maria Gini	University of Minnesota, USA
Lapo Governi	University of Florence, Italy
Wang Han	Nanyang Technological University, Singapore
Takamasa, Koshizen	Honda Research Institute Japan Co. Ltd., Japan
Ravi, Kothari	IBM Research - India, India
Roumen Kirilov Kountchev	Technical University of Sofia, Bulgaria
Hamid Krim	North Carolina State University, USA
Chan-Su Lee	Yeungnam University, South Korea
Pedro U., Lima	Instituto Superior Técnico, Puerto Rico
Mario Mata	Universidad Europea de Madrid, Spain
Zbigniew Michalewicz	University of Adelaide, Australia
Luis Moreno	University of Carlos III, Spain
Ali, K. N. Nahvi	Toosi University of Technology, Iran
Jayanthi Ranjan	Institute of Management Technology (IMT), India
Yassine Ruichek	University of Technology of Belfort-Montbéliard, France
Ashok Samal	University of Nebraska-Lincoln, USA
Ishwar Sethi	Oakland University, USA
M. Hassan, Shafazand	Shahid Bahonar University of Kerman, Iran
Rajeev, Sharma	Pennsylvania State University, USA

Yu, Sun	Michigan State University, USA
Kok Kiong, Tan	National University of Singapore, Singapore
Christopher R., Wren	Mitsubishi Electric Research Laboratories, USA
Yeon-Mo, Kumoh Yang	National Institute of Technology, South Korea
Mohammed, Yeasin	University of Memphis, USA
Anthony Yezzi	Georgia Institute of Technology, USA
Peng-YengYin	National Chi Nan University, Taiwan

Contents

About the Editors

Prof. (Dr.) Fred Paas is a professor and chair of the research group Educational and Developmental Psychology at the Department of Psychology, Education and Child Studies of Erasmus University Rotterdam in the Netherlands. He is also a professorial fellow at the School of Education, University of New South Wales, Sydney, Australia. He holds a master's degree (1988) in human movement sciences and a Ph.D. degree (1993) in instructional technology. His research focuses on the design of effective and efficient learning environments by applying contemporary, multidisciplinary scientific knowledge about the human cognitive system. He has published more than 300 articles and book chapters and he was recognized as the world's most productive author in the five best journals in the field of educational psychology. Since 2015, he has been the editor-in-chief of the journal Educational Psychology Review, which is the number one journal in the field of educational psychology. He is also on the editorial board of several other renowned journals, such as the Journal of Educational Psychology, Educational Technology Research and Development, and the International Journal of Environmental Research and Public Health

Prof. (Dr.) Srikanta Patnaik is the director of IIMT, Bhubaneswar. He has received his Ph. D. (Engineering) on Computational Intelligence from Jadavpur University, India, in 1999 and supervised 34 Ph. D. theses and more than 60 M. Tech theses in the area of machine intelligence, soft computing applications, and re-engineering. Dr. Patnaik has published more than 100 research papers in international journals and conference proceedings. He is author of 2 text books and edited 82 books and few invited book chapters, published by leading international publisher like Springer-Verlag, Kluwer Academic, etc. He is the editor-in-chief of International Journal of Information and Communication Technology and International Journal of Computational Vision and Robotics published from Inderscience Publishing House, England, and also editor-in-chief of Book Series on "Modeling and Optimization in Science and Technology" published from Springer, Germany.

Prof. (Dr.) Taosheng Wang is a doctor, professor of Applied Economics (International Trade Direction), specialist of General Office of the State Council, director of Chinese Society of World Economy, vice chairman of the Expert Committee of China Association of Trade in Service, master tutor. He is the dean of School of Business of Hunan International Economics University, chief expert of the provincial-level social science key research base. He is also foregoer of provincial-level key construction course "International Trade," provincial-level teaching team "International Trade," provincial-level characteristic specialty "International Economics and Trade." His main research areas are international economic and trade theory and policy; institutional innovation and international trade; international competitive advantage and competitive strategy.

Learning Analytics and Curriculum Design

The Study of Computer-Assisted Technology on the Quality of English Writing-Take Automatic Evaluation System as an Example

Tianrong Lu(✉)

Dalian Neusoft University of Information, No. 8 Software Park Road, Dalian, Liaoning, China
lutianrong@neusoft.edu.cn

Abstract. This study explores the efficacy of an Automatic Evaluation System (AES) in enhancing English writing skills among university freshmen in Dalian City. Focusing on two classes of 50 students each, with comparable academic backgrounds, the research was designed around a 10-week experiment involving pre-tests, a mid-term evaluation, and post-tests aligned with the College English Test Band 4 (CET-4) standards. The primary objective was to assess the impact of AES feedback on improving students' writing proficiency. Utilizing SPSS 26.0 for statistical analysis, the research findings indicate a significant improvement in writing scores for students exposed to the AES, compared to those in the control group. Despite the positive outcomes, the study also highlights limitations, such as the experiment's short duration and the small sample size, which might affect the generalizability of the results. Furthermore, while the AES effectively identified specific errors, its feedback on essay structure and content requires enhancement. Future research is recommended to extend the duration and expand the sample size, including a more diverse demographic to substantiate the findings. The study underscores the potential of automated evaluation tools in supporting English language instruction but calls for improvements in offering holistic feedback.

Keywords: computer-assisted technology · automatic evaluation system · English writing

1 Introduction

Writing, as one of the primary means of communication in English, shows the creativity in students' cognition and thought processes. In the context of college English writing instruction, teachers have observed that many students have a weak foundation in English writing, with common issues such as spelling mistakes, tense confusion, grammatical errors, repetitive content, and a lack of innovation in ideas. Additionally, large class sizes and lengthy essays often result in a significant volume of work for teachers to grade, making the grading process a substantial challenge. Implementing an automatic evaluation system (AES) in writing classes can assist teachers by reducing grading time, identifying specific errors, offering suggestions for improvement, and monitoring and evaluating the writing process, thereby encouraging students to effectively revise their essays.

© The Author(s), under exclusive license to Springer Nature Switzerland AG 2025
F. Paas et al. (Eds.): EduTA 2024, LAIS 51, pp. 3–13, 2025.
https://doi.org/10.1007/978-3-031-95252-4_1

2 Literature Review

Shermis (2003) noted that automatic evaluation systems are based on the process writing approach and constructivist theory, utilizing computer technology for evaluation and feedback on writing [1]. Khoii & Doroudian (2014) mentioned that feedback from automatic evaluation systems is generated through computer programs that review essays and then analyze the content using qualitative and quantitative methods to produce written evaluations [2].

The development of automatic evaluation systems has undergone significant evolution since its inception in 1966 in the United States [3]. This whole developing process can be categorized into three distinct periods: First, the initial phase. The initial phase began in the 1960s with the creation of the first automatic scoring system, Project Essay Grader (PEG), by Ellis Page and others in the United States. PEG was designed for scoring writing in large-scale examinations [4]. It employed natural language processing techniques to extract surface linguistic features from essays that directly reflect the quality of the writing, such as essay length, word length, and punctuation count [5]. Second, the expansion phase. The second phase saw the development of more sophisticated systems like E-rater by ETS and IntelliMetric by Vantage Learning in the 1990s [6]. These systems, similar to PEG, were built on scoring models developed from large corpora of essays graded by human raters [7]. Moreover, they expanded the range of features considered in the evaluation process to include not just lexical and syntactical elements, but also aspects of discourse content and structure [8]. Third, the diversification phase. The early years of the 21st century introduced a diversification in the types of automatic scoring engines available, including "My Access! Criterion" and "Writing Roadmap" [9]. Beyond merely providing scores for student essays, these evaluation systems assessed specific dimensions of the text, such as content, organization, vocabulary, grammar, and formatting standards. They also offered personalized feedback based on these assessments [10].

Empirical research on automatic evaluation systems mainly fall into two categories, focusing on their impact on writing quality and learning outcomes. First, intra-group comparisons. This approach involves comparing pre-test and post-test performances within the same group of students using an automatic scoring system. Attali (2004) observed significant improvements in style, grammar, language use, and overall scores on writing assignments in U.S. writing classes after the application of the Criterion system compared to initial drafts [11]. Hoon (2010), using My Access! as the research platform, found that most students acknowledged the system positively, believing that feedback from the automatic scoring system had a beneficial effect on their writing skills. Teachers involved in the experiment also found the system easy to use, with timely feedback that contributed positively to the development of writing instruction models [12]. Second, inter-group comparisons. To address the limitations of intra-group comparisons, researchers have conducted studies that divide students into experimental and control groups to examine whether automatic scoring systems enhance writing quality [13]. Rich et al. (2013) conducted an empirical study using Writing RoadMap 2.0 (WRM2.0) and found that students' test scores significantly improved after using the WRM2.0 software, generally outperforming students who had not used the software [14]. Stevenson and Phakiti (2014) indicated that the impact of automatic scoring systems on writing scores was moderate, which might be influenced by students' language proficiency and their first language [15].

3 Research Design

3.1 Research Objects and Objectives

This research is centered on two freshman classes at a university in Dalian City, with each class comprising 50 students. These freshmen have acquired the ability to write complete English essays during their high school education. Being new to university life, they are more receptive to new teaching approaches. Additionally, establishing a strong base in English writing is crucial for their subsequent English language studies. These students have comparable scores and performance levels from the national college entrance examination (Gaokao) and receive instruction from the same teacher, who allocates the same number of hours to each class weekly, to maximize the reliability of the experiment. The primary question this study aims to address is whether an automatic evaluation system can improve the quality of students' English writing.

3.2 Research Materials and Tools

This study utilized the essay questions from three sets of the College English Test Band 4(CET-4) paper conducted in December 2022 as test subjects. Initially, students from two classes were asked to complete a pre-test writing task (i.e., Writing Test Question 1), with the topic being "The role of physical exercise in achieving success at college." During the implementation phase of the research, the experimental and control groups each undertook three writing tasks (i.e., Writing Test Question 2), with this round's topic being "The importance of developing a healthy lifestyle." Finally, at the conclusion of the study, students from both classes completed a writing task (i.e., Writing Test Question 3), with the post-test essay topic being "The necessity of developing social skills." It is important to note that the reason for selecting essay questions from the same session of the College English Test Band 4(CET-4) was as follows: firstly, because the exam questions are jointly devised by English experts, their reliability and validity are relatively high, making them suitable for assessing students' writing abilities; secondly, to ensure the fairness of the examination, the difficulty level of the different sets of questions in the same session of the College English Test Band 4(CET-4)is essentially consistent, reducing the uncertainty in the research and maximizing the reliability of the experimental data, making it an appropriate choice for the test subjects.

The instrument used in the study is the Pigai Website, created by Beijing Ciwang Technology Co. LTD. Utilizing corpus and cloud computing technologies, Pigai assesses students' essays by measuring how closely they align with a standard corpus. This enables the immediate generation of scores, feedback, and analysis of the essays' content, thereby facilitating students' writing skill enhancement through self-directed practice. Pigai offers a variety of features, of which four are particularly noteworthy:

First, online essay correction systems leverage Natural Language Processing (NLP) technology for the automatic detection of grammar, spelling, structure, and other aspects of writing. The platform offers specific suggestions for improvement, such as identifying issues with logical coherence, grammatical errors, and monotony in sentence structure (see Fig. 1 for details). It specifically analyzes the appropriate use of key words and phrases in essays. Based on this analysis of word usage, the system provides concrete vocabulary replacement suggestions to facilitate the improvement of vocabulary skills.

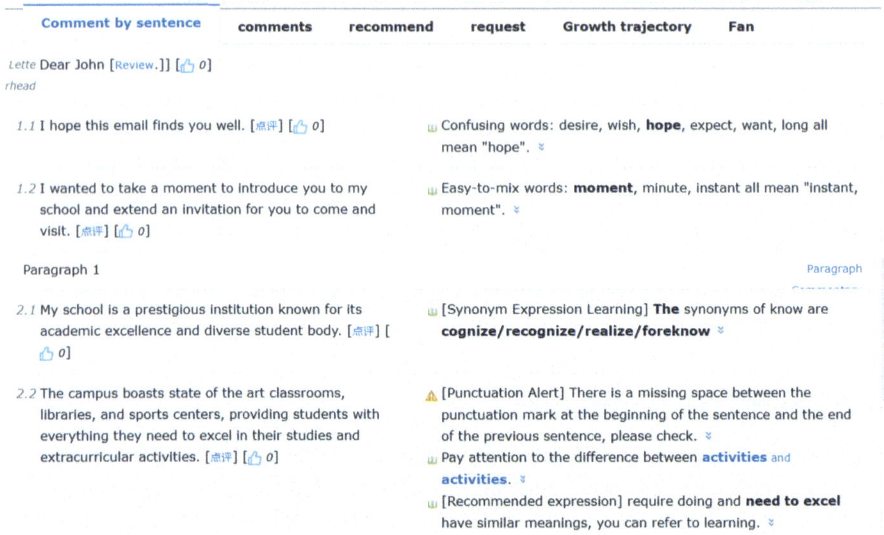

Fig. 1. The specific suggestions for improvement by sentences

Second, the platform assesses the type and complexity of vocabulary used in student essays, providing a richness index as shown in the illustration (see Fig. 2 for details). It highlights frequently occurring vocabulary, helping students recognize words that may be overused. This vocabulary index evaluation aims to improve students' lexical resource by encouraging them to explore and utilize a wider range of vocabulary in their writing, thus avoiding repetition and promoting more sophisticated expression.

Third, the system meticulously categorizes error types in student essays, breaking them down into categories such as grammatical errors, spelling errors, and punctuation mistakes, while providing frequency statistics to help students identify common mistakes (see Fig. 3 for details). This approach not only aids in identifying and understanding recurrent errors but also in tracking improvement over time, providing a comprehensive feedback loop that supports ongoing learning and development in writing skills.

Fourth, As shown in the diagram (see Fig. 4 for details), by monitoring key data points during the learning process, such as average word length, vocabulary richness, vocabulary difficulty, average sentence length, clause density, and the number of clauses, the system can identify areas where the user may be weak (for example, in vocabulary, grammar, sentence structure, etc.).

Fig. 2. The richness index of the type and complexity of vocabulary used

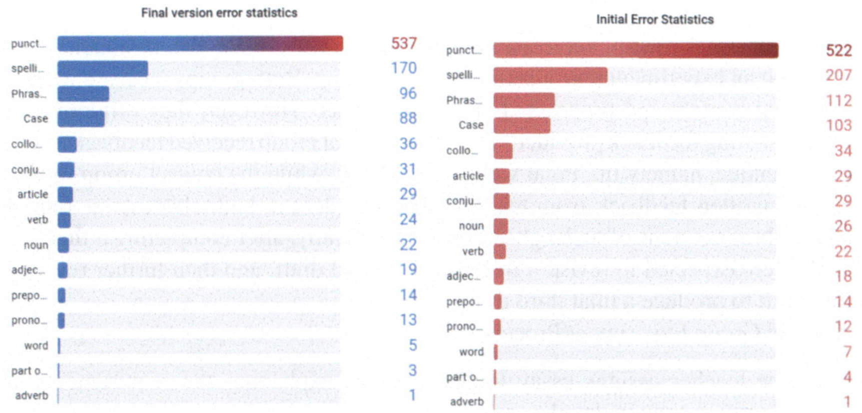

Fig. 3. The frequency statistics of error types

3.3 Research Methods and Process

This study employs a test method, spanning 10 weeks across three phases, with a total of five writing tests conducted.

Phase 1: Preliminary stage of the study

The study selects College English Test Band 4 (CET-4) writing questions for the pre-test, asking students from both the experimental and control groups to complete the pre-test writing tasks. These tasks are graded by teachers according to the CET-4 writing evaluation criteria. Following this, the English writing scores of students from both classes at the pre-test stage are statistically analyzed using SPSS 26.0. This analysis aims to determine whether there is a significant difference in English writing abilities between the two groups of students before the study begins.

Phase 2: Mid-term of the study

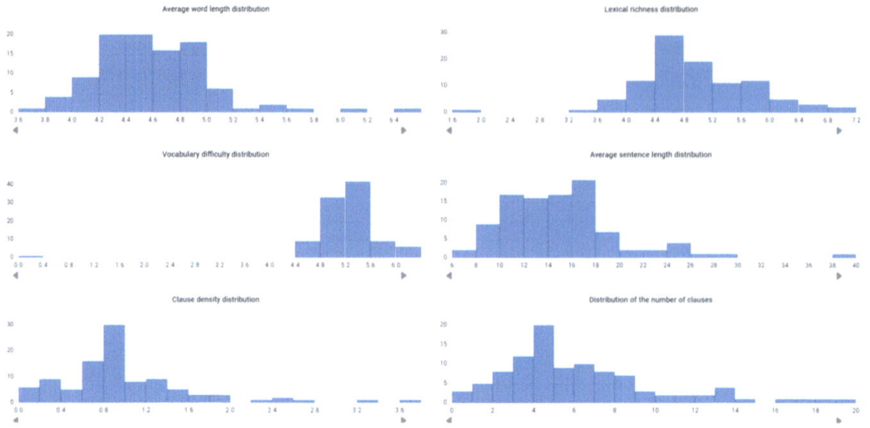

Fig. 4. The key data points of student's writing

During the experiment, two classes with similar levels of English proficiency were divided into an experimental group and a control group. Both groups received identical instructional content and progressed at the same pace. However, they differed in the method of writing feedback provided. The experimental group received feedback through a new technique, namely the Pigai website feedback, while the control group received traditional written feedback from teachers. Each student in both classes was required to submit three versions of an English essay. Specifically, after completing a first draft, students were expected to revise it to create a second draft, and then further revise the second draft to produce a final third draft.

Control Group Experiment Process:

Step 1. The teacher assigns essay topics in class, explaining the requirements and designated writing time in detail.
Step 2. Students complete their essays in class within 30 min and submit their first drafts to the teacher.
Step 3. The teacher grades the students' essays according to the College English Test Band 4 (CET-4) essay scoring criteria. After grading, the teacher returns the essays to the students, requesting that they revise their work.
Step 4. Students revise their essays based on the teacher's feedback within 30 min, and then submit their second drafts to the teacher.
Step 5. The teacher reviews and grades the second drafts, still using the CET-4 essay scoring criteria.
Step 6. The teacher again distributes the graded essays to the students, asking them to make further revisions, with another 30-min period allocated for this final revision before collecting the final drafts.
Step 7. The teacher reviews and grades the final drafts, continuing to apply the CET-4 essay scoring criteria, and then distributes the graded essays back to the students.
Step 8. The teacher summarizes the three sets of essays submitted by the entire class.

Experimental Group Experiment Process:

Step 1. The teacher timely assigns essay topics in class, providing detailed explanations of the essay requirements and the allotted writing time.
Step 2. Students write their essays within a maximum of 30 min and submit their first drafts to the Pigai system.
Step 3. The Pigai system evaluates the essays, offering scores, comments, sentence-by-sentence reviews, and suggestions for knowledge expansion.
Step 4. Students revise their essays based on the suggestions from the Pigai system within 30 min, and submit the second drafts.
Step 5. The Pigai system again provides scores, comments, sentence-by-sentence reviews, and suggestions for knowledge expansion.
Step 6. Following the Pigai system's advice, students revise their essays within 30 min, and submit the third drafts.
Step 7. The Pigai system provides a final evaluation including scores, comments, detailed reviews, and expansion suggestions.
Step 8. The teacher summarizes the three sets of essays submitted by the entire class.

Phase 3: Late-Stage Research

After the 10-week experiment concluded, in the third phase, this study conducted a writing test on students from both classes using College English Test Band 4 (CET-4) writing test question, which is of a similar difficulty level to the pre-test. The grading teachers and the scoring standards remained the same as in the pre-test. The teachers used SPSS26.0 software to statistically analyze and compare the post-test scores to evaluate whether the essay feedback model provided by the automatic evaluation system effectively improved the English writing skills and to analyze whether there was a change in the quality of the students' writing.

4 Research Findings and Discussion

In this research, the scores before and after the experiment for both the control group and the experimental group are analyzed and compared. The analysis involves applying independent samples t-test and paired samples t-test on the data collected from these groups, utilizing SPSS version 26.0. The outcomes are detailed below:

Table 1 reveals that the independent samples t-test performed on the pre-experiment data shows the control group's average score before the experiment at 73.94, while the experimental group's average score stands at 74.12. This indicates that the experimental group's average is marginally higher than that of the control group. However, the significance level ($Sig = 0.851 > 0.05$) suggests no meaningful difference in the pre-experiment scores of the experimental and control groups. Therefore, prior to the experiment, the writing abilities of the two groups were similar, rendering them appropriate candidates for this study.

Regarding the scores after the experiment (refer to Table 2 for more information), the average score for the control group is 75.42, in contrast to the experimental group's average score of 79.98. With a Sig. (two-tailed) value of 0.004, falling below the threshold of 0.01, a significant difference is evident in the post-experiment scores of the two groups. This highlights the distinct benefit of employing the automatic evaluation system in the experimental group.

Table 1. Independent Samples T-test on the Pre-experiment Data

			n	Mean		S. D.			Std. Error Mean	
Pre-experiment total score		Experimental Class	50	74.12		11.909			1.684	
		Control Class	50	73.94		11.650			1.648	
		Levene's Test for Equality of Variances		t-test for Equality of Means						
		F	Sig.	t	df	Sig. (2-tailed)	Mean Difference	Std. Error Difference	95% Confidence Interval of the Difference	
									Lower	Upper
Pre-experiment total score	Equal variances assumed	.035	.851	.076	98	.939	.180	2.356	−4.496	4.856
	Equal variances not assumed			.076	97.953	.939	.180	2.356	−4.496	4.856

Table 2. Independent Samples T-test on the Post-experiment Data

			n	Mean		S. D.			Std. Error Mean	
Post-experiment total score		Experimental Class	50	79.98		10.139			1.434	
		Control Class	50	75.42		13.501			1.909	
		Levene's Test for Equality of Variances		t-test for Equality of Means						
		F	Sig.	t	df	Sig. (2-tailed)	Mean Difference	Std. Error Difference	95% Confidence Interval of the Difference	
									Lower	Upper
Pre-experiment total score	Equal variances assumed	8.890	.004	1.910	98	.059	4.560	2.388	−.179	9.299
	Equal variances not assumed			1.910	90.930	.059	4.560	2.388	−.183	9.303

To determine if there is a statistically significant enhancement in scores from before to after the experiment within each group, paired samples t-tests were conducted on both the pre-experiment and post-experiment scores for the control and experimental groups separately.

Based on Table 3, it can be seen that the mean score of the experimental class before the experiment is 74.12, and the mean score after the experiment is 79.98, representing

Table 3. Paired Samples T-test on the Pre-experiment and Post-experiment Data

		Mean	n	S. D.	Std. Error Mean
Pair 1	Experimental Class post-experiment	79.98	50	10.139	1.434
	Experimental Class pre-experiment	74.12	50	11.909	1.684
Pair 2	Control Class post-experiment	75.42	50	13.501	1.909
	Control Class pre-experiment	73.94	50	11.650	1.648

		Paired Differences					t	df	Sig. (2-tailed)
					95% Confidence Interval of the Difference				
		Mean	Std. Deviation	Std. Error Mean	Lower	Upper			
Pair 1 Experimental Class	Post-experiment & Pre-experiment	5.860	2.642	.374	5.109	6.611	15.684	49	.000
Pair 2 Control Class	Post-experiment & Pre-experiment	1.480	2.131	.301	.847	2.086	4.911	49	.000

an average increase of 5.86. The standard deviation decreased from 11.909 to 10.139, indicating an overall improvement in the writing scores of the students in the experimental class and a reduced variation compared to the pre-experiment. In the control class, the mean score before the experiment is 73.94, and the mean score after the experiment is 75.42. The standard deviation increased from 11.650 to 13.501, indicating a slight improvement in the writing scores of the students in the control class, but with a wider score distribution.

The paired comparison, as depicted in Table 3, reveals that the significance (Sig) value (two-tailed) for both the experimental and control classes before and after the experiment is 0.000, which falls below the threshold of 0.01. This demonstrates a significant change in scores for both groups. However, the experimental class experienced a notably more substantial and statistically significant score enhancement compared to the control class, underscoring the significant impact of employing an automatic evaluation system in enhancing students' writing proficiency.

5 Conclusion

5.1 Main Findings

Using an automated grading system for feedback has shown to positively impact students' English writing skills. Through a 10-week writing training on the correction network, the average score of student essays increased from 74.12 to 79.98. A comparison of the pre- and post-test results of the experimental class revealed varying degrees of improvement in students' essays after receiving feedback from the automatic evaluation system. Furthermore, this system effectively addressed the issue of grading English essays, significantly enhancing the efficiency of teachers' grading work. It alleviated some of the teachers' workload and provided them with detailed data on students' growth trajectories, exemplary essays, comprehensive comments, and the entire class's essay performance. This enabled teachers, as guides and motivators, to better understand the overall situation, supporting them in more effectively improving students' English writing skills.

5.2 Research Insights

Firstly, from the perspective of teachers, it is essential for them to continually update their English writing pedagogy and teaching methods. Shifting from traditional teaching methods and exploring new correction techniques not only facilitate the renewal of teaching philosophies but also enhance students' writing skills. Secondly, from the students' standpoint, it is crucial to engage in extensive practice and repeated revisions. They should pay close attention to the suggested revisions and continuously correct mistakes and improve their essays based on these recommendations. This process helps minimize errors in vocabulary, grammar, and details, thereby improving the quality of their English essays. Lastly, regarding the application of automatic evaluation systems, although the system has had a positive impact on English essay instruction, there are still areas for improvement. Investigations have shown that the system's feedback often focuses on specific details of the essay rather than offering comprehensive evaluations. Therefore, future research and development should aim to enhance the system's feedback on the structure and content of essays, making it more scientific and holistic.

5.3 Limitations and Future Directions

Firstly, the duration of the experiment was relatively short. This teaching experiment lasted for three months, with only one English writing class per week, indicating that the writing training provided was insufficient. Effective writing training requires a longer period to observe tangible results. Extending the duration of the experiment would allow students to make full use of the correction network to revise their essays multiple times. Secondly, the sample size of the experiment was limited. The experiment involved only two classes, resulting in a small and limited range of essay samples. To draw more general conclusions, it is necessary to increase the number of essay samples and expand the sample range. Future research should target students of varying English proficiency levels and from different schools and regions to arrive at more rigorous experimental conclusions.

References

1. Shermis, M.D., Burstein, J.C.: Automated Essay Scoring: A Cross-Disciplinary Perspective. Lawrence Erlbaum Associates Inc. (2003)
2. Khoii, R., Doroudian, A.: Automated scoring of EFL learners' written performance: a torture or a blessing? In: Conference Proceedings. ICT for Language Learning, no. 6, pp. 5146–5155 (2014)
3. Chen, H., Lee, J.Y.: Impact of automated essay scoring on EFL students' writing quality and writing anxiety: insights from a mixed-methods study. Comput. Assist. Lang. Learn. **5–6**, 586–610 (2022)
4. Garcia, P., Thompson, L.: Advancements in artificial intelligence for writing assistance: a review of automatic evaluation systems. J. Artif. Intell. Res. **71**, 455–489 (2023)
5. Nguyen, H., Boles, W.: Automated feedback in ESL writing: exploring students' perceptions. Comput. Educ. **143**, 103675 (2020)
6. O'Neill, S., Russell, A.M.: Machine learning and the future of writing assessment: analyzing the capabilities of current automated evaluation tools. Assess. Eval. High. Educ. **01**, 142–158 (2023)
7. Kumar, V., Stracke, E.: Feedback in automated writing evaluation: implications for second language writing. TESOL Q. **01**, 103–132 (2021)
8. Attali, Y.: Exploring feedback and revision features of criterion. Natl. Counc. Meas. Educ. **04**, 12–16 (2004)
9. Li, M., Sotiriadou, P.: The effect of automated writing evaluation on writing competence in EFL contexts: a meta-analysis. Educ. Res. Rev. **34**, 100379 (2022)
10. Patel, R., Dexter, S.: AI in education: the promise and perils of automated essay scoring in English language learning. J. Lang. Learn. Technol. **01**, 1–18 (2022)
11. Attali, Y., Bridgeman, B., Trapani, C.: Performance of a generic approach in essay scoring. J. Technol. Learn. Assess. (10), 1–19 (2010)
12. Hoon, T.: Online automated essay assessment: potentials for writing development, 9 August 2010. http://ausweb.scu.edu.au/aw06/papers/refereed/tan3/paper.html
13. Ahmed, S., Perez, R.M.: Evaluating the effectiveness of automated feedback in writing instruction: a longitudinal study. J. Educ. Technol. Soc. **02**, 108–122 (2021)
14. Rich, C.S., Schneider, M.C, D'Brot, J.M.: Applications of automated essay evaluation in west Virginia (2013)
15. Stevenson, M., Phakiti, A.: The effects of computer-generated feedback on the quality of writing. Assess. Writ. **19**, 51–60 (2014)

Digital Literacy of College Foreign Language Teachers in the Context of Digital Education

Yali Qiang(✉)

Guangdong University of Science and Technology, Dongguan, China
254894038@qq.com

Abstract. Under the background of digital education, this study mainly investigates foreign language teachers' digital literacy and explores strategies for improving the digital literacy of them. This study focuses on 208 foreign language teachers in colleges and explores their digital literacy from five dimensions, which are digital awareness, utilization and development of digital resource, digital teaching skills, empowering learners, and professional development. It has found that college foreign language teachers' digital literacy is at a high level. There exists no significant difference in digital literacy of foreign language teachers with different sexes, educational attainment and academic rank, but there is significant difference among foreign language teachers of different age groups. Based on the research results, corresponding improvement strategies are proposed to provide reference for the improvement of digital literacy among foreign language teachers in colleges.

Keyword: Digital Literacy · Digital Education · Foreign Language Teachers in Universities

1 Introduction

In the times of digital transformation in higher education, digital technology has exerted a profound impact on the field of education. As the backbone of foreign language education, digital literacy has been an essential core quality for foreign language teachers. Digital literacy not only involves the application abilities of basic information technology, but also includes the competence to effectively blend digital technology into foreign language teaching and innovate teaching methods with technology, which is crucial for foreign language teachers to boost the quality and effectiveness of foreign language teaching, promote students' language acquisition, and fit in with the educational technology environment. The Ministry of Education in China issued the *Teacher Digital Literacy* standard in 2022, which aimed to promote Chinese education digitalization strategies and cultivate teacher teams with digital ability. At the 2024 World Digital Education Conference, boosting teachers' digital literacy and ability is one of the parallel forum topics. At present, there is an increasing amount of research on teachers' digital literacy, but there is still room for further strengthening of research on foreign language

© The Author(s), under exclusive license to Springer Nature Switzerland AG 2025
F. Paas et al. (Eds.): EduTA 2024, LAIS 51, pp. 14–26, 2025.
https://doi.org/10.1007/978-3-031-95252-4_2

teachers' digital literacy [1]. On the basis of the perspective of professional development among foreign language teachers in universities, through questionnaire survey, this study analyzes the present situations of digital literacy among foreign language teachers in universities, and explores practical and feasible improvement strategies, aiming to provide ideas for the research on foreign language teachers' digital transformation of higher education and digital literacy.

2 Digital Literacy

In 1994, Yoram Eshet Alkalai, a professor at the Open University of Israel, first proposed the five major frames of "digital literacy", namely, recreation literacy, image literacy, information literacy, branch literacy, and socio emotional literacy [2]. "Real-time thinking literacy" was newly added to the frame in 2012. In 2017, the European Union formulated *EU Digital Literacy Framework for Educator* in the field of education, which systematically described teachers' digital literacy from three dimensions, professional development, teaching competence, and student empowerment. Innovative practices in foreign language teaching have been carried out based on the perspective of digital literacy [3].

In 2019, the State Council issued *China Education Modernization 2035*, which has proposed the strategic task of "accelerating educational reform in the information age". In the sector standard of *Teacher Digital Literacy*, issued by the Chinese Ministry of Education, "teacher digital literacy" is defined as "the consciousness, competence, and obligation that teachers possess to take full advantage of digital technology to obtain, handle, apply, administrate, and assess digital information and resources, find, analyze, and tackle educational and teaching problems, innovate, transform, and take full advantage of teaching activities [4]. Meanwhile, the *Teacher Digital Literacy* divides the digital literacy framework into five dimensions, which are digital consciousness, digital technology knowledge and skills, digital applications, digital social obligation, and professional development, which form an organic whole that is interrelated and mutually reinforcing, jointly determining the levels of teachers' digital literacy [5]. The digital awareness of teachers is the foundation for enhancing digital literacy, which determines their cognitive level and attitude towards digital education, and is a prerequisite for teachers to carry out education and instruction in the digital times; Digital technology knowledge and skills is the foundation for teachers to achieve the deep integration of digital technology with education, which directly affects whether teachers can effectively apply digital technology to instruction practice and educational reform; Digital applications involve how teachers use digital technology to innovate, transform, and take full advantage of educational and instructional activities, which is the core goal for teachers to achieve digital teaching; Digital social responsibility emphasizes the social responsibilities that teachers should undertake in the context of the digital age, such as protecting student privacy and maintaining network security; Professional development is to facilitate teachers to better adapt to the high-quality development needs of education in the digital times through continuous learning and practice.

In the process of promoting education digitization, the overall teaching environment of most universities in China has undergone tremendous changes. The teachers can use a series of hardware devices such as interactive large screens, electronic whiteboards, and intelligent all-in-one machines, as well as practical software tools such as smart attendance, classroom interaction, discussion, and teaching recording systems. In addition, some digital resources such as micro lecture videos, virtual experiments, and digital textbooks have also emerged, all of which require teachers to directly use and participate in the development of these digital educational resources.

3 Digital Literacy of Foreign Language Teachers

The understanding and attitude of foreign language teachers towards information technology will affect their digital teaching practice, among which the proficiency of technical tools is an important influencing factor. Before the official publication of *Teacher Digital Literacy*, related research mostly focused on teachers' information literacy. The research on college foreign language teachers' information literacy mainly focused on the connotation of information literacy of foreign language education, the current situation, requirements, and training paths of college foreign language teachers' information literacy [6].

After the publication of *Teacher Digital Literacy*, Lai Chun et al. constructed a digital literacy framework for foreign language teachers, which is divided into four main dimensions: basic digital literacy, teaching literacy of integrated digital technology resource, professional development literacy, and social moral literacy [7]. Hu Jiehui et al. studied the beliefs and practical characteristics of digital literacy among foreign language teachers, as well as their relationship, and pointed out that digital assessment and the transformation of artificial intelligence foreign language teaching models are weak links in digital applications. The relationship between digital beliefs and practices of foreign language teachers is influenced by various factors such as their subject education philosophy, proficiency in technology application, and emotional experience [8].

The improvement of teachers' digital literacy will bring many changes to classroom teaching, such as personalized and more flexible teaching methods, and more efficient teaching evaluation. Teachers are organizers of teaching activities, guides and participants in learning activities. Teachers are the key to achieving deep combination of information technology with teaching processes and contents under digital conditions. Only with sound digital literacy can teachers build a blended digital teaching environment, skillfully utilize intelligent technology to carry out classroom teaching activities, and create high-quality smart teaching models.

Currently, more and more university foreign language teachers are beginning to realize the importance and value of digital teaching. They gradually realize that digital teaching can not only enrich teaching methods and improve teaching effectiveness, but also meet the personalized and diverse learning needs of students [9]. Many foreign language teachers have already utilized various digital teaching tools, such as multimedia teaching software, online teaching platforms, virtual laboratories, as well as digital teaching resources, large language models, and so on.

Although digital teaching has been widely applied and achieved certain results in university foreign language teaching, it also faces some challenges and problems. For example, some teachers lack digital literacy and find it difficult to effectively apply digital teaching tools and resources; The incomplete construction of digital teaching facilities in some schools has affected the development of digital teaching; Some students lack self-learning and self-management abilities, making it difficult for them to fit in with the learning needs of digital instruction.

4 Research Design

The research design encompasses research objective, research questions, research hypothesis, research subject, and research instrument.

4.1 Research Objective

This study focuses closely on the issue of digital literacy among college foreign language teachers, taking foreign language teachers in Dongguan City, Guangdong Province as the research object. Through questionnaire survey, the present situations of digital literacy among foreign language teachers in universities under the background of digital education is studied and analyzed, and targeted strategies are proposed to provide reference for the enhancement of digital literacy among foreign language teachers in universities.

4.2 Research Questions

Specifically, the following questions will be answered in this research:

1. What is the level of digital literacy of college foreign language teachers in terms of:
   1.1 digital awareness;
   1.2 utilization and development of digital resource;
   1.3 digital teaching skills;
   1.4 empowering learners; and
   1.5 professional development?
2. Is there a significant difference in digital literacy of college foreign language teachers when grouped in terms of:
   2.1 sex;
   2.2 age;
   2.3 educational attainment; and
   2.4 academic rank?
3. Based on the findings of the study, what strategies might be proposed to increase the digital literacy of foreign language teachers in universities?

4.3 Research Hypothesis

The null hypothesis to be tested in this research at 0.05 level of significance is as follows:

Ho: There is no significant difference in digital literacy of foreign language teachers in universities when they are grouped in terms of sex, educational attainment, age, and academic rank.

4.4 Research Subject

This study adopted a random sampling method and carried out a questionnaire survey in four universities in Dongguan city, Guangdong Province. The subject of this survey was the teachers engaged in foreign language teaching in colleges. The total of 210 questionnaires were distributed through Wenjuanxing, and duplicates, regular filling, and questionnaires with a time of less than two minutes were cleared out. Finally, 208 valid questionnaires were achieved, with 99% effective rate. The sample data shows that the sex, age, educational attainment, and academic rank of the research subjects in this survey have a certain proportion, and the distribution of the survey subjects is relatively reasonable. The following Table 1 is the basic information of the sample.

Table 1. The Basic Information of the Subject

	Type	Number	Percentage
Sex	Male	59	28.15%
	Female	149	71.85%
Age	Under 26	16	7.6%
	26–35	67	32.54%
	36–45	63	30.15%
	46–55	36	17.31%
	Over 55	26	12.4%
Educational attainment	Bachelor's degree holder	17	8.36%
	Master's degree holder	165	79.25%
	Doctoral degree holder	26	12.39%
Academic rank	Senior rank	62	29.95%
	Intermediate rank	115	55.36%
	Junior rank	11	5.24%
	Ungraded	20	9.45%

4.5 Research Instrument

In the guidance of *Teacher Digital Literacy*, released by the Ministry of Education in China in December 2022 and *European Framework for the Digital Competence of Educators* released in 2017, this study synthesizes the structural system of teacher digital literacy and combines it with the characteristics of foreign language education in Chinese universities. A self-designed survey questionnaire about the present level of digital literacy among college foreign language teachers is developed. This questionnaire is separated into two parts, basic information and digital literacy status. Basic information comprises the teachers' sex, age, educational attainment, and academic rank. The literacy status section consists of 24 questions, including five abilities, digital awareness, utilization and development of digital resource, digital teaching skills, empowering learners, and professional development.

The questionnaire is with a Likert five point scale, which ranges from "completely disagree" to "completely agree", with scores ranging from one to five. The higher is the grade, the higher is the teacher's digital literacy. In terms of questionnaire reliability, seen in Table 2, the Cronbach's Alpha coefficient of the total questionnaire is 0.977, and the reliability of each subscale ranges from 0.866 to 0.920, indicating high reliability and good internal consistency of the scale. This study uses correlation analysis to test the construct validity. The correlation coefficients between each dimension of the questionnaire and the overall value range from 0.792 to 0.887, indicating a significant correlation between each dimension and the overall questionnaire. The correlation coefficients between each dimension of the questionnaire range from 0.456 to 0.549, showing a moderate correlation. The scale has good construct validity.

Table 2. The Results of Reliability Test

Instrument	Cronbach's Alpha	Interpretation
Digital awareness	0.914	Excellent
Utilization and development of digital resource	0.913	Excellent
Digital teaching skills	0.866	Excellent
Empowering learners	0.876	Excellent
Professional development	0.920	Excellent

5 Research Results

This study applies SPSS 26.0 statistical software to analyze the collected data and employs descriptive statistical analysis to understand the general status of digital literacy of foreign language teachers in universities. ANOVA and T-test are adopted for different sex, age, educational attainment, and academic rank.

5.1 Overall Status of Digital Literacy of Foreign Language Teachers

After analyzing the collected data, the results of descriptive statistics are showed in Table 3.

Table 3 presents that the status of foreign language teachers' digital literacy of has grand mean of 4.185 and is interpreted at high level, which means that the college foreign language teachers do well in digital competence. From the perspective of specific literacy, the average score of digital awareness dimension is the highest, reaching 4.415 points, indicating that the overall awareness of digital teaching application among foreign language teachers in universities is relatively high. In addition to teachers' awareness of

Table 3. The Mean Value and Standard Deviation of Digital Literacy of Foreign Language Teachers

Dimension	Mean	St. Deviation	Rank
1. Digital awareness	4.415	0.610	1
2. Utilization and development of digital resource	3.753	0.491	5
3. Digital teaching skills	4.197	0.732	4
4. Empowering learners	4.316	0.650	2
5. Professional development	4.207	0.701	3
Grand mean	**4.185**	**0.578**	

digital teaching, the dimensions with scores from high to low are empowering learners, professional development, digital teaching skills, and utilization and development of digital resource, respectively, 4.316, 4.207, 4.197, and 3.753. The competence to apply and develop digital resources is currently the weakest section in the digital literacy of foreign language teachers in universities, and only this dimension is lower than the average score, indicating that the lack of this part of literacy seriously reduces the overall digital literacy level of teachers.

5.2 Difference in Foreign Language Teachers' Digital Literacy When Grouped According to Profile Variables

To further understand whether there are significant differences in digital literacy among foreign language teachers in universities in terms of sex, age, educational attainment, and academic rank, this research uses independent sample t-test and one-way analysis of variance to make an analysis of the data results.

5.2.1 Sex

An independent sample t-test is made for foreign language teachers with different genders in universities, and the result is shown in the following table.

From Table 4, it can be found that the average grade for female teachers is 4.219, while the one for male teachers is 4.151. Female teachers have slightly higher digital literacy than male teachers, but the difference has not reached a significant level. In terms of various dimensions, female teachers have a higher ability to empower learners than males and reach a significant level ($P = 0.042 < 0.05$). Male teachers in all other dimensions have slightly higher abilities in "utilization and development of digital resource" compared to female teachers, but the difference between the two groups did not reach a significant level.

5.2.2 Age

This study conducts a one-way analysis of variance on foreign language teachers from different age groups in universities. The results have showed that there are significant

Table 4. Difference in Digital Literacy of Foreign Language Teachers in Terms of Sex

Dimension	Sex	Mean	t-value	P-value
Digital awareness	Male	4.369	−1.355	0.176
	Female	4.459		
Utilization and development of digital resource	Male	3.769	0.440	0.660
	Female	3.745		
Digital teaching skills	Male	4.168	−0.759	0.448
	Female	4.229		
Empowering learners	Male	4.256	−1.843*	0.042
	Female	4.367		
Professional development	Male	4.156	−1.278	0.202
	Female	4.254		
Overall	Male	4.151	1.061	0.289
	Female	4.219		

differences in the dimensions of "digital awareness" ($F = 2.408$, $P = 0.049 < 0.05$) and "professional development" ($F = 2.411$, $P = 0.047 < 0.05$) among teachers of different age groups, and there is an extremely significant difference in the dimension of "empowering learners" ($F = 3.437$, $P = 0.009 < 0.01$). The least significant difference (LSD) method is used to conduct further inter group post hoc tests on the dimensions that have achieved significant differences, and only the significant difference data above the 0.05 level are retained which is seen in Table 5.

From Table 5, it can be seen that teachers in the 45–55 age group have significantly higher digital competence of "digital awareness" and "empowering learners" than those in the 36–45 age group and the over 55 age group. At the same time, teachers in the "professional development" dimension are also significantly higher than those in the over 55 group; The 26–35 age group is significantly higher in the dimension of "empowering learners" than the 36–45 and over 55 age groups, and significantly higher in the dimension of "professional development" than the over 55 age group.

5.2.3 Educational Attainment and Academic Rank

This study conducts a one-way analysis of variance about college foreign language teachers' the digital literacy with different educational attainment and academic rank. The results register that there is no significant difference in the overall level and F-test of the five sub dimensions of digital literacy among foreign language teachers with different educational attainment and academic rank, indicating that the digital literacy status of foreign language teachers with different educational attainment and academic rank are equivalent.

From Fig. 1, it can be seen that the average scores of teachers with different academic rank show the same trend in each dimension. As their academic ranks increase, their

Table 5. Difference in Digital Literacy of Foreign Language Teachers in Terms of Age

Dimension	Age	Mean	F-value	P-value	Post hoc tests
Digital awareness	Under 26	4.381	2.408*	0.049	D > C D > E
	26–35	4.464			
	36–45	4.323			
	46–55	4.556			
	Over 55	4.178			
Empowering learners	Under 26	4.200	3.437**	0.009	B > C B > E D > C D > E
	26–35	4.386			
	36–45	4.214			
	46–55	4.465			
	Over 55	3.933			
Professional development	Under 26	4.143	2.411*	0.047	B > E D > E
	26–35	4.267			
	36–45	4.128			
	46–55	4.335			
	Over 55	3.833			

Legend: B refers to26–35 years old; C refers to 36–45 years old; D refers to 46–55 years old; E refers to over 55 years old.

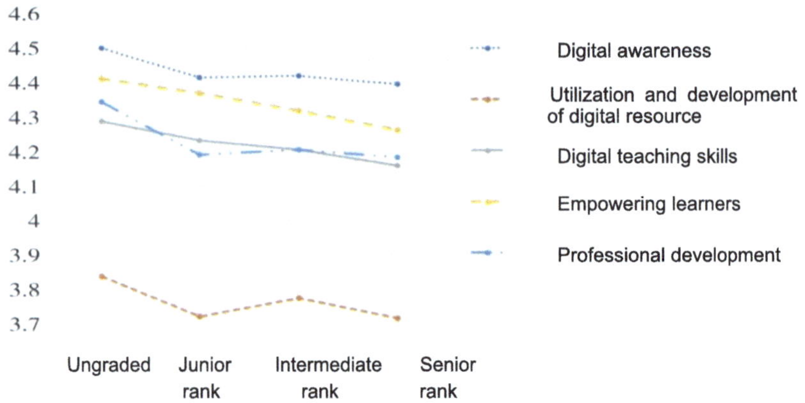

Fig. 1. Difference in in Digital Literacy of Foreign Language Teachers in Terms of Academic Rank

digital literacy gradually decreases. This is similar to the trend of digital literacy among foreign language teachers in different age groups in universities. As there is a certain correlation between the evaluation of academic rank and the age of teachers, it also indirectly demonstrates the internal consistency of the data in this study.

6 Research Conclusions and Improvement Strategies

This part encompasses research conclusions and improvement strategies for digital literacy of foreign language teachers in universities.

6.1 Research Conclusions

Firstly, the overall digital literacy level of foreign language teachers in Dongguan's universities is at a high level. Among them, the awareness dimension of digital teaching application scores the highest, while the competence to apply and develop digital resources performs the worst, and digital teaching skills are relatively weak. This indicates that college foreign language teachers have established a positive attitude and awareness towards the application of digital teaching in the teaching process, but there is still a lack of practical operational skills.

Secondly, there is no gender difference in the overall performance level of digital literacy of foreign language teachers in universities, but female teachers have slightly higher overall competence in quality literacy than male teachers, and their ability to empower learners is remarkably higher than the male teachers. Among the other four dimensions, three dimensions also score higher than male teachers.

Thirdly, there are significant differences in digital literacy of college foreign language teachers with different age groups. Teachers aged 45–55 have the highest digital literacy, followed by those aged 26–35, while teachers aged over 55 have the weakest digital literacy. Teachers aged 36–45 also have relatively weak digital literacy, showing an overall trend of "first rising, then falling, then improving, and finally falling".

Fourthly, there is no significant difference in digital literacy of foreign language teachers in universities with different educational attainment and academic rank.

6.2 Strategies for Enhancing Digital Literacy

6.2.1 Strengthening Digital Literacy and Promoting the Internalization of Digital Literacy

The results of this study indicate that current foreign language teachers in universities have a high awareness of digital teaching and fully recognize the importance of digital technology in teaching. Therefore, foreign language teachers in universities need to further strengthen their digital literacy, internalize their digital literacy, and make it a "meta-literacy" for teachers.

Foreign language teachers should pay timely attention to the latest developments in the sector of educational technology both domestically and internationally, adhere to the habit of improvement, actively participate in digital literacy training and practice, and strive to enhance their ability to apply digital technology. Foreign language teachers should actively embrace the transformation, updating, and rational optimization of educational resources through smart technology, actively promote digital applications in classroom teaching, teaching evaluation, collaborative teaching and research, and professional development, explore the deep combination path of digital technology with foreign language teaching, and deeply develop the potential of new technologies to boost the quality and efficiency of the present instruction. In the process of applying digital technology for foreign language teaching, foreign language teachers must have a firm belief in perseverance and overcoming difficulties, have the courage to continuously learn digital knowledge in unfamiliar fields, master the competence to apply digital tools and resources, overcome the problems and challenges faced in educational digitalization practice, enhance adaptability to the digital instructional environment, and promote the transformation and innovation of digital foreign language teaching models.

6.2.2 Enhancing Teachers' Practical Skills in "Digit + teaching"

In the times of digital education, teaching is no longer a simple combination of technology and instructional methods, but a fusion innovation of technology with instruction with a more complex learning environment [10]. The survey results of this study indicate that current foreign language teachers in universities have a positive attitude and awareness towards the integration of digital technology into teaching. However, their ability to use and develop digital resources and digital teaching skills is significantly lacking, and there is ample room for improvement in teachers' digital teaching practical skills. Therefore, strengthening the practical skills of "digit + teaching" is a key link in improving college foreign language teachers' overall digital literacy.

Firstly, universities need to improve their digital teaching software and hardware support. Universities need to build digital and intelligent teaching facilities and environments such as smart classrooms, and build or introduce digital teaching platforms to encourage teachers to construct and use digital course resources on the platforms. On this basis, a supporting incentive system should be established to provide policy guidance for teachers to enhance their digital teaching skills. Secondly, universities should organize specialized training on digital teaching skills for teachers, which includes multi-dimensional and comprehensive training on digital teaching skills from the aspects of digital teaching resource construction, teaching environment construction, teaching activity process design, and teaching evaluation innovation. Teachers can improve their specialized abilities in weak areas. Finally, universities ought to stimulate teachers to actively participate in the practice and research of digital instruction, such as blended online and offline teaching models, to enhance their digital teaching skills in the instructional process and deepen their theoretical understanding in the research process of digital teaching.

6.2.3 Improving Digital Teaching Competence and Cultivating Teaching Literacy of Digital Resource

Digital competence is a fundamental need for teacher literacy in the digital times. Teacher digital competence means the traits exhibited by teachers in using digital technology to impart knowledge in teaching, which involve motivation, attitude or values, digital teaching skills, digital knowledge, digital evaluation and reflection abilities, etc. [11]. Its connotation mainly includes constructing digital knowledge and forming good digital literacy, capable of developing digital resources, solving digital teaching problems, digital communication and collaboration, and promoting the deep combination of digital technology and curriculum teaching, and building a digital teaching community to promote professional growth.

In the times of digital education, foreign language instruction requires teachers to apply digital technology instruments and resources to organize personalized instructional activities, and fully utilize online learning platforms such as Super Star and Wisdom Tree to provide customized learning content and activities for students of different courses. The improvement of teachers' digital literacy can help make foreign language teaching more contemporary, and the blended online and offline teaching greatly can expand the field of foreign language learning. By utilizing massive multimedia network resources, online teaching platform collaboration tools, and social media platforms to create a virtual real integrated learning space, teachers can interact with students online at any time after class, view students' questions and reply, thumb up, or watch, and also post topics to students online.

Teachers should cultivate the competence to adopt digital technology to online foreign language teaching, not only providing students with static teaching content and resources, but more importantly, the ability to carefully organize all the teaching content of a course, design and record it into logical micro lessons, and create high-level PPT courseware. Teachers should provide students with course resources including multimedia materials such as sound, images, videos, and dynamic illustrations, and upload them to the course platform for students to download and share, and learn at any time. By reforming traditional teaching models and encouraging students to change their learning methods, foreign language learning can better meet the personalized needs of "digital natives".

6.2.4 Establishing a Lifelong Learning Community Where Digital Technology and Professional Skills Support Each Other

There is no significant correlation between the level of digital literacy and educational attainment and academic rank of college foreign language teachers, fully demonstrating that the formation of digital literacy is not related to the knowledge and skills already possessed by teachers, but rather a professional literacy oriented towards the future, as the British historian, Arnold Toynbee once said, when civilization no longer responds creatively to new challenges, the wheel of progress will slip off their axis [12]. To cope with future technological changes, foreign language teachers in universities must continuously reconstruct their professional skills and achieve lifelong learning in the process of professional development. The learning of digital technology and the improvement of professional skills should be an alternating and mutually reinforcing process. Mastering digital technology can not only enhance teachers' comprehensive abilities and digital teaching skills, but also provide teachers with rich learning paths and resources. The continuous improvement of professional skills will drive teachers to consciously seek higher digital literacy to arm them. Only when two technical skills support and coexist with each other can high-quality development of profession of college foreign language teachers be achieved.

7 Conclusion

The digital technology revolution has produced new challenges and opportunities to foreign language teachers, and enhancing their digital literacy is a key aspect of it. On the basis of this, this research adopts a questionnaire survey research method to explore the current development status and improvement strategies of digital literacy of foreign language teachers. This study is a preliminary attempt to apply the national standard of *Teacher Digital Literacy* to the practice of language education. It is hoped that the research results can provide useful insights for increasing the digital literacy of foreign language teachers and assisting in the digital transformation of foreign languages. However, this study only focuses on college foreign language teachers in Dongguan, which has certain limitations. In addition, it is also possible to explore in depth the promotion strategies of digital assessment and artificial intelligence foreign language teaching models, and facilitate the widespread application of digital teaching in the sector of foreign language education.

References

1. Zhang, J.: EFL teachers' digital literacy: the role of contextual factors in their literacy development. Front. Psychol. **14**, 86–91 (2023)
2. Wu, D., Gui, X.J.: Teacher digital literacy: connotation, standards, and evaluation. Electron. Educ. Res. **8**, 108–114 (2023)
3. Falloon, G.: From digital literacy to digital competence: the teacher digital competency framework. Educ. Technol. Res. Dev. **68**, 2449–2472 (2020)
4. Wu, J.Q., Ren, F.X., Li, M.: Teacher digital competence: connotation, evolutionary path, and framework construction. Heilongjiang High. Educ. Res. **9**, 83–90 (2021)
5. Zheng, X.D., Ma, Y.F., Yue, T.Y.: EU teacher digital competence framework: a new guide for the development of technological innovation teacher. Electron. Educ. Res. **2**, 121–128 (2021)
6. Jiang, L., Yu, S., Zhao, Y.: Teacher engagement with digital multimodal composing in a Chinese tertiary EFL curriculum. Lang. Teach. Res. **4**, 613–632 (2021)
7. Lai, C., Lv, B., Gong, Y.: Teacher digital literacy framework in foreign language teaching activity design. Foreign Lang. World **3**, 31–38 (2023)
8. Hu, J.H., Zhang, T.F.: Research on the beliefs and practices of digital literacy of foreign language teachers in Chinese universities. Foreign Lang. Foreign Lang. Teach. **5**, 73–85 (2023)
9. Zhao, L.Z.: Improvement of digital literacy and digitalization of foreign language teaching practice for college foreign language teachers. J. Changchun Inst. Technol. (Soc. Sci. Ed.) **2**, 104–109 (2024)
10. Chen, J., Li, D.L.: Development of technical teaching content knowledge for college English teachers from the perspective of intermediary regulation. Foreign Lang. Foreign Lang. Teach. **5**, 22–32 (2020)
11. Wang, L., Bu, Z.T., Wang, B.R.: Research on digital literacy in international language education: review and prospect. Foreign Lang. World **3**, 82–89 (2021)
12. Farjon, D., Smith, A., Voogt, J.: Technology integration of pre-service teachers explained by attitudes and beliefs, competency, access, and experience. Comput. Educ. **130**, 81–93 (2019)

Exploration of English Multi-interactive Teaching Mode Based on Computer Information Technology

Yaying Lao(✉)

Guangdong Polytechnic of Science and Technology, Guangzhou, Guangdong, China
1228108523@qq.com

Abstract. To further enhance the quality and standard of college English education, it is necessary to fully utilize modern and specialized teaching methods, leveraging computer information technology to establish a multi-interactive teaching model that promotes improved teaching outcomes in college English. This study, by analyzing the advantages of the prevalence of information technology and online resources, constructs a multi-interactive teaching model encompassing various aspects such as instructional guidance, curriculum design, teaching practice, and teaching evaluation, with information technology serving as the primary teaching tool. Taking 118 second-year English majors from University R as experimental subjects, 58 students were assigned to the experimental group, where the proposed multi-interactive teaching model was implemented, while the remaining 60 students served as the control group, receiving traditional teaching methods. By comparing the final exam scores between the two groups, the effectiveness of the information technology-based multi-interactive teaching model on college English teaching outcomes was validated. Through the comparison of scores between the two classes, it was found that the average score of the experimental group increased by 15.25 points compared to the control group, with a lower degree of dispersion in students' average scores, indicating more stable performance. The English multi-interactive teaching mode based on computer information technology can significantly improve the quality of college English teaching.

Keywords: College English · Information Technology · Diversified Interactive Teaching · Multi-Interactive Teaching · Empirical Analysis · Model Building

1 Introduction

With the development of computer information technology, technical and innovative support has been provided for modern teaching innovations. In the context of the digital development and transformation of the educational environment, it is imperative to enhance the quality of college English teaching and foster students' practical abilities in English language interaction and communication. Teachers need to fully grasp the crucial opportunity of integrating current computer information technology into college English

© The Author(s), under exclusive license to Springer Nature Switzerland AG 2025
F. Paas et al. (Eds.): EduTA 2024, LAIS 51, pp. 27–36, 2025.
https://doi.org/10.1007/978-3-031-95252-4_3

classrooms, clarify the objective connections between teaching elements, and achieve a deep integration of information technology with English language teaching, thereby constructing a teaching model that aligns with the improvement of college students' language skills [1]. Computer information technology has become a research hotspots in related fields both domestically and internationally. Utilizing its advantages to construct diverse interactive teaching models that are in line with the characteristics of English teaching.

Current research abroad utilizing computer information technology to enhance English language teaching primarily focuses on innovating language teaching methodologies through information technology. For instance, foreign scholars such as Prateet Baskota [2] and others have explored the significance of applying CMC (Computer-Mediated Communication) and AI (Artificial Intelligence) in English education, revealing the potential of computer technology in language acquisition. Jemerson Eugenio [3] and his team, addressing the deficiencies in linguistic representation, heuristic approaches, and instrumental language in current English language teaching, designed an interactive English communication enhancement technology module leveraging information technology. This technology simulates authentic language environments to exercise students' English conversation skills. Furthermore, Amir Reza Rahimi [4] and his colleagues have emphasized the necessity and logic of integrating artificial intelligence technology into English language teaching, thereby enhancing the potential for more innovative language teaching methods to contribute to the innovative development of language itself.

Information technology has been popularized and applied in English teaching in China, especially the application of various interactive teaching models. For example, domestic scholar Wang Yanmin [5] analyzed the path of information technology for smart teaching construction in her research, emphasizing the importance of information technology in strengthening teacher-student connections. Zhang Tingting [6], in her study, highlighted the role of information technology in revolutionizing teaching methods in the information age, elucidating the efficiency of blended online and offline teaching and the significance of multi-dimensional teaching evaluation, providing a reference for the application of information technology in language teaching. Additionally, scholars such as Zhou Yishu [7] comprehensively analyzed the advantages of computer information technology and constructed a teaching model that deeply integrates language teaching with information technology, offering a robust academic reference for enhancing the effectiveness of English language teaching.

Based on current research findings, while there is a relatively abundant body of research on the application of information technology in language teaching both domestically and internationally, empirical studies validating the effectiveness of information technology in enhancing college English language skills are relatively scarce. Therefore, this study aims to analyze the advantages of information technology in the English language teaching context, construct a multi-interactive teaching model that aligns with the improvement of college students' English skills, and conduct an empirical analysis with 118 second-year students from University R as the research subjects to verify the effectiveness of the model.

2 Theoretical Foundation of the Multi-interactive Teaching Model Based on Information Technology

Driven by the rapid development of information technology, this paper constructs a multi-dimensional interactive English teaching model based on computer technology. By analyzing the limitations of English classrooms, combined with constructivist theory and human-computer interaction theory, a three-dimensional interactive teaching framework integrating intelligent platform, virtual reality, and big data analysis is, and its effect on improving students' language application ability and cross-cultural communication ability is verified with empirical cases. In practice, online classrooms, interactive software, virtual laboratories, personalized learning platforms and other ways are often used to design real-time teaching activities for students, so that they can participate in a rich variety of interactive learning processes with the help of a variety of tools and means [8].

2.1 Constructivist Learning Theory

This model is theoretically grounded in Constructivism Theory and Cognitive Learning Theory. Constructivist theory emphasizes that learning is a process in which students actively construct knowledge and are the core participants in the learning process. The process of constructing meaningful learning involves diverse, communicative, and collaborative interactions, fostering a multi-layered and multi-faceted teaching model [8]. Information technology provides abundant resources and tools for creating such situations, enabling students to explore and practice in diverse learning environments, and encouraging learners to actively construct knowledge in specific contexts. Enable students to transcend time and space limitations, communicate and collaborate with classmates and teachers from different regions, thereby promoting the internalization of knowledge and the improvement of abilities through interaction.

2.2 Cognitive Learning Theory

On the other hand, Cognitive Learning Theory advocates a learner-centered approach to teaching, where teachers stimulate students' motivation and initiative, enhancing their observational skills, abilities to analyze and synthesize learning content, ultimately fostering the substantial acquisition of new knowledge and skills [9]. The multi-interactive teaching model based on information technology, rooted in Cognitive Learning Theory and Constructivism Theory, positions teachers as facilitators and supervisors of learning, further emphasizing students' proactive exploration of knowledge and skills.

3 Construction of a Diversified Teaching Model Based on Information Technology

3.1 Framework Design of the Diversified Teaching Model Based on Information Technology

The multi-interactive college English teaching model based on information technology emphasizes five dimensions in curriculum design: "tiered + categorized" teaching guidance, "compulsory + elective" course structure, "online + offline" project-based teaching, "in-class + out-of-class" teaching practice, and a "process + outcome" diversified evaluation mechanism [10]. These five dimensions are synergistically integrated to construct the multi-interactive teaching model (as shown in Fig. 1), closely aligning with the English language teaching objectives of enhancing language proficiency and application skills through the design of a diversified and interactive teaching model.

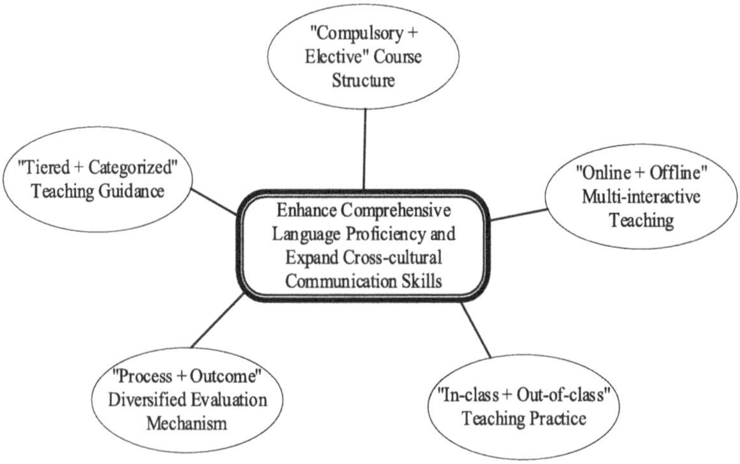

Fig. 1. Construction Framework of Diversified Interactive Teaching Mode Based on Information Technology

The five core elements shown in Fig. 1 cover the whole chain [11] of college English teaching from teaching guidance to curriculum setting, teaching practice to teaching evaluation. In this process, the teachers carefully select the teaching methods and strategies that match the students'; language ability according to the teaching guiding principles of classification and classification, aiming to effectively improve the teaching quality and optimize the learning effect. In addition, the model also skillfully integrates the advantages of compulsory and elective courses, online and offline teaching, with the help of advanced information technology, to build a teaching platform to promote the interaction between teachers and students. This platform greatly expands the space-time boundary of traditional classroom teaching for the implementation of diversified interactive teaching [12]. By making the multiple interaction mode deeply rooted in each link of teaching——whether it is the process, result or practical activities inside and outside the class, supplemented by comprehensive supervision and evaluation mechanism and rich interactive practice platform inside and outside the class, this mode effectively promotes the improvement of students'; comprehensive language application ability.

3.2 Practical Approaches to Implementing the Multi-interactive Teaching Model Based on Information Technology

Driven by information technology, the core of the multi-interactive teaching mode is to realize the seamless integration of language teaching and information technology, and based on this, to build a learning platform that can promote students' multiple interaction. The specific implementation path is illustrated as shown in Fig. 2. Concretely, during the college English teaching process, information technology is utilized to organize the usage of the English teaching platform in a rational manner. Adhering to the current college English teaching guidelines, information technology and online teaching platforms are employed to facilitate students' multi-interactive learning. By leveraging information technology, specific teaching tasks are designed, and learning projects that encourage student-student and student-teacher interactions are assigned. This ensures that information technology is seamlessly integrated into the multi-interactive teaching model, offering opportunities for interaction and enhancing students' awareness of leveraging information technology to improve their English skills [13].

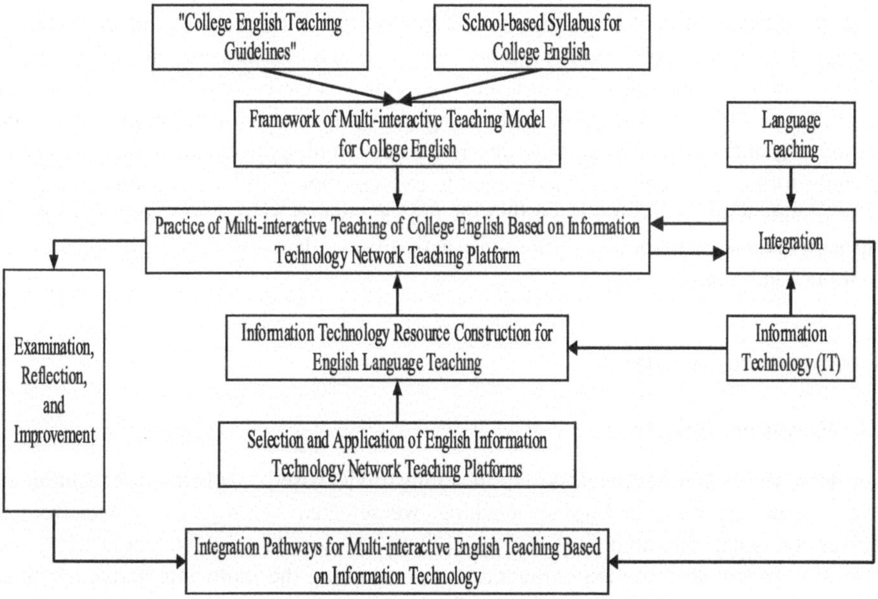

Fig. 2. Practical Approaches to Implementing the Teaching Model

3.2.1 Interest Exploration Activities for Designing Language Based on Information Technology

Integrating information technology to design exploratory activities for college students that can stimulate their exploration of language and spontaneous language learning. Teachers use information technology to create scenarios that are close to the real language environment for students, allowing for planned and purposeful communication in a virtual language environment. Actively utilizing online learning platforms such as MOOCs, Coursera, edX, etc. to provide students with rich learning resources, and utilizing VR technology to simulate language environments, such as virtual travel to English speaking countries. For example, in teaching activities aimed at improving students' listening skills, VR virtual interactive technology can be used to click on characters in animations to have conversations with them, which not only exercises students' listening skills but also enhances their oral expression abilities. At the same time, enable students to enhance their interest and motivation in English learning through virtual interaction.

3.2.2 Building Diverse Collaborative Exploration Methods

Based on information technology, a diversified collaborative inquiry learning model is constructed. Through online knowledge lectures, teachers provide students with language skills, refined learning subjects, and knowledge dimensions for diversified collaborative inquiry. By utilizing information technology to design diverse collaborative exploration activities for students, we aim to enhance their English proficiency and cultivate their teamwork spirit. Firstly, by utilizing online collaboration platforms such as Google Docs, Office 365, or other online collaboration tools, students can collaborate in real-time to edit documents, discuss issues, and share resources, improving the efficiency of student resource exchange and enhancing their information technology application abilities. In addition, using computer information technology to establish online English learning forums, students are encouraged to express their opinions, ask questions, and solve doubts with each other regarding the topics covered by teachers' online lectures. Through diverse collaborative exploration activities, students' writing skills and critical thinking can be improved.

4 Empirical Analysis

4.1 Research Subjects

In order to verify the effectiveness of the teaching model based on information technology proposed in this study in English teaching, we selected 118 s year students from R University as experimental subjects. Specifically, 58 second-year students from Class A01 were assigned to the experimental group, adopting the multi-interactive teaching model proposed in this study, and their final exam scores served as the experimental data. Meanwhile, the final exam scores of 60 students from Class A08 were used as the control group data. Notably, the two classes had comparable average final scores in college English, and their English interaction abilities fell within the same range.

4.2 Practical Process

The students in Class A01 were instructed following the information technology-based teaching model. Teachers published the semester's teaching schedule and provided learning task plans for students via the internet, reducing the blindness in their learning. Additionally, utilizing the remote education platform enabled by computer information technology, teachers disseminated teaching-related resources, offering a platform for teacher-student interaction. Through the group function available on the online teaching platform, roles such as group leader, recorder, resource manager, and spokesperson were clearly assigned to enhance interactive communication among students. Furthermore, teachers were no longer constrained by textbooks during the teaching process but leveraged information technology to access abundant teaching content from the information network, enriching and diversifying the teaching materials provided to students [13]. In the practice of the teaching model, teachers could enrich their own cultural knowledge and professional competence using information technology, adopting heuristic teaching methods to fully embody students' central role in learning [14]. In contrast, students in the control group, Class A08, were taught using traditional methods and analyzed traditional textbooks. The final exam scores of both classes served as the quantitative basis for validating the effectiveness of the multi-interactive teaching model.

4.3 Results and Analysis

The data of the final grades, homework completion and class performance of the two classes were collected to verify the effectiveness of multi-interactive English teaching methods based on information technology.

(1) Comparative analysis of students' final results with different teaching methods

The final test results are detailed in Table 1.

Table 1. Descriptive Statistics of Final Exam Scores for Experimental and Control Groups

Category	Mean	standard deviation (S)	Average score ($\overline{X}$)	Z
Control Group	58	7.97	61.34	2.15
Experimental Group	60	7.31	76.59	

Test the difference in the average value of the class English proficiency test from two independent sample spaces to judge whether the difference of their respective representative population is obvious. The Z calculation formula is:

$$Z = \frac{\bar{X_1} - \bar{X_2}}{S_{\bar{X_1} - \bar{X_2}}} \quad (1)$$

Comparing the z value of the two independent sample means obtained with the theoretical z value, infer the probability of occurrence, The relationship between Z value and difference is shown in Table 2.

Table 2. Degree of difference in Z test

Z absolute value	P	Difference degree
≥2.58	≤0.01	highly significant
≥1.96	≤0.05	notable
<1.96	>0.05	quiet

According to the analysis, the performance of the students in the experimental group has improved significantly. Compared with the final scores of the A08 class using traditional teaching methods, the final scores of the students in the experimental group are 15.25 points higher, and the standard deviation of the final scores in the experimental group is lower, indicating that the degree of dispersion is low and the stability of the student performance is higher. However, the average score of the control group is low, and the standard deviation is large, indicating that the dispersion of students' grades is large, and the high scores and low scores are quite different. Moreover, the standard error of the mean in the experimental group is lower, indicating that the accuracy and reliability of the student achievement data sample are higher [15]. Therefore, we know that the integration of information technology into the multiple interactive teaching mode can significantly improve the quality of college English teaching, which plays a very important role in promoting the smooth development of college English teaching.

(2) Comparative analysis of students' homework completion in different teaching methods

The proportion of class A08 in Class A01 is higher than that of Class A 08, indicating that the information technology teaching mode helps students to better complete the homework tasks. See Table 3 for details.

Table 3. Statistical results of work completion of experimental and control groups

Category	Job completion (excellent rate)	
	Submit on time	Excellent homework
Control Group	85.00%	95.00%
Experimental Group	50.00%	70.00%

(3) Different teaching methods of students' classroom performance

The number of active speeches and group discussion activity of class A01 are higher than that of class A08, key data comparison as shown in Table 4. It indicate that the information technology teaching mode can improve students' participation in class. See Table 5 for details.

Table 4. The Key data comparison

	Experimental Group	Control Group	P value
Classroom engagement	82.3%	47.6%	<0.01
Improvement of oral fluency	29.7%	12.4%	<0.05

Table 5. Statistical results of work completion of experimental and control groups

Category	Class performance (engagement)	
	Speak actively	Group discussion activity
Control Group	50 Times/week	centre
Experimental Group	80 Times/week	high

5 Conclusion

This mode not only greatly stimulates students 'interest in learning, enhances their participation in class, but also effectively enhances students' comprehensive ability. By comparing the data of 2 groups, We can find that the English scores of the students in the experimental group were 15.25 points higher than that of the control group. This result strongly proved that the multi-interactive teaching mode can significantly improve the students' English scores and show better performance in maintaining the stability of their scores. Accordingly, this study comes to the conclusion through empirical test: integrating information technology into college English teaching and constructing a multi-interactive teaching mode is an effective way to improve the quality of college English teaching and cultivate students' ability of applying English in practice. Looking forward to the future, with the innovation of education concept, Interactive college English teaching mode, relying on the power of information technology, shows a broader development potential. By continuously optimizing this teaching model and deepening the integration of information technology and English teaching, we can significantly enhance students' learning experience, stimulate their innovative thinking, and improve their cross-cultural communication skills.

References

1. Nian, X.: Exploration of multi-interactive English teaching model based on computer information technology. Overseas Engl. (5), 79–81 (2024)

2. Baskota, P., Poudel, T.: Artificial intelligence and computer-mediated communication: the text analysis and undergrad's class observation. Discov. Educ. **3**(1), 131 (2024)
3. Eugenio, J., Raymundo, J.: Language functions in ESL classrooms: basis for the enhancement of English conversational skills of junior high school students. Int. J. Linguist. Lit. Transl. **7**(7), 107–116 (2024)
4. Rahimi, R.A., Pavón, S.A.: The role of ChatGPT readiness in shaping language teachers' language teaching innovation and meeting accountability: a bisymmetric approach. Comput. Educ. Artif. Intell., 100258 (2024)
5. Wang, Y.: Research on the construction of multi-interactive smart teaching of English in open education under the background of "internet + education". Overseas Engl. (8), 166–168 (2024)
6. Zhang, T.: Research on the online and offline blended teaching mode of "advanced English" under the cloud class + BOPPPS model. Overseas Engl. (Lower) (10), 189–190 (2021)
7. Zhou, Y., Yan, L., Xin, T.: Construction and practice of project-based teaching model of college English based on multi-interaction. High. Educ. Lett. **9**(13), 100–104 (2023)
8. Alzeer, S., Thompson, P.: Toward a tool for evaluating corpus-based word lists for use in English language teaching contexts. Appl. Corpus Linguist. **4**(3), 100098 (2024)
9. Ma, H.: Research on multi-interactive teaching model of primary school English under the information technology environment. Engl. Teach. **22**(13), 42–44, 54 (2022)
10. Javad, M.E.B., Saeed, K.: The effect of corpus-assisted language teaching on academic collocation acquisition by Iranian advanced EFL learners. J. Appl. Res. High. Educ. **16**(4), 1188–1213 (2024)
11. Yang, L.: Research on multi-interactive teaching model of college English under the information technology environment. Wen Cun Yue Kan (17), 130–132 (2024)
12. Ammari, M.R., Ahmad, A.R.W., Abushunar, S.M., et al.: Faculty members' perspective on employing Youtube in teaching English for ESL undergraduate students. J. Lang. Teach. Res. **15**(4), 1313–1322 (2024)
13. Luo, L.: Analysis of "multi-interactive blended" translation teaching strategy based on information technology. Overseas Engl. (3), 148–150 (2023)
14. Jafari, A., Sharifa, J.: In-service training for improving the quality of English language teaching in Zanzibar's secondary schools. Qual. Assur. Educ. **32**(3), 371–386 (2024)
15. Lai, Y.: The value and countermeasures of constructing a "multi-interactive" teaching model in high school English teaching in the information age. Kong Zhong Mei Yu (9), 53–55 (2023)

The Personalized Recommendation System for TESOL Learners Based on Particle Swarm Optimization Algorithm

Yuwei Xi(✉)

Faculty of Humanities and Social Sciences, Tyne and Wear NE1 7RU, UK
xiywuk2019@163.com

Abstract. This paper presents a personalized recommendation system for TESOL learners based on Particle Swarm Optimization (PSO). The system aims to enhance the learning experience by offering tailored recommendations that optimize various metrics such as precision, recall, F1-score, diversity, novelty, relevance, accuracy, user satisfaction, and recommendation quality. The proposed PSO-based approach effectively improves the recommendation results by addressing the multi-objective optimization challenges inherent in personalized learning. Through a series of experiments, the PSO algorithm demonstrated superior performance over traditional collaborative filtering and content-based filtering methods, particularly in terms of diversity and novelty. The results highlight the potential of PSO in improving the relevance and effectiveness of recommendations in educational contexts, providing TESOL learners with more personalized, diverse, and valuable content. This research contributes to the development of advanced recommendation techniques that can be applied to other domains requiring personalized content delivery.

Keywords: Particle Swarm Optimization · Personalized Recommendation System · TESOL · Multi-objective Optimization · Learning Recommendation

1 Introduction

1.1 Research Background and Importance

With the rapid development of information technology, personalized learning has gradually become an important research direction in the field of education. TESOL (Teaching English to Speakers of Other Languages), as a language teaching model widely used worldwide, aims to provide non-native learners with a tailored English learning experience [1]. Traditional TESOL teaching often relies on direct guidance from teachers and unified teaching materials, which is difficult to meet the differences in learning progress, interests and needs of different learners. In order to solve this problem, the research and application of personalized recommendation systems has become one of the key ways to improve the teaching effect of TESOL. Personalized recommendation systems automatically recommend suitable learning resources and activities by analyzing students'

© The Author(s), under exclusive license to Springer Nature Switzerland AG 2025
F. Paas et al. (Eds.): EduTA 2024, LAIS 51, pp. 37–47, 2025.
https://doi.org/10.1007/978-3-031-95252-4_4

historical learning behaviors, interest preferences and learning progress, thereby helping learners obtain the most valuable content at the right time and optimize learning effects [2].

In the last years, recommendation systems have gained quite remarkable achievements in such well-known fields as e-commerce, social media, and online education. The studies on personalized recommendation systems for learners of TESOL remain very limited, particularly on using optimization algorithms to enhance accuracy, diversity, and personal touch in recommendations [3]. Various challenges still exist. As an intelligent algorithm widely used in multi-objective optimization problems, PSO has become an important tool for improving the performance of recommendation systems with its strong global search ability and adaptability. PSO can optimize multi-dimensional objective functions by simulating the behavior of particle groups in nature, thereby achieving more accurate and personalized recommendations. Therefore, to do research about personal recommendation systems for TESOL learners based on PSO will have far-reaching theoretical significance and be of practical importance, meanwhile, providing a new efficient idea for solving problems that occur about education in practice [4].

1.2 Research Objectives

The main objective of this study is to design and implement a personalized recommendation system for TESOL learners based on the particle swarm optimization algorithm, aiming to improve the accuracy, personalization and diversity of the recommendation system through multi-dimensional performance optimization. First, this study analyzes the personalized needs of TESOL learners and combines learning behavior analysis to construct a recommendation system framework suitable for this field. Secondly, this study uses the particle swarm optimization algorithm to comprehensively optimize multiple objectives in the recommendation process, including the relevance, diversity, and novelty of the recommendations, to ensure that the recommendation results can fully meet the personalized needs of learners. Through the design and implementation of the system, this paper not only explores the application potential of PSO in educational recommendation systems, but also verifies the advantages of the algorithm in improving recommendation effects and user satisfaction through a series of experiments. In addition, this study also analyzes the differences between the PSO-based recommendation system and common methods such as traditional collaborative filtering algorithms and content-based recommendation algorithms in different evaluation indicators through comparative experiments, and evaluates the actual effect of the PSO algorithm in personalized recommendation.

2 Related Technology and Research Review

2.1 Research Progress in Personalized Recommendation for TESOL Learners

In recent years, with the vigorous development of online education, personalized recommendation technology has gradually been applied to the field of TESOL (Teaching English to Speakers of Other Languages). The personalized needs of TESOL learners are usually manifested in multi-dimensional differences such as learning progress, interest preferences, and language proficiency. Therefore, the traditional one-size-fits-all teaching method is difficult to effectively meet the needs of each learner. For this reason, a personalized recommendation system came into being, which can make recommendations based on learners' historical behaviors, learning goals and interests to improve their learning efficiency and interest. In recent years, researchers have proposed a variety of personalized recommendation models for TESOL learners [5]. These models mainly rely on learners' learning data, including learning time, post-class feedback, participation in activities, etc. By analyzing these data, intelligent recommendations are made for content suitable for learners' current learning stage and goals.

At present, personalized recommendation systems for TESOL learners mainly focus on collaborative filtering, content-based recommendation, and hybrid recommendation methods. Collaborative filtering methods recommend content that other learners like by analyzing the behavior of similar learners; content-based recommendation focuses on the similarity between learners and learning content; and hybrid recommendation methods combine multiple recommendation strategies to improve recommendation effects in different dimensions. However, these methods still have some limitations. For instance, the collaborative filter is susceptible to data sparse while content-based recommendations could repeat themselves and get boring; development to overcome which problems can gradually introduce a Particle Swarm optimization algorithm, such an optimally powerful device in learning and research by developing Multiple objective optimizations to optimize personalized recommender systems on improving recommender results better with multiple results being both effective and representative of diversity, it might do more fully reveal many problems and more issues may need consideration because [6].

2.2 Overview of Particle Swarm Optimization Algorithm and Its Application in Recommendation System

Particle Swarm Optimization (PSO) algorithm is a global optimization algorithm that simulates the foraging behavior of bird flocks in nature. It was proposed by Kennedy and Eberhart in 1995. PSO algorithm simulates the process of a group of particles moving in the search space to find the optimal solution of the problem. Each particle represents a potential solution and adjusts its movement direction according to its own experience and the experience of the group, so that the entire group tends to the optimal solution. PSO has strong global search capabilities, can avoid local optimal solutions, and has low computational complexity, which makes it widely used in many optimization problems [7].

In recommendation systems, PSO algorithms are widely used to optimize multiple objectives of recommendation results. For example, in personalized recommendations, PSO can simultaneously optimize multiple objectives such as accuracy, diversity, and novelty of recommendations, thereby avoiding the recommendation system from relying too much on a single performance indicator [8]. The multi-objective optimization characteristics of PSO enable it to fill in the gaps in traditional recommendation methods (such as collaborative filtering and content-based recommendations) and improve the overall performance of recommendation systems. PSO can effectively improve the quality of recommendation by constantly adjusting the recommendation parameters and finding the optimal parameter combination, while avoiding the problems caused by overfitting and data sparsity. In addition, the adaptability of the PSO algorithm enables it to dynamically adjust the recommendation strategy according to the needs and behaviors of different users, providing more personalized learning resource recommendations. This makes PSO have important application prospects in personalized recommendation systems for TESOL learners [9].

2.3 Limitations and Challenges of Existing Personalized Recommendation Systems

Although personalized recommendation systems have achieved remarkable application results in various fields, especially in education, existing recommendation systems still face many limitations and challenges. First, it is still a key problem for personalized recommendation systems: in data sparsity, there are an enormous number of users with relatively few behavioral data, and thus too hard for the recommendation system to capture the user's true preference, which influences the recommendation performance. There are huge differences in the learning habit, interest, and needs of the learners, especially in personalized recommendations of TESOL learners, and the problem of data sparsity is prominent. Thus, traditional methods of collaborative filtering and content-based recommendation hardly work effectively [10].

In this regard, most personalized recommendation systems, to date, are over-dependent on the historical data without considering the dynamic change in user needs.

In education, the need and preference of a learner may evolve with time, while traditional recommendation systems are usually static. Determining the recommended content is quite difficult to adjust in real time according to the learners' feedback. Therefore, this is another challenge in the development of a system that can learn in real time and dynamically adjusts the recommended content. Moreover, the urgent optimization of multi-objective issues in recommendation systems is also a necessity. Personalized recommendation is not limited to improving accuracy but is a consideration of diversity, novelty, and learning efficiency in multiple dimensions. How to balance these goals and avoid the simplification of recommended content while improving quality is an important direction worth researching.

The PSO algorithm effectively solves these problems by introducing a multi-objective optimization method. It optimizes multiple objectives of the PSO algorithm, enhances performance in multiple dimensions of the recommendation system, overcomes limitations of traditional methods, and provides new ideas for the development of personalized recommendation systems.

3 Design and Implementation of Recommendation System Based on Particle Swarm Optimization Algorithm

3.1 Requirements Analysis and Framework Design of Personalized Recommendation System

With the development of information technology, personalized recommendation system has become a key tool to improve user experience. The personalized recommendation system for TESOL (Teaching English to Speakers of Other Languages) learners aims to recommend the most suitable learning resources for them based on their learning habits, interests, progress and other characteristics. TESOL learners face the diversity of content and the complexity of knowledge points in the learning process. Therefore, an effective personalized recommendation system should have the following basic requirements (Table 1).

Table 1. Characteristics of effective personalized recommendation system

Requirement	Description
User Profile Modeling	Collect data on the learner's basic information, learning history, preferences, and habits to build a personalized profile
Resource Recommendation	Recommend learning materials based on the learner's progress, interests, and learning needs
Real-time Updates	Continuously update the learner's profile and adjust recommendations as the learner progresses through their studies
Accuracy and Relevance	Ensure that the recommended resources are highly relevant to the learner's learning style and current level
Diversity in Recommendations	Provide a variety of learning materials to prevent monotony and broaden the scope of the learner's knowledge
Feedback Incorporation	Adjust recommendations based on the learner's feedback to ensure that the system is always improving

3.2 Application and Improvement of Particle Swarm Optimization Algorithm

The application of particle swarm optimization (PSO) algorithm in personalized recommendation system is mainly to optimize the accuracy, relevance and diversity of recommendation results. The following are several key formulas of PSO algorithm, which help us adjust the search position of particles in the recommendation system to obtain the best recommendation results.

3.2.1 Particle Speed Update Formula

The core of particle swarm optimization algorithm lies in the continuous update of particle position and speed, the formula is as follows:

$$v_i^{t+1} = wv_i^t + c_1r_1(p_i^t - x_i^t) + c_2r_2(g^t - x_i^t) \tag{1}$$

v_i^t represents the velocity of particle i at the t th moment; x_i^t represents the position of particle i at the t th moment; w: inertia weight, controlling the influence of particle velocity; c_1, c_2 are individual and social learning factors respectively; r_1, r_2 are random numbers, usually in the interval [0, 1]; p_i^t is the best historical position of particle i at the t th moment; g^t is the global best position of all particle groups.

3.2.2 Fitness Function Formula

In the personalized recommendation system, the particle swarm optimization algorithm evaluates the quality of the recommendation results through the fitness function. The fitness function usually considers the accuracy and diversity of the recommendation, and the formula is as follows:

$$F = \alpha \cdot \text{Accuracy}(R, H) + \beta \cdot \text{Diversity}(R) \tag{2}$$

F is the fitness value of the particle; α, β are weight coefficients, which are used to adjust the contribution of accuracy and diversity respectively; Accuracy(R, H) is the matching degree between the recommended resource set RRR and the learner's historical behavior set H; Diversity(R) is the diversity of the recommended resource set R.

3.2.3 Constraint Formula

In order to ensure that the recommendation results of the recommendation system meet the actual needs, the particle swarm optimization algorithm also needs to consider some constraints. For example, suppose we need to limit the number of resources recommended each time to $N_{\max}$, we can use the following constraint formula:

$$\sum_{i=1}^{N} r_i \leq N_{\max} \tag{3}$$

r_i is the recommendation weight of resource i; $N_{\max}$ is the maximum number of resources recommended each time.

This constraint ensures that the recommendation system will not recommend more than a preset number of resources to learners at one time, thereby avoiding information overload and maintaining the accuracy of the recommendations.

3.3 System Implementation and Functional Modules

The implementation of this system is based on the above design framework, which mainly includes core modules such as data collection, feature extraction, recommendation generation and feedback optimization (as shown in Fig. 1).

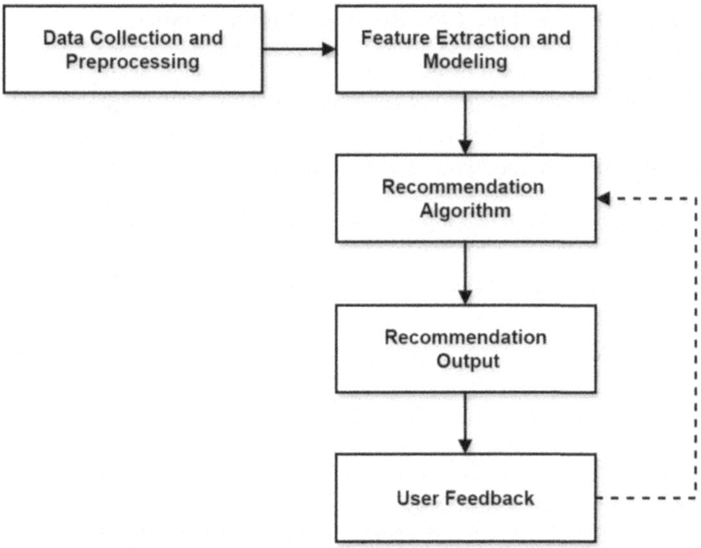

Fig. 1. Core Information Flow of the Personalized Recommendation System

Figure 1 illustrates the core information flow of the personalized recommendation system. It starts with data collection and preprocessing, followed by feature extraction, and ends with the recommendation output being shown to the user. The system continuously refines the recommendation algorithm based on user feedback, creating a feedback loop for optimization.

4 Experiment and Result Analysis

4.1 Experimental Design and Dataset Introduction

This experiment is still based on the TESOL learner dataset, which includes learner behavior data, learning progress and feedback data. Different from the basic experiment, the current experiment introduces more complex evaluation criteria and explores the convergence characteristics and the impact of iterations in the system optimization process. We use the particle swarm optimization algorithm (PSO) for experiments and compare it with other common recommendation algorithms.

The dataset used in this experiment comes from the actual learning records of TESOL learners, including learners' historical behavior data (e.g., completed courses, resource usage and feedback information) and learners' personal preferences. The dataset contains real user data from multiple educational platforms and learning systems, covering the learning activity records of more than 5,000 learners. Each learner's record includes the usage and feedback of at least 20 learning resources, and the total amount of the dataset reaches more than 100,000. In order to ensure the fairness and accuracy of the experiment, we cleaned and preprocessed the data, removed the noise data, and ensured the reliability of the data.

4.2 Performance Evaluation of the Recommendation System

In order to conduct an in-depth evaluation of the performance of the recommendation system, we not only used conventional precision, recall and F1 score, but also explored indicators such as the algorithm convergence characteristics and the relationship between the number of iterations in the optimization process.

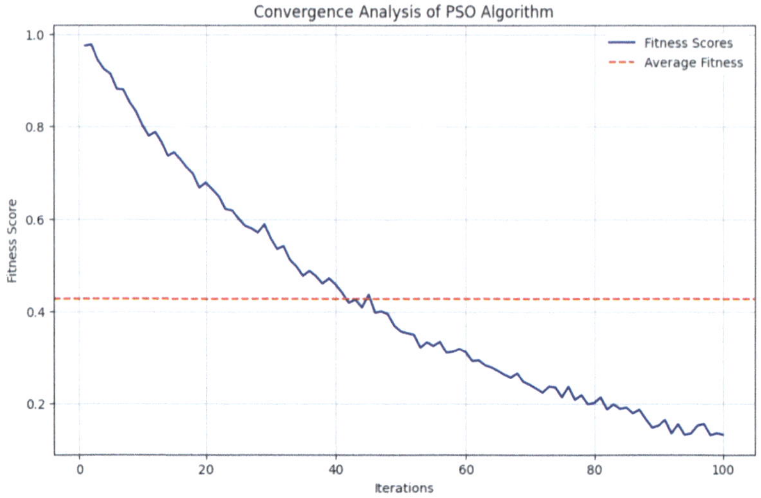

Fig. 2. Convergence analysis of the PSO algorithm

Figure 2 shows the convergence characteristics of the particle swarm optimization (PSO) algorithm during the iteration process. As the number of iterations increases, the fitness score gradually stabilizes, indicating that the system gradually finds the optimal solution. The red dotted line represents the average value of the fitness, showing the performance fluctuation and final stability of the algorithm during the convergence process.

Figure 3 shows the relationship between recommendation accuracy and iteration number. As the number of iterations increases, the recommendation accuracy gradually improves, showing the optimization effect of the particle swarm optimization algorithm on the recommendation results in the continuous adjustment process.

Figure 4 shows the performance of different recommendation algorithms (PSO algorithm, collaborative filtering algorithm and content-based recommendation algorithm) in multiple dimensions such as precision, recall rate, F1 score, diversity, novelty, relevance, accuracy, user satisfaction and recommendation quality. According to the actual algorithm characteristics, the PSO algorithm performs well in multi-dimensional optimization, but collaborative filtering performs better in precision and recall rate. The content-based recommendation system performs well in recommendation relevance and novelty, but is slightly inferior in diversity and user satisfaction.

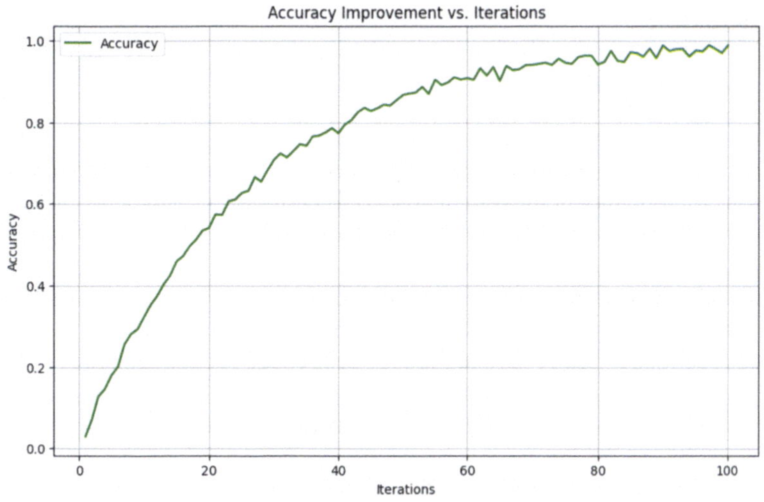

Fig. 3. Relationship between the number of iterations and recommendation accuracy

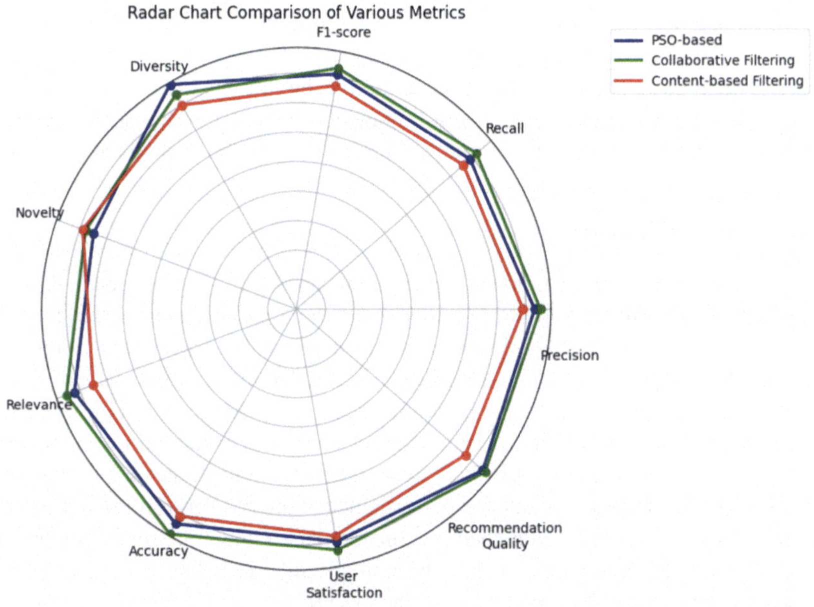

Fig. 4. Radar Chart Comparison of Various Metrics

5 Conclusion

5.1 Main Conclusions

This study proposes a personalized recommendation system for TESOL learners based on the particle swarm optimization (PSO) algorithm, and experimentally compares and analyzes the performance of the system in multiple performance indicators such as precision, recall rate, F1 score, and diversity. By designing and implementing a comprehensive personalized recommendation system and combining the PSO algorithm to optimize the recommendation results, this study proves the effectiveness and superiority of the particle swarm optimization algorithm in the field of personalized recommendation. Compared with traditional collaborative filtering and content-based recommendation algorithms, the PSO algorithm has improved performance in multiple dimensions, especially in diversity and novelty indicators, showing its advantages in multi-objective optimization problems.

The experimental results show that the recommendation system based on PSO optimization not only achieves good results in standard evaluation indicators, but also can provide more personalized and accurate recommendation results, thereby improving the learning effect and satisfaction of TESOL learners. By comparing and analyzing the performance of different algorithms, this study further verifies the wide application potential of PSO in recommendation systems and provides new ideas and directions for the development of personalized recommendation technology in the future. In addition, the personalized recommendation system framework designed in this study has strong scalability and can be applied and promoted in recommendation systems in other fields.

5.2 Research Limitations

Although the personalized recommendation system based on PSO proposed in this study has achieved good results in multiple dimensions, there are still some limitations and shortcomings. First, the TESOL learner dataset used in the experiment is relatively small in scale, and the diversity and coverage of the data are limited, which may not fully reflect the complexity in real application scenarios. The needs and preferences of different learners vary greatly in actual situations, so the experimental results based on the current data set alone may not fully represent the personalized needs of all learners. In order to more accurately evaluate the performance of the PSO algorithm in the recommendation system, future research needs to consider larger and more representative data sets, especially multi-dimensional information involving user behavior data, to improve the robustness and generalization ability of the system.

Secondly, although the PSO algorithm has shown a high effect in optimizing the recommendation results, its computational complexity is high, especially when processing large-scale data, it may face performance bottlenecks. In addition, the parameter setting of the PSO algorithm has a great influence on the final optimization results, but there is currently a lack of effective automatic parameter adjustment mechanism. Therefore, how to further optimize the parameter selection of the PSO algorithm and reduce its computational complexity to improve the efficiency of the recommendation system is still an urgent problem to be solved.

5.3 Future Research Directions

In the future, many aspects can be further expanded and deepened. First, deep learning technology can be combined with the particle swarm optimization algorithm to explore more intelligent recommendation systems. Deep learning is able to mine nonlinear relationships and complex patterns of data better, especially in dealing with large-scale and complex data sets, providing more accurate prediction results. In particular, the deep learning model combined with the PSO algorithm will be able to further improve the accuracy and diversity of the recommendation results while ensuring personalized recommendations.

Another important direction of research is the real-time and adaptability of the personalized recommendation system. The current recommendation system is mostly based on offline training and batch processing mode, which may have a certain lag in application scenarios that require real-time recommendations, such as TESOL online learning platform.

In the future, online learning- and incrementally updated-based PSO optimization algorithms could be further researched so that the dynamic adjustment and real-time optimization of the recommendation system adaptively occur through the activities of the users themselves, the feedback from their usage, to further provide recommendations flexibly and timely.

References

1. Huang, J., Zhang, X.: Intelligent recommendation method for offline course resources tax law based on chaos particle swarm optimization algorithm. J. Intell. Fuzzy Syst. **45**, 10603–10617 (2023)
2. Zhang, L.: An IoT-based English translation and teaching using particle swarm optimization and neural network algorithm. Soft. Comput. **27**, 14431–14450 (2023)
3. Li, L.: Analysis of syntactic complexity and semantic coherence of academic English writing based on particle swarm optimization. Mob. Inf. Syst. (2022)
4. Li, R.: State of health and charge estimation based on adaptive boosting integrated with particle swarm optimization/support vector machine (AdaBoost-PSO-SVM) model for lithium-ion batteries. Int. J. Electrochem. Sci. (2022)
5. You, G., Shiue, Y., Yeh, W., Chen, X., Chen, C.: A weighted ensemble learning algorithm based on diversity using a novel particle swarm optimization approach. Algorithms **13**, 255 (2020)
6. Xu, H., Bao, S., Zhang, X., Liu, S., Jing, W., Ji, Y.: Blood glucose prediction method based on particle swarm optimization and model fusion. Diagnostics **12** (2022)
7. Cha, Y.: Learning resource recommendation method based on particle swarm optimization algorithm. J. Comput. Appl. (2014)
8. Zervoudakis, K., Mastrothanasis, K., Tsafarakis, S.: Forming automatic groups of learners using particle swarm optimization for applications of differentiated instruction. Comput. Appl. Eng. Educ. **28**, 282–292 (2019)
9. Adebayo, O.S., Aziz, N.A.: Static code analysis of permission-based features for android malware classification using apriori algorithm with particle swarm optimization (2015)
10. Salih, S.A., Rashid, T.A.: Learner performance-based behavior optimization algorithm: a functional case study. In: International Conference on Interactive Collaborative Robotics (2022)

Research on User Demands of Online Open Courses

Li Yang, YanFang He(✉), and YongChao Xie

Hunan Railway Professional Technology College, Zhuzhou, China
53610285@qq.com

Abstract. Obtaining the user demands of online open courses is the primary task of successfully constructing the quality evaluation index system of online open courses. To tap the real demands of users, Indicators of user demands and the weight of each indicator of online open courses are all determined based on teacher-users and student-users. This paper focuses on using the KJ method and data analysis to obtain the user demands of online open courses.

Keywords: Online Open Courses · User Demands · Weight of Demand · KJ Method · Data Analysis

1 Introduction

Wu Yan, Director-General of the Department of Higher Education of the Ministry of Education, said, "Vigorously promoting the construction, use and learning of online open courses is a key move to achieve the quality of higher education in China, especially the quality of talent training" [1]. In June 2016, the "Guiding Opinions on Deepening Education and Teaching Reform in Universities Affiliated by Central Departments" clearly requires the universities under the Ministry to vigorously promote the construction of online open courses and provide special funds and policy guarantees [2]. Since the launch of the first batch of national quality online open courses in 2017, many undergraduate and higher vocational courses have been recognized as national and provincial quality online open courses every year. Among them, the number of high-quality online open courses in higher vocational schools is also increasing year by year. For example, in 2020, 99 national excellent online open courses in higher vocational education have been identified [3].

The quality evaluation of online open courses is an important part of course construction work, which effectively guarantees the smooth implementation of course construction projects. In April 2015, the Ministry of Education's "Opinions on Strengthening the Construction, Application and Management of Online Open Courses in Colleges and Universities" proposed that colleges and universities should be encouraged to formulate teaching quality certification standards for online open courses, and the online open courses recognized by the school should be incorporated into training programs and teaching plans [4]. While the country is vigorously promoting the construction of online open courses and the number of courses is growing rapidly, learners have higher and

© The Author(s), under exclusive license to Springer Nature Switzerland AG 2025
F. Paas et al. (Eds.): EduTA 2024, LAIS 51, pp. 48–58, 2025.
https://doi.org/10.1007/978-3-031-95252-4_5

higher requirements for the quality of online open courses. Speeding up the formulation of quality evaluation methods for online open courses and improving the construction of the quality evaluation system have become the top priorities for the development of online open courses.

There are many studies on the quality evaluation of online open courses at home and abroad, but there are few studies on the evaluation of online open courses from the perspective of users. Accurately obtaining the demands of online open course users is the key to successfully building an online open course quality evaluation index system based on user experience. This paper will use a variety of methods to obtain the demands table of online open course users and prepare for the subsequent quality evaluation research of online open courses.

2 Research Status

Scholars have conducted extensive research on the user demands of online open courses.For example, Li Xuelan [1] divides the users of online open courses into three major value subjects: managers, teachers and students. Then, the demands information of the three major value subjects is analyzed by means of literature analysis, questionnaire survey and other methods, and the demands of the value subjects are decomposed layer by layer by using the KJ method. Finally, the AHP method is used to obtain the weight of each demand, and the online course quality improvement strategy based on the demands of the value subject is obtained. Zhang Jianfei [2] divided the users of online open courses into students, peer teachers, teaching administrators, and experts. A multi-role and multi-dimensional evaluation framework have been established including 3 first-level evaluation indicators, 11 second-level evaluation indicators, and 34 third-level evaluation indicators. However, the weights of indicators at all levels have not been determined. Zhao Xiangqin [3] considers programmers in different positions, teachers of computer science, and students of computer and design departments as users of the programming language online learning platform. First, the basic demands of users for the online learning platform are obtained. Then, the functional requirements and weights of 26 kinds in 5 categories including responsive design, teaching teachers, teaching mode, teaching form and course service are obtained based on the Kano model method and the Better-Worse coefficient method. This provides a reference for the design and optimization of the programming language online learning platform. In order to analyze the real demands of the army for the function of the bullet cabinet, Yongliang [4] first analyzed the customer demands based on the KJ method, then built the KANO model and conducted a questionnaire survey to obtain customer demands data. Finally, the better-Worse coefficient method is used to analyze the data to obtain the demands table of the bullet cabinet function, which provides a reference for manufacturers to design the function of the bullet cabinet. From the research results of scholars, it can be seen that the users of online open courses can be basically divided into three categories: students, teachers and managers. Among them, the demands of students and teachers are closely related to the quality of course teaching, while the demands of managers have a smaller weight in the demands table of course quality [5, 6]. Therefore, this paper will take teachers and students as the main users of online open courses and build the demands table of online open courses based on the demands of the two major users, so as to provide demands data for subsequent research on the quality evaluation of online open courses.

3 Analysis of User Demands Based on KJ Method

The KJ method (also known as an Affinity Diagram) was proposed by the Japanese scholar Jiro Kawakita, which is a method of sorting out cluttered information [7, 8]. The basic steps of the KJ method are as follows:

① Information collection: Collect a large amount of fragmented information through meeting records, surveys, brainstorming and other methods. This information can be opinions, phenomena, suggestions, etc.
② Record information: Write each piece of information on a piece of paper. The information should be short and clear for subsequent understanding and processing.
③ Classification forms affinity groups: this information is classified by discussion method, and similar or related information is grouped into a group, which is an affinity group.
④ Name and label categories: Take a subject name for each affinity group and label it. The subject name of the affinity group should be concise and reflect the core content of the affinity group.
⑤ Form an affinity graph: all affinity groups are put together to form an affinity graph, which reflects the views, problems and relationships of each affinity group.
⑥ Analysis and summary: According to the affinity map formed in step 5, further analysis is conducted to identify the root cause of the problem, discover potential solutions, or provide a basis for subsequent decisions.

Many literatures show that using the KJ method to analyze user demands can effectively organize information and help designers clarify design ideas. It is a very useful requirement analysis method. In this paper, the KJ method is used to analyze the user demands of online open courses, and the process is shown in Fig. 1. First, select the appropriate user type and number of users to form the interviewee. Then by reading a lot of relevant literature, determine the user's original demand for online open courses. Obtaining the original needs of users from the literature can reduce the workload of research. Next, users are instructed to put forward various requirements for open online courses based on their own needs. The process is based on brainstorming. According to KJ method, the user needs are integrated and classified to form an affinity chart. The final affinity chart is a list of the needs of online open courses' users.

3.1 Identify Users of Online Open Courses

Teacher users of online open courses: refer to teachers who need to use the online open course platform to teach, mostly peer teachers. This can be a teacher from the same school as the curriculum development school or from a different school.

Student users of online open courses: refer to students who need to use the online open course platform for learning, including current students and social students, as well as new teachers.

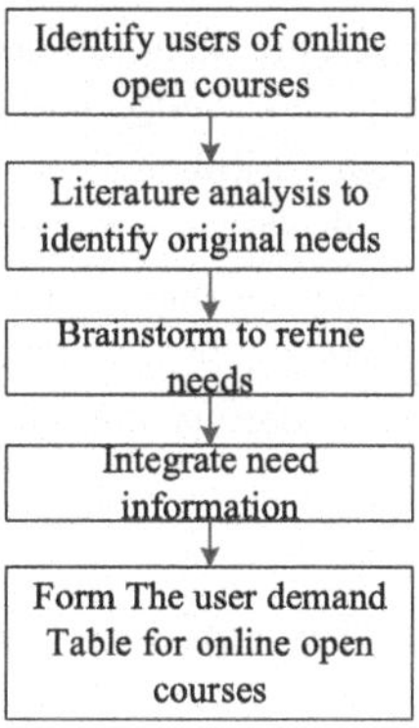

Fig. 1. User demands analysis process of online open courses based on KJ method

3.2 Literature Analysis to Obtain the Original User Demands Table

On the Chinese CNKI academic platform, search for references with keywords of "online open courses", "quality evaluation", and "demands analysis", and select 20 representative literatures as research samples. Using the literature analysis method, the user demands research in the reference literature is extracted and summarized, and the original demands table of online open course users is obtained as shown in Table 1. In the table, according to different user types, the needs are divided into two categories: teacher needs and student needs. Teacher needs are divided into six categories: course resources, teaching methods, interaction, course design, assessment method, teaching objectives. Student needs are divided into 8 categories, such as course resources, course design, course team, teaching methods, interaction, assessment method, learning interest, learning purpose. These requirements are the preliminary research results of researchers, and using these research results as the research basis of this subject can greatly shorten the research cycle.

It needs to be emphasized that the mainstream online open course platforms in China are mature (such as iCourse.com, Smart Vocational Education Platform, etc. [9, 10]), so user demands for platform functions are not investigated when researching.

3.3 Brainstorm to Refine the User Requirement Table

In order to have a deeper understanding of the demands of online course users, we organized 30 student representatives and 8 teacher representatives from different majors and grades in our school to conduct face-to-face brainstorming to obtain more detailed demands for online courses. Supplement Table 1 by adding the demands shown in the column "Brainstorm Supplement". For example, Teachers users supplemented "The video time is about 5 to 10 min, which can provide assistance for offline teaching; provide assistance for offline teaching; Teachers can master historical learning records from past discussion areas or Q&A areas; Knowledge design serves offline teaching; chapter titles are set reasonably; knowledge points can be found quickly; Objective subject; Highlight ideological and political goals; have expansion goals." Students users supplemented

"The resources of the same knowledge point should not be too many, and the content of the resources should not be repeated; Knowledge is easy to understand and interesting; Teachers have a passion for class, preferably with a sense of intimacy and a sense of humor; Adopt different teaching methods according to the characteristics of different knowledge points; No malicious or boring comments; There are test questions during the video playback; the test questions are few and precise; Set suspense to stimulate interest in learning; Set suspense to stimulate interest in learning; Timely issuance of certificates".

3.4 Integration of User Demands Information

It can be seen from Table 1 that many demands of teachers and student users are similar or repeated and need to be further integrated. The integration process is roughly merging synonyms, removing duplicate demands, and standardizing the terms of requirements. For example, teachers and students' demands for course resources include "rich course resources", "mainly video resources", and "clear and smooth video". The demands of the two can be integrated into "rich course resources, video resources mainly, and high video quality".

The KJ method is used to integrate the demands in Table 2, and the 25 secondary demands indicators of online open course users shown in Table 2 are obtained. The secondary demands for course design are "clear teaching goals, Reasonable teaching progress, Reasonable and progressive knowledge structure, Practical and understandable knowledge points, have expanded knowledge". The secondary demands for course resources are "Abundant resources but not redundant, mainly video resources, Clear and smooth video, how teachers appear". The secondary demands for teaching methods are " Various teaching methods, combining theory and practice, Teaching process embedded in video". The secondary demands for interaction are "Timely teacher Q&A, Active communication area, No malicious swiping in the comment area". The secondary demands for assessment method are "Moderate difficulty and reasonable assessment test questions, Reasonable grade composition, Test questions based on objective questions". The secondary demands for course team are "good teacher image, teachers have passion, High professional level of teachers, Teachers spread positive energy". The secondary demands for learning interest are "arouse interest in learning, achieve learning goals, Get the certificate in time".

3.5 Determination of the Weight of User Demands Indicators

A complete user demands table includes not only the user's demands indicators, but also the weight of each demand indicator in order to better design or evaluate whether the product meets the user's requirements. Most scholars use the expert scoring method to determine the weight of each demand indicator [11–15]. As mentioned in the preface; in order to construct an online open course quality evaluation index system based on user demands, it is necessary to obtain the real demands of course users. Therefore, the weight of user demands should not be determined by experts but should be determined by teachers and students.

Table 1. Demands table of online open course users

User Type	Demands Indicators	Specific Description	Brainstorm Supplement
Teacher	Course Resources	Rich teaching resources; micro-course resources as the main;	The video time is about 5 to 10 min, which can provide assistance for offline teaching;
	teaching methods	Teaching methods are diverse;	Preferably project-based or modular teaching
	Interaction	Teachers can post discussion questions and answer questions;	Teachers can master historical learning records from past discussion areas or Q&A areas
	Course Design	The course content is reasonable; the knowledge points are small but precise, easy to understand and practical	Knowledge design serves offline teaching; chapter titles are set reasonably; knowledge points can be found quickly
	Assessment method	A variety of assessment methods to assess the effectiveness of online learning;	Objective subject
	Teaching Objectives	The teaching objectives are clear and in line with the knowledge objectives, skills objectives and literacy objectives of the talent training plan; teaching has process control and result control;	Highlight ideological and political goals; have expansion goals;
Students	Course Resources	The teaching resources are abundant; the resources are mainly video, animation, etc.; the time is moderate; the video is clear and smooth;	The resources of the same knowledge point should not be too many, and the content of the resources should not be repeated
	Course Design	The design of knowledge points is step-by-step, with difficulty levels and extended learning; the knowledge is novel and practical;	Knowledge is easy to understand and interesting
	Course Team	Teachers with good image, standard Mandarin, decent dress and high professional level;	Teachers have a passion for class, preferably with a sense of intimacy and a sense of humor
	Teaching methods	Diversified teaching methods; combination of theory and practice;	Adopt different teaching methods according to the characteristics of different knowledge points
	Interaction	Teachers answer questions timely; the communication area is active and positive;	No malicious or boring comments
	Assessment method	Reasonable test questions and reasonable score composition	There are test questions during the video playback; the test questions are few and precise
	Learning interest	arouse interest in learning	Set suspense to stimulate interest in learning
	learning purpose	Achieve knowledge and skills goals; earn certificates	Timely issuance of certificates

Use the Questionnaire Star platform to set up questionnaires and distribute questionnaires to course users. To scientifically obtain the ranking of the importance of each indicator by users, all the questions in the questionnaire adopt the ranking question type.

Table 2. Secondary demands table of online open course users

Primary needs	Secondary needs	Number
Course Design A1	Clear teaching goals	A11
	Reasonable teaching progress	A12
	Reasonable and progressive knowledge structure	A13
	Practical and understandable knowledge points	A14
	Have expanded knowledge	A15
Course Resources A2	Abundant resources but not redundant	A21
	Mainly video resources	A22
	Clear and smooth video	A23
	How teachers appear	A24
Teaching methods A3	Various teaching methods	A31
	Combining theory and practice	A32
	Teaching process embedded in video	A33
Interaction A4	Timely teacher Q&A	A41
	Active communication area	A42
	No malicious swiping in the comment area	A43
Assessment method A5	Moderate difficulty and reasonable assessment test questions	A51
	Reasonable grade composition	A52
	Test questions based on objective questions	A53
Course Team A6	Good teacher image	A61
	Teachers have passion	A62
	High professional level of teachers	A63
	Teachers spread positive energy	A64
Learning interest A7	Arouse interest in learning	A71
	Achieve learning goals	A72
	Get the certificate in time	A73

Too many questions and too many options will wear down the patience of the respondents, so the number of questions is designed to be 8, and the options to be sorted for each question are controlled within 3 to 5.

Through statistical analysis of the 69 questionnaires collected, the weight of the importance of the primary demands and the weight of the importance of the secondary demands to the primary demands are determined. Figure 2 is the user scoring results of the importance of the primary demands. Figure 3 shows the user scoring results of the secondary demands of "course design".

The average score for each option in Figs. 2 and 3 is obtained using formula (1):

$$S = (\Sigma f \times W)/n \tag{1}$$

Among them, S represents the average score of a single option, f represents the frequency, W represents the single weight, and n represents the number of people who answered this question.

Using formula (2), the final importance weight of the secondary indicator can be obtained.

$$Z = X * Y \tag{2}$$

where X is the weight of the importance of the primary needs, Y is the weight of the importance of the secondary demands to the primary needs, Z is the final importance weight of the secondary indicator.

For example, in Fig. 2 the score of "course design" is 4.43. In Fig. 3 the importance score of "clear teaching objectives" in "course design" is 3.78. According to formula (2), the final importance weight of "clear teaching objectives" in all demands is 4.43 * 3.78 = 16.7454.

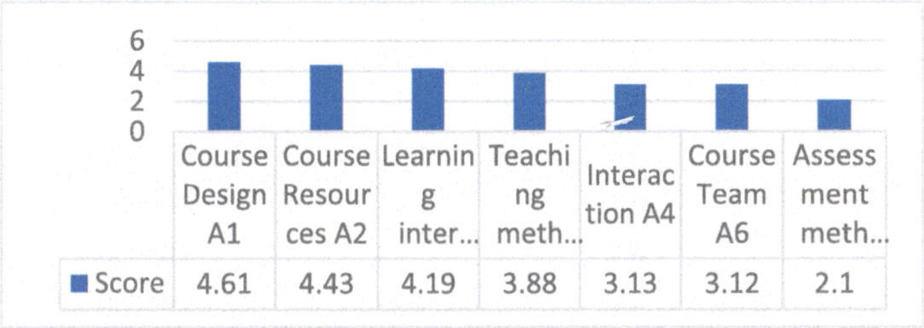

Fig. 2. User scoring results of the importance of primary demands

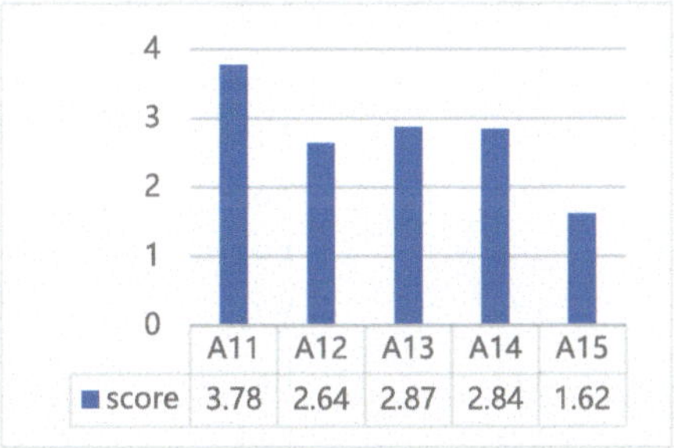

Fig. 3. User scoring results of the secondary requirements "Course Design"

Finally, the user demands table of online open course is obtained, as shown in Table 3. The secondary requirements are sorted according to the score. The higher the score, the higher the ranking, the higher the importance. The top three demands are "clear teaching objectives", "Reasonable and progressive knowledge structure", and "abundant resources but not redundant".

Table 3. User demands table of online open course

Primary demands	Weight of primary demands	Secondary demands	Weight of secondary demands relative to primary demands	Weight of secondary demands	Sort
A1	4.43	A11	3.78	16.74540	1
		A12	2.64	11.69520	5
		A13	2.87	12.71410	3
		A14	2.84	12.58120	4
		A15	1.62	7.17660	16
A2	4.61	A21	3.19	14.70590	2
		A22	2.13	9.81930	8
		A23	2.19	10.09590	7
		A24	1.77	8.15970	11
A3	3.88	A31	2.42	9.38960	9
		A32	2.1	8.14800	12
		A33	1.12	4.34560	23
A4	3.13	A41	2.45	7.66850	14
		A42	1.93	6.04090	19
		A43	1.25	3.91250	24
A5	2.1	A51	2.39	5.01900	21
		A52	1.99	4.17900	18
		A53	1.22	2.56200	25
A6	3.12	A61	2.52	7.86240	13
		A62	2.16	6.73920	17
		A63	2.68	8.36160	10
		A64	1.93	6.02160	20
A7	4.19	A71	2.71	11.35490	6
		A72	1.81	7.58390	15
		A73	1.14	4.77660	22

4 Conclusion

Accurately obtaining the demands of online open course users is the premise of designing and evaluating the quality of an online open course. This paper adopts a variety of methods to study the user demands of online open courses. Firstly, the original demands of teachers and students are obtained by literature analysis and brainstorming. Then the secondary user demands table is sorted out by the KJ (affinity graph) method. Finally, the weight of the importance of each demand is obtained by using the questionnaire survey method and the user scoring method. The final user demands table of online open course consists of two levels of user demands and the weight of the importance of each demand.

From Table 3, it can be seen that teachers and student users have the highest expectations for the demands of "clear teaching objectives", "Reasonable and progressive knowledge structure", and "abundant resources but not redundant". This calls for significant attention from course developers and course evaluators. "Clear teaching objectives" requires course developers to be familiar with the vocational positions and abilities of the course, and to clarify the knowledge objectives, ability objectives and literacy objectives of the course. Only in this way can the overall design of the course be carried out. "Reasonable and progressive knowledge structure" requires course developers to reconstruct and design the course content according to the degree of difficulty. "Abundant resources but not redundant" means that the teaching resources are rich and cover all the knowledge and ability target points. However, to reduce the burden of learners, it is not recommended that teaching resources be redundant, which will reduce the enthusiasm of learners.

The research results of this paper will provide data reference for the team's subsequent research on the quality evaluation of online open courses.

Acknowledgment. This work was supported by planning special project of Hunan Province Education Science in 2024 (No. XJK24BZJ018).

References

1. Li, M., Chen, W., Pei, Z., Cao, Z.: Analysis of the value subject demands of online open courses in colleges and universities based on the quality house (QFD). J. Hubei Norm. Univ. (Philos. Soc. Sci.) **41**(06), 100–108 (2021)
2. Zhang, J., Hu, Y., Ji, X.: Research on the quality evaluation system of online open courses–based on the sharing platform of high-quality online open courses in Zhejiang Province. J. Continuing High. Educ. **30**(06), 19–24+37 (2017)
3. Zhao, X.: Functional requirements analysis of programming language online learning platform based on Kano model. Packag. Eng. **2**, 1–11 (2022)
4. He, Y.: Research on the function design of bullet cabinet based on KJ method and KANO model. Packag. Eng. **3**, 1–12 (2022)
5. Yang, X., Zhou, H., Zhou, X.: Review of domestic online course quality accreditation research. e-Educ. Res. **40**(06), 50–57 (2019)
6. Zhao, X., Zhou, Y., Linyun, Q., Zhang, Z.: Teaching quality evaluation of college physics MOOC based on fuzzy comprehensive evaluation method. Phys. Eng. **28**(5), 28–38 (2018)
7. Merisotis, J.P., Phipps, R.A.: Quality on the line: benchmarks for success in internet-based distance education (2017). http://www.ihep.org/sites/default/files/uploads/docs/pubs/qualityontheline.pdf
8. Khan, B.H.: Web-based training, pp. 76–95. Educational Technology Publications, USA (2001)
9. Britain, S, Liber, O.: A framework for pedagogical evaluation of virtual learning environments. http://www.jtap.ac.uk/reports/htm/jtap-041.html,2017

10. Qin, J.: Research on The Quality Evaluation of Entrepreneurship Education in Colleges and Universities Based on QFD. Tianjin University, Tianjin (2009)
11. Li, Q., Liu, N.: Research on MOOC quality assurance system. Open Educ. Res. **5**, 66–73 (2015)
12. Qiu, J., Yufang, O.: Research on the construction and application of MOOC quality evaluation index system. High. Educ. Dev. Eval. **5**, 72–81 (2015)
13. Huang, R., Ren, Y.: The Theory and Practice of Promoting High-Quality Educational Resources Sharing Through Informatization. Higher Education Press, Beijing (2017)
14. Dong, Y., Huang, R., Li, X.: Development and revision of "quality certification standards for online courses" Res. Electron. Educ. **6**, 65–70 (2003)
15. Tong, X., Jia, X.: Exploration on the construction of MOOC quality evaluation system. China Distance Educ. **5**, 63–71 (2017)

Design and Implementation of College Oral English Teaching System Based on Web Technology

Mingyi Ling(✉)

College of Humanities and Social Sciences, Sanya Aviation and Tourism College, Sanya 572000, Hainan, China
zzqklmyx@163.com

Abstract. Online oral English teaching needs to combine intelligent algorithms for data processing to promote the intelligent effect of the oral English teaching system. The purpose of this paper is to improve the efficiency of the oral English teaching system by combining web technology. Moreover, this paper proposes a new scheme for server status and request information, which can reduce the response time. In addition, this paper constructs an oral English teaching system based on the improved algorithm, and verifies the effect of the model proposed through simulation experiments and comparative experiments. Through experimental analysis, the oral English teaching system based on web technology has a faster data processing effect and can play an important role in oral English teaching.

Keywords: Web · English · oral · teaching system

1 Introduction

Teachers can give full play to the advantages of online teaching and effectively explain the teaching content in online English teaching. Meanwhile, the performance of teachers in online English classes is also more active and active. At the same time, students can ask questions that they don't understand in class, and teachers can also understand students' mastery of knowledge by asking questions.

However, compared to traditional classroom teaching, classroom interaction in online teaching mode is more frequent, and students will face more problems in learning. How to enable students to actively participate in learning in online classrooms, and how to make the effectiveness of online teaching the same or even better than traditional classroom teaching, are issues that every online educator must consider [1].

Before conducting online teaching, teachers need to choose suitable online teaching software in advance. These software have strong interactivity and fun, and students will have great interest in the learning process. In addition, there are various types of remote education learning software, and teachers need to choose according to the teaching environment and their own situation in their region. For example, teachers can

© The Author(s), under exclusive license to Springer Nature Switzerland AG 2025
F. Paas et al. (Eds.): EduTA 2024, LAIS 51, pp. 59–68, 2025.
https://doi.org/10.1007/978-3-031-95252-4_6

search for online education platform resources online or directly search for schools or institutions in their city [2].

In online teaching, teachers should use more online learning software and make full use of the online course resources provided by online platforms. Teachers can provide remote explanations for students to watch online. In this way, students can choose to learn according to the teaching content. In this way, students can record the problems they encounter and their own ideas at any time, and teachers can also use the software's online broadcasting function for learning discussions [3].

During the broadcasting process, students can interact by recording videos or voice, which can directly reflect their learning outcomes. In a learning discussion, teachers can discuss a certain issue, but due to the large number of people and students speaking more, in order not to affect the learning progress of other students, teachers can record the entire discussion process and play it out during the broadcast, allowing other students to participate. This function can not only help other students answer questions and record solutions to problems, but also help teachers timely grasp the learning situation of students, avoiding confusion and anxiety caused by students thinking alone. It is also conducive to stimulating students' interest and enthusiasm for knowledge learning, allowing them to participate more actively in online classrooms [4].

Due to the fact that online teaching is carried out through online platforms, it is necessary to attach importance to student participation and learning outcomes in the course. In online teaching, students need to learn relevant knowledge and acquire it through teaching videos and other means, which are taught by teachers. However, in this situation, students' learning methods are passive and they are unable to actively participate in the classroom, which leads to unsatisfactory teaching outcomes. Therefore, teachers should pay attention to mobilizing students' enthusiasm in teaching [5].

Teachers need to carefully design questions. When designing problems, teachers need to analyze the content of the textbook as comprehensively and deeply as possible, and students can be divided into small groups for discussion. Secondly, teachers should create a relaxed and harmonious classroom atmosphere. Finally, when teaching knowledge points, teachers should pay attention to appropriately dispersing and transferring the content of classroom teaching, and avoid focusing their attention on a particular knowledge point. By adopting these methods, teachers can not only fully mobilize the enthusiasm of students to participate in teaching activities, but also enhance their initiative and enthusiasm for learning [6].

Practical research has found that in the full-time online teaching mode, students' mastery of the knowledge they have learned is not very ideal. Teachers can use these issues to promote students' learning and improve their learning efficiency. The low participation and enthusiasm of students in online teaching is an important reason for the poor teaching effectiveness. Through the setting of these questions, arouse students' interest in the classroom content, thereby improving teaching effectiveness. For example, when explaining the relevant knowledge points of organic matter, teachers can design some questions for students to answer, such as having students orally explain the calculation method of organic matter unsaturation, or having a few classmates write silently on the blackboard. This can not only increase students' understanding and understanding of the teaching content, but also stimulate their interest and enthusiasm for learning [7].

In order to enhance teaching effectiveness, in online teaching, teachers should try to increase classroom interaction as much as possible and strengthen student subjectivity. For example, in the classroom, having students discuss a topic; Students complete a simple exercise based on the teacher's explanation; Assign some small tasks for students to complete, etc. [8]. These can enhance communication and interaction, as well as between students, enabling students to gain more information through communication with teachers. The improvement of student subjectivity can promote teachers to better carry out teaching work. The online teaching mode improved learning efficiency. However, this teaching model also has some drawbacks, and teachers need to continuously improve and perfect it in practice, so that this online teaching can better serve the education industry [9].

At present, in online teaching, the supervision of student learning mainly relies on teachers monitoring the learning situation of students. More and more people are starting to learn through live streaming, and the number of students learning through the internet is constantly increasing, making it difficult for teachers to effectively supervise these students [10]. In addition, it is difficult for students to maintain a good learning state in online teaching because their self-directed learning ability is relatively poor, and they are unable to persist in learning in this teaching mode [11].

Therefore, in the online teaching mode, schools need to fully attach importance to supervising and managing students, and communicate and contact parents in a timely manner. In addition, schools should regularly assess and evaluate the teaching ability and level of teachers.

2 Web Load Balancing in Oral English Education System

This paper tests the performance of the server from both dynamic and static aspects. This study specially screens and extracts some parameters and data that can further show the quality of the server hardware configuration.

To ensure good server performance, it is necessary to reduce the number of connections to the server as much as possible. It indicates that the server is overloaded. The request load of the server j can be determined by formula (1):

$$\mathrm{Load[j]} = \sum\nolimits_{i=1}^{N} \mathrm{R(i)[j]^{*}w[i]} \tag{1}$$

If $R(i)[j]$ means that request $R(i)$ is connected to server j, then the request payload for server j is formula (2):

$$Load[j] = \sum\nolimits_{i=l}^{N} R(i)[j]^{*}w[i]w + [k] \tag{2}$$

If request $R(i)$ has been processed by server j, the request payload for server j is:

$$\mathrm{Load[j]} = \mathrm{Load[j]} - \mathrm{w[i]} \tag{3}$$

Next, this paper explains the judgment method of load balancing: This method extracts all the contents of the cluster server, and then compiles a record array.

$$L = \alpha \times \frac{\lambda l_i \times C_i}{\lambda l_M \times C_m} + \beta \times \frac{\lambda 2_i \times M_i}{\lambda 2_M \times M_M} + \gamma \times \frac{N_i}{N_M} + \theta \times \frac{R_i}{R_M} \tag{4}$$

In formula (4), λl_M represents the CPU processing power, λl represents the CPU processing power, λl_i represents the processing power of i CPUs, $\lambda 2_M$ represents the memory parameter of the reference server, $\lambda 2$ represents the memory parameter, and $\lambda 2_i$ represents the memory parameter of the i-th server. The CPU usage is represented by C, C_i represents CPU usage of the i-th server, then C_M is the base server CPU usage. The memory usage is represented by M, so M_i represents server memory of the i-th server and M_M represents base server memory usage. R represents the request load, the hard disk transmission amount of the reference server is represented by R_M, and the hard disk transmission amount of the i-th server is represented by R_i. N represents network traffic, reference server network traffic is represented by λl_M. N_M, and network traffic of the i-th server is represented by N_i. γ represents the weight of the network comparison, α represents the weight of the CPU comparison, β represents the weight of the memory comparison, and θ represents the weight of the load comparison.

When comparing server CPU, load performance, memory, and network, the dynamic load and static parameters of hardware need to be carefully considered. Based on the list of cluster servers, it was found that the existing servers with the lightest load face the following challenges: firstly, if two heavier loads are placed on the same server, it is highly likely to reduce the response efficiency of server clients; The second issue is that server load calculation is continuous, which can to some extent affect server performance. In the face of the above difficulties, the minimum load strategy can be adopted, and a random probability model can be selected. By calculating and comparing weights, the probability space of each server can be reasonably set. Assuming the allocator receives a new request task, calculate a random number between [0, 1]. If the relative weight of the server is small, its probability space must be relatively large, and more request tasks can be allocated with a higher probability of allocation. Each forwarding belongs to an independent event, and the final forwarding will not interfere with it. Similarly, performing another forwarding operation will not affect each other. Therefore, it can improve the forwarding effect and ensure the stability of the server, which can further reflect the optimized effect.

3 System Construction and Test

3.1 System Design

The entire architecture can be divided into three services: file service, video service and Web service from the perspective of service framework. End users can directly access these three services, as shown in Fig. 1. There are MySQL database service cluster, Tomcat application service cluster and storage in the back-end of Web service architecture. The static file access service in the Tomcat application service is in a detached state and reaches the user terminal via the file service.

Based on the classification of the above experimental process, this study proposes the basic process of Web3D oral English inquiry learning (as shown in Fig. 2): the platform or educator presents the research topic to the learner, borrows technology to build a learning situation that the learner is interested in, then learners independently put forward hypotheses for the question and design the experimental plan, and then explore and verify the answers to the question, and finally draw conclusions and conduct self-evaluation and

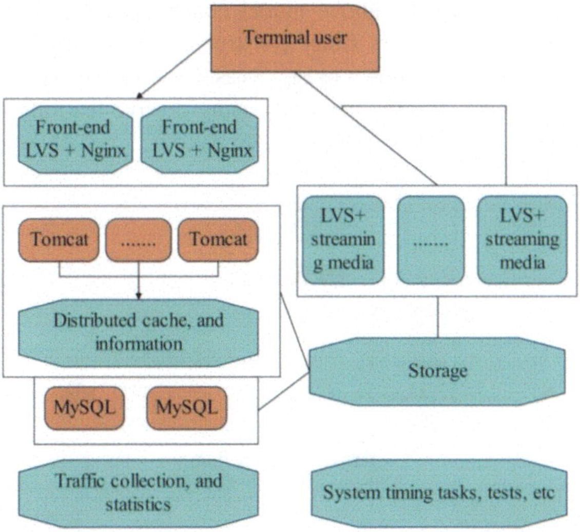

Fig. 1. Overall architecture of the Web platform.

reflection. In addition, the inquiry-based learning process not only needs to stimulate learners' thirst for knowledge, but also has certain learning effects. Therefore, pre-test and post-test links are added to understand the learners' experimental preparation knowledge and knowledge understanding and construction. Considering that the virtual experiment is not only suitable for the classroom environment with explanations, but also suitable for the autonomous learning environment, the knowledge preview link is added.

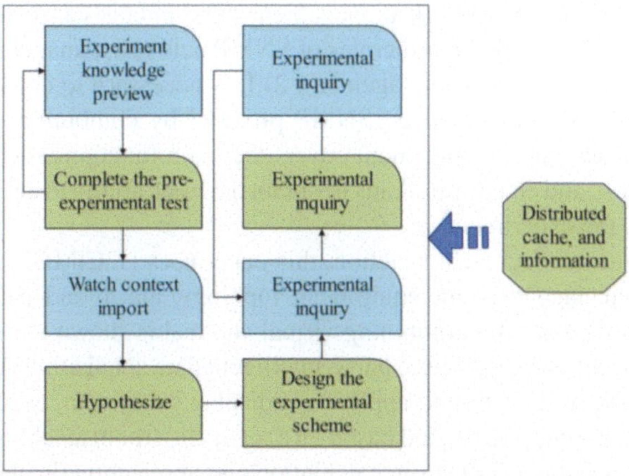

Fig. 2. The basic process of Web3D virtual oral English inquiry learning.

Based on the elements of the above model construction and the analysis of the inquiry-based learning process, this study constructs the learning behavior model of Web3D oral English inquiry (as shown in Fig. 3) as the infrastructure for collecting and recording learners' inquiry learning activities using diverse learning resources in the learning environment.

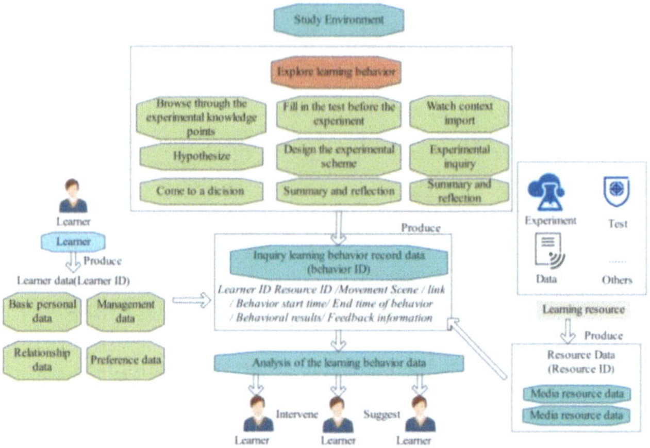

Fig. 3. Learning behavior model of Web3D oral English inquiry.

As shown in Fig. 4. The overall framework of the platform adopts a hierarchical design idea to realize the independence of different levels and ensure the high stability, practicability and scalability of the system.

To comprehensively promote the innovation of network management teaching means and the construction of training platform, the following methods and means are adopted: 1) The implementation of network management experiments can enable students to deeply understand and verify the principles of SNMP network management, and master the specific process of network management. 2) It is necessary to construct a network management model system based on SNMP protocol by combining virtual and real means, so that the network management covers the main functions such as equipment, topology, and alarm, and enables students to obtain the basic ability to effectively manage the network after the experiment.

For the network management function, this paper uses H3CiMC network management system to manage network equipment, topology, alarm and performance. The logical network structure after combining virtual and real is shown in Fig. 5.

The system mainly includes three parts: intelligent data acquisition subsystem, intelligent data analysis subsystem and application display subsystem, as shown in Fig. 6. The three subsystems are constructed based on a set of classroom teaching analysis index system, and complete data collection, calculation and application display respectively.

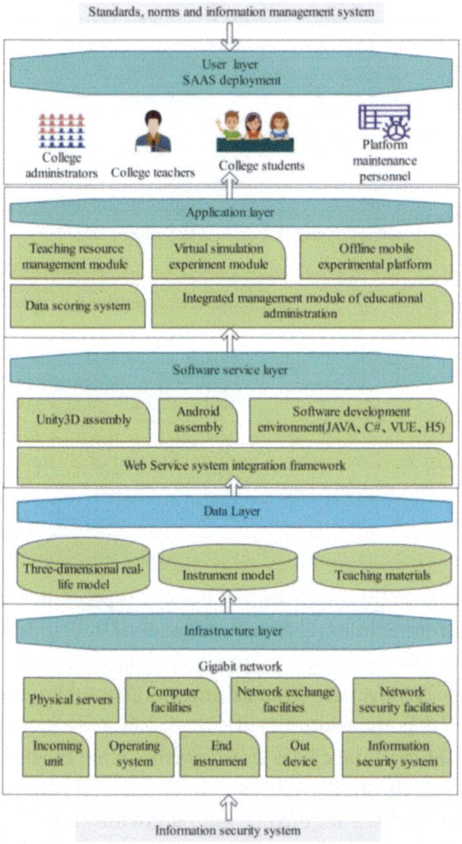

Fig. 4. Architecture of network teaching platform.

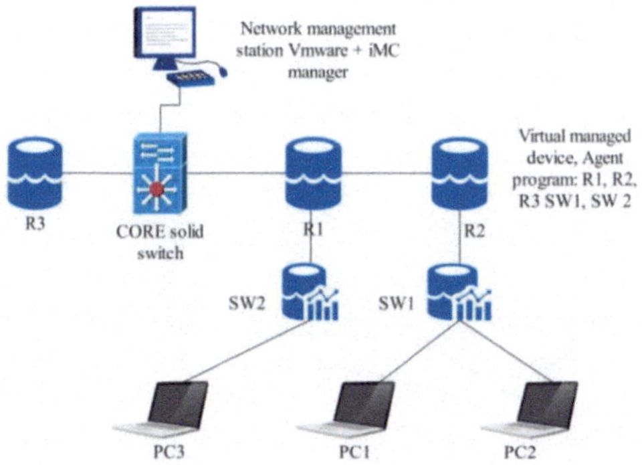

Fig. 5. Logical topology of virtual-real network management experimental model.

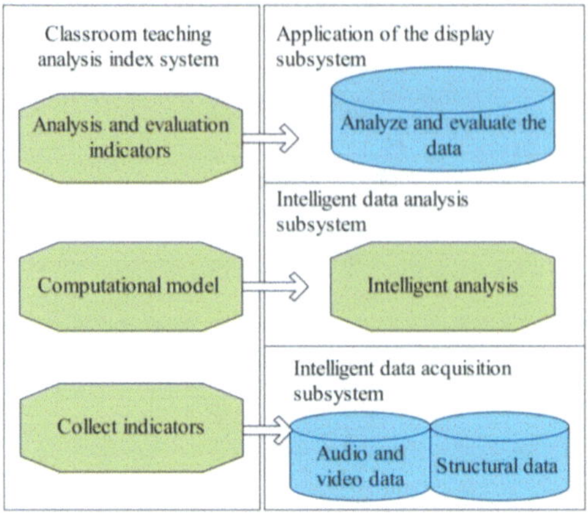

Fig. 6. Logical architecture diagram of System

3.2 Experiments

To verify the effectiveness of this algorithm, the cache replacement algorithm written in c language is used. The average interval time parameter of the guide is set 0.5. Considering the time limit of visiting the web page, a greater weight is given to the last visit time. At the same time, the previous average interval time is also taken into account, and 2/5 of the data block is selected as the training set.

Considering the locality of the cache space, the data that the user may use is extracted from the server in advance through prediction, and replaced in the cache space to greatly reduce the user's broadband waste, ensure the bandwidth index results, and accelerate network access. In this paper, the prefetch threshold is set to 5, and the performance of the proposed cache replacement algorithm is evaluated by analyzing the wastage of broadband under different prefetch thresholds.

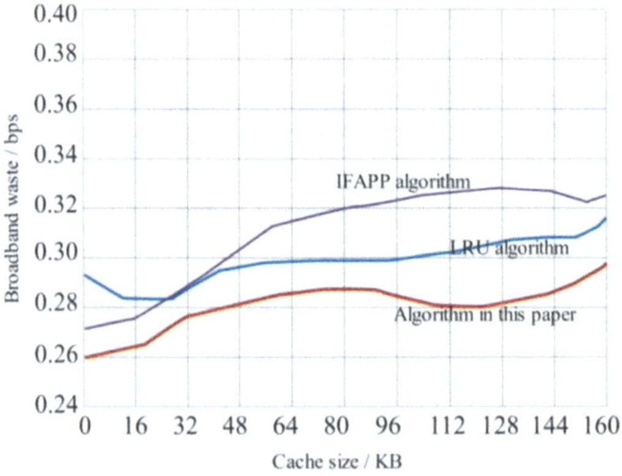

Fig. 7. Bandwidth waste.

In Fig. 7, the broadband waste of the method proposed is low. When the cache size is 160 KB, the broadband waste is 0.28 bps, while the broadband waste of the LRu algorithm and the IFAPP algorithm is 0.3 bps and 0.33 bps respectively. This is because the method proposed uses PageRank algorithm to equalize the average cost of web site access. By enhancing the autonomy of edge nodes, it improves the control effect of Http protocol header, reduces the average cost of web site access, and ensures the bandwidth index result.

Two classes with the same initial oral English scores are used to conduct an experimental teaching for a term. The experimental class adopts the system proposed in this paper for oral English teaching, and the control class adopts the traditional oral English teaching model. Moreover, each class has 20 students. Comparison of oral English scores is shown in Table 1.

Table 1. Comparison of oral English scores.

Num	Experimental group	Control group	Num	Experimental group	Control group
1	93	87	11	86	67
2	93	87	12	84	66
3	92	86	13	83	65
4	92	83	14	82	63
5	92	83	15	81	60
6	91	78	16	81	58
7	91	73	17	80	57
8	91	73	18	79	50
9	90	71	19	78	50
10	89	68	20	78	49

In Table 1, the oral English scores in the experimental class are generally higher than those in the control class after the experiment. The college oral English teaching system based on Web technology can play an important role in college oral English teaching and effectively improve the students' oral English level.

4 Conclusion

This paper constructs a set of oral English classroom teaching system suitable for the intelligent teaching environment of college education, and realizes a digital classroom teaching analysis system. The system can intelligently collect and analyze the whole process data of teachers' behavior and students' behavior without teachers and students' perception. Moreover, the system realizes the accompanying data collection of the whole process of the classroom. Based on artificial intelligence analysis technology, it can collect and analyze data in multiple dimensions to help teachers carry out classroom teaching analysis and precise teaching research based on specific data, so as to carry out evidence-based in-depth teaching reflection and teaching iteration design. It has certain theoretical and practical value for enriching the evaluation theory of oral English teaching and exploring the analysis method of digital classroom.

References

1. Sun, Z., Anbarasan, M., Praveen Kumar, D.J.C.I.: Design of online intelligent English teaching platform based on artificial intelligence techniques. Comput. Intell. **37**(3), 1166–1180 (2021)
2. Fitria, T.N.: Teaching English through online learning system during Covid-19 pandemic. Pedagogy J. Engl. Lang. Teach. **8**(2), 11–21 (2020)
3. Chen, Y.: [Retracted] college English teaching quality evaluation system based on information fusion and optimized RBF neural network decision algorithm. J. Sens. **2021**(1), 6178569–6178580 (2021)
4. Sun, R., Zhang, H., Li, J., Zhao, J., Dong, P.: Assessment-for-learning teaching mode based on interactive teaching approach in college English. Int. J. Emerg. Technol. Learn. (iJET) **15**(21), 24–39 (2020)
5. Sholihah, H.I.A., Hidayat, A.W., Srinawati, W., Syakhrani, A.W., Khasanah, K.: What linguistics advice on teaching English as a foreign language learning using blended learning system. Linguist. Cult. Rev. **5**(1), 342–351 (2021)
6. Wang, L., Xu, Z.: An English learning system based on mobile edge computing constructs a wireless distance teaching environment. Mob. Inf. Syst. **2021**(1), 2718859–2718870 (2021)
7. Rostami, F., Yousefi, M.H.: Iranian novice English teachers' agency construction: the complexity dynamic/system perspective. Asian Pac. J. Second Foreign Lang. Educ. **5**(1), 4–14 (2020)
8. Chen, S.: Design of internet of things online oral English teaching platform based on long-term and short-term memory network. Int. J. Contin. Eng. Edu. Life Long Learn. **31**(1), 104–118 (2021)
9. Hakim, M.A.R., Serasi, R., Revola, Y., Adnan, N.I., Astari, A.R.N.: The reconstruction of blended learning: teaching model for full online learning system: a study on English education department in islamic universities in Indonesia. Int. J. Learn. Teach. Educ. Res. **22**(4), 1–20 (2023)
10. Shang, W.L.: Application of machine learning and internet of things techniques in evaluation of English teaching effect in colleges. Comput. Intell. Neurosci. **2022**(1), 7643006–7643016 (2022)
11. Li, M.: Multidimensional analysis and evaluation of college English teaching quality based on an artificial intelligence model. J. Sens. **2022**(1), 1314736–1314748 (2022)

Research on Multi-interactive Teaching Model of College English Listening and Speaking Based on Intelligent Terminal

Meng Zheng(✉)

Department of Foreign Languages, Chengdu Neusoft University, Chengdu 611700, China
zm887625@163.com

Abstract. At present, online English listening and speaking teaching needs to strengthen the interaction between teachers and students, the interaction mode needs to be innovated, and the research level needs to be deepened. In order to extract the local and global information of the speech spectrum, better capture the spectrum structure of speech signals, and realize speech enhancement in low SNR and unfamiliar environments, this paper proposes a time-frequency domain speech enhancement network based on parallel multi-attention mechanism and codec structure. Experimental results show that compared with classic and mainstream models, the model proposed in this paper has achieved better enhancement effect. Moreover, the model proposed in this paper is effective in English listening and speaking teaching. Therefore, multi-interactive English teaching provides a new way of thinking for listening class, and it improves listening from the perspective of communication and interaction, and creates a real scene of listening and speaking to improve students' comprehensive English application ability.

Keywords: intelligent terminal · English · listening and speaking · multiple interaction

1 Introduction

Under the background of the information age, higher education is actively optimizing and innovating the teaching concepts and teaching models of various disciplines. For a long time, college English education has focused on cultivating students' English practical ability and exploring more scientific, good and effective teaching models.

With the rapid development of digitization, networking, and intelligence, technical support has been provided to create efficient smart classrooms, creating a smart, flexible, and convenient teaching environment, thereby improving the teaching experience and optimizing the learning process [1].

In daily teaching, teachers gradually improve their ability to use smart teaching platforms, closely connecting pre class, in class, and after class, fully optimizing digital resources, enhancing their teaching wisdom, innovation, and practical application abilities. This lays a practical foundation for the transformation of future scientific research

© The Author(s), under exclusive license to Springer Nature Switzerland AG 2025
F. Paas et al. (Eds.): EduTA 2024, LAIS 51, pp. 69–81, 2025.
https://doi.org/10.1007/978-3-031-95252-4_7

achievements. In the classroom, students are able to participate in classroom activities with full enthusiasm. They actively share, showcase, listen, watch, and solve doubts, achieving "reducing time and increasing efficiency" in their learning. And it promotes the improvement of students' self-learning ability. Teachers use the Honghe blackboard with an intelligent teaching platform for teaching, which combines universality and practicality; Intelligent podium paired with central control platform. Synchronize control of interactive display device screens using a miniature "blackboard" to achieve efficient and convenient classroom operations. The smart teaching platform can provide real-time and intelligent data feedback on students' listening and speaking abilities throughout the entire process [2]. The entire evaluation process is more diverse, objective, and specific. Teachers only need to focus on reviewing the results, analyzing and understanding each student's errors and shortcomings based on the information feedback from the data. Then provide personalized guidance to students in a targeted manner. So as to help teachers accurately manage the classroom [3].

The intelligent teaching platform has the functions of screen broadcasting, preview feedback, audio and video teaching, student monitoring, classroom testing, quick answer, file distribution and collection, Internet access, video live broadcast, remote command, classroom scoring, etc. Based on these functions, teachers can personalize teaching activities for students, including following reading training, dictation training, fill in the blank exercises, repeat listening, listening selection, reading aloud training, oral expression, group discussions, dubbing, etc. Teachers distribute materials before class. Students browse, preview, and annotate key and difficult points. The intelligent teaching platform intelligently analyzes and displays the marked key and difficult points in the form of a list, and then automatically matches the knowledge point courseware for teachers to explain in class. In class, the teacher leads Word text from outside [4]. Focusing on general English, the platform automatically generates audio for students to practice reading. The intelligent teaching platform automatically calculates the results of reading in the form of charts and displays grades, allowing students to engage in self listening, mutual listening, and peer evaluation; Teachers import the Word text of TED speeches from external sources, set playback intervals and times, and students practice dictation. The intelligent teaching platform automatically calculates and displays the details of the dictation results report. Teachers evaluate and record grades: teachers issue oral expression content requirements and organize group discussions, while students engage in oral expression training. The intelligent teaching platform comprehensively evaluates the completeness, accuracy, rhythm, and fluency of oral expression, and then teachers evaluate, provide feedback, and record grades; After class, teachers use intelligent teaching platforms to reorganize test papers, and students complete listening and speaking tests [5].

The smart teaching platform provides strong technical support for the reform and innovation of audio-visual and oral courses, breaking the previous limitations of listening exercises, oral dialogue exercises, and single evaluation methods that were only limited to objective questions. The smart teaching platform provides a large amount of online resources as auxiliary materials for textbooks, achieving a reliance on textbooks and increasing the integration of ideological and political elements, cross-cultural communication, and traditional Chinese cultural elements. The smart teaching platform has built a bridge for integrated learning before, during, and after class. The smart teaching platform provides real-time learning feedback and evaluation, achieving the scientificity and impartiality of process assessment [6].

Teachers fully utilize various advanced information technologies and abundant teaching resources to create a student-centered smart teaching environment through online teaching management platforms, creating conditions for resource construction, environmental construction, and connotation construction. Under the guidance of the "output oriented approach", set up integrated in class and out of class resources and create a good interactive learning environment [7]. Teachers use smart teaching platforms to develop multimedia teaching resources, including teaching videos, audio materials, online exercises, and simulation training, to guide students in listening and speaking exercises such as listening and reading, follow-up reading, oral testing, and debate, forming a diverse interactive teaching mode of student student interaction, teacher-student interaction, and human-machine interaction [8].

Teachers use smart teaching platforms to simulate academic seminars. They learn how to interact with others during the seminar, how to articulate their own views, and how to identify details, record key points, and seize the appropriate time to ask questions during academic lectures [9]. The teaching teacher first explains the key points to pay attention to, teaches students key sentence patterns, and then allows students to watch relevant academic videos in the multimedia classroom through the internet. When encountering details that need attention, the teacher stops to explain and guides students to take good notes. In addition, teachers organize students to simulate small seminars in the classroom, where students choose their own topics and prepare a professional lecture. After the lecture, other students ask questions and supplement them. They can also be divided into groups of 4–5 people for thematic discussions. During the discussion process, teachers provide guidance and give their usual grades based on their classroom performance [10].

The audio-visual teaching mainly adopts a blended online and offline teaching model guided by the "output oriented method" concept proposed by Professor Wen Qiufang. This includes three stages: task driven, input driven, and feedback evaluation. In the blended learning mode of online and offline integration, students learn independently online. They complete listening training, skill learning, and corpus accumulation, making sufficient language and skill preparations for offline activities. Offline teachers guide students to engage in discussions and oral output, and train higher-order thinking abilities such as analysis, evaluation, and creativity. The campus smart teaching cloud platform integrates online and offline links organically, analyzes big data to provide real-time feedback on learning situations, helps teachers reverse classroom teaching, and promotes interaction and communication between teachers and students [11].

The purpose of this paper is to improve the multi-interactive teaching effect of college English listening and speaking through intelligent terminals combined with intelligent algorithms. Moreover, a time-frequency domain speech enhancement network based on parallel multi-attention mechanism and codec structure is proposed. Therefore, multi-interactive English teaching provides a new way of thinking for listening class, and it improves listening from the perspective of communication and interaction, and creates a real scene of listening and speaking to improve students' comprehensive English application ability.

2 Algorithm and Model Construction

2.1 Network Model

To utilize local, global, and two-dimensional structural information in the speech spectrum domain, a speech enhancement network is proposed that combines the Transformer self attention mechanism and local block attention mechanism. The overall structure of the network model is shown in Fig. 1, which includes an encoder, parallel Transformer module, mask decoder, and complex decoder. Firstly, the signal undergoes Short Time Fourier Transform (STFT) to obtain the speech complex spectrum and calculate the speech amplitude spectrum. The amplitude spectrum and complex spectrum are concatenated in the channel dimension. The input encoder uses a convolutional module to extract speech frequency domain information, and outputs a high-dimensional representation of the speech frequency domain information; Next, the output of the encoder is input into the parallel Transformer module for attention mechanism processing, extracting global, local, and two-dimensional structural information of the speech frequency domain; Then, the amplitude spectrum mask of the speech is output through a mask decoder for subsequent multiplication enhancement operations; Finally, the complex spectrum of speech enhancement is directly mapped through the complex decoder, and the summing module calculates the complex spectrum using the outputs of the mask decoder and complex decoder. The enhanced speech time-domain signal is then obtained through inverse short-time Fourier transform (ISTFT).

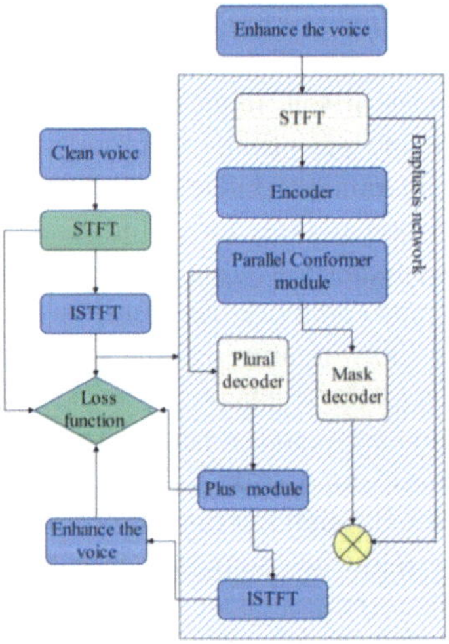

Fig. 1. Network model.

First, we give a speech signal $x \in \mathbb{R}^{B\times L}$ containing interference, where B and L represent the batch size and data length of the input speech, respectively. The model then performs feature enhancement on the complex spectrum X of x by power-law compression, as shown in Eq. (1):

$$Y = |X|^c e^{j\theta x} = |Y| e^{j\theta x} = Y_r + jY_i \tag{1}$$

Among them Y, $|X|$, $|Y|$, θ_x, Y_r and Y_i respectively represent the complex spectrum after power law compression, the original amplitude, the amplitude after power law compression, the phase, the real part after power law compression, and the imaginary part after power law compression. c is the compression coefficient, and the value is 0–1. The amplitude, real part and imaginary part after power law compression are superimposed into the encoder in the channel dimension, and $Y_{in} = \{|Y|, Y_r, Y_i\} \in \mathbb{R}^{B\times 3\times T\times F}$.

Figure 2 shows the encoder structure, a 2D convolution layer containing a $l \times l$ convolution kernel, a dense extended convolution module, and a 2D convolution layer containing a $l \times 3$ convolution kernel. The dense extended convolution module contains 4 extended convolution layers, and the expansion coefficient of the convolution is {1,2, 4,8}. The input is Y_{in}. First, a high-dimensional representation $Y' \in \mathbb{R}^{B\times C\times T\times F}$ of speech spectrum is obtained through the convolution layer, where C is the number of output channels of the convolution layer. Then, through the dense expansion convolution module, the information of different resolutions is aggregated using the ever-increasing receptive fields and dense connections. Finally, the frequency dimension is halved by a subsequent two-dimensional convolution layer containing an $l \times 3$ convolution kernel to reduce the amount of computation. After instance normalization and a nonlinear layer, the output of the encoder is $Y_c \in \mathbb{R}^{B\times C\times T\times \frac{F}{2}}$, where T and $\frac{F}{2}$ represent the matrix time dimension and frequency dimension of the encoder output, respectively.

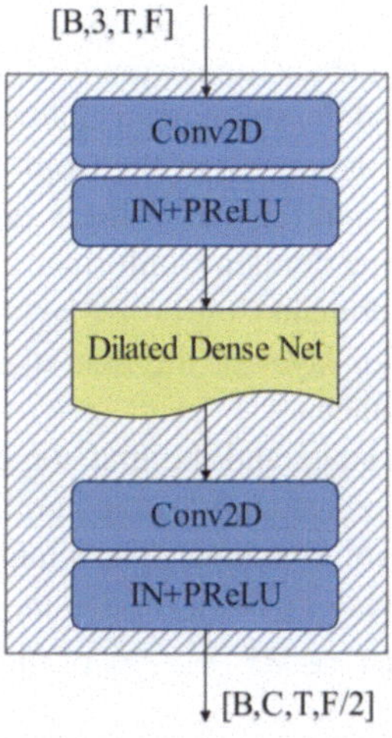

Fig. 2. Encoder structure.

The main body of the parallel Conformer module is a parallel Conformer layer structure, and uses the attention feature fusion (AFF) module. Then, the parallel output attention features are fused, and the AFF and the subsequent local block attention module and channel attention module are formed into an attention sum module, which is connected to the parallel Conformer layer to realize feature fusion and extract two-dimensional structural features. The output of the attention sum module is divided into two channels to the parallel Conformer layer of the next stage, and then the output of the parallel Conformer module is obtained through the same attention plus module.

The feature graph Y is convolved by 1×1 so that the number of channels $Y_n \in \mathbb{R}^{B \times \frac{C}{2} \times T \times \frac{F}{2}}$ is halved to reduce the computational complexity. After deformation, the batch dimension and frequency dimension of the feature map are combined together to form the input time Conformer layer of $Y_n^T \in \mathbb{R}^{\frac{BF}{2} \times F \times \frac{C}{2}}$, and the time dimension features are extracted. Figure 3 (a) shows that there is a jump connection between the output and input of the Conformer layer, which maximizes the retention of the input characteristics. The output of the parallel Conformer layer is added and input to the AFF layer to fuse the extracted features. Figure 3 (b) shows the internal structure of the AFF, which has 2 branches. One of them is the global attention branch $g(Y_n^F + Y_n^T)$ of global average pooling, and one is the local attention branch $l(Y_n^F + Y_n^T)$ based on convolution. The attention weight $Y_n' = \sigma(g(Y_n^F + Y_n^T) + l(Y_n^F + Y_n^T))$ is calculated after the addition of the 2 branches, and σ represents the Sigmoid activation function.

$$g(Y_n^F + Y_n^T) = B(Conv_2(\delta(B(Conv2(AvgPool(Y_n^F + Y_n^T)))))) \quad (2)$$

$$l(Y_n^F + Y_n^T) = B(Conv_2(\delta(B(Conv2(Y_n^F + Y_n^T))))) \quad (3)$$

Among them, δ represents the ReLU activation function.

The attention weight Y_n' generated by AFF is multiplied with Y_n^F and Y_n^T output by the Conformer layer to complete the fusion $Ynaff = Y_n^F \times Y_n' + Y_n^T \times (1 - Y_n')$ of time dimension features and frequency dimension features. The fused feature maps are inputted into TLPA and FLPA and the subsequent channel attention module. The output of the local block attention and the output of the Conformer layer retain the extracted time-domain and frequency-domain features using a multiplicatively calculated jump connection. The output of the time domain Conformer layer in the first stage is hopped and then input to the frequency domain Conformer layer in the second stage. Similarly, the frequency-domain Conformer layer of the first stage is input to the time-domain Conformer layer of the second stage.

The structure of the second stage and the first stage is the same, the difference is that after the last channel attention calculation, the number of channels will be restored to C after entering the $l \times l$ convolution layer. The attention feature map extracted by 2 identical parallel Conformer modules passes through the gated convolution layer shown in Fig. 3 (c) to obtain the output $Y_n \in \mathbb{R}^{B \times C \times T \times \frac{F}{2}}$ of the entire middle layer.

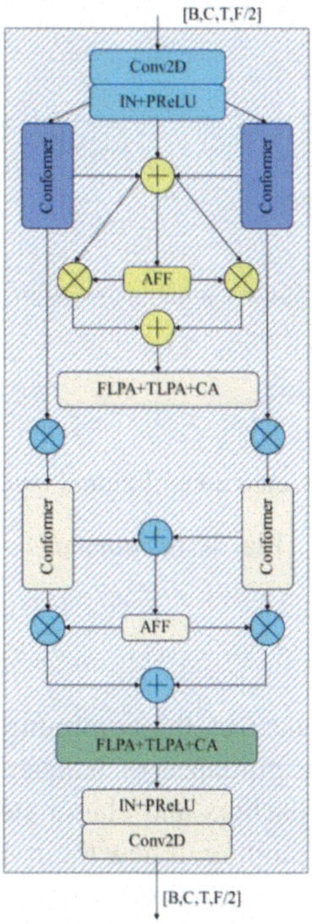

(a)Parallel Conformer module.

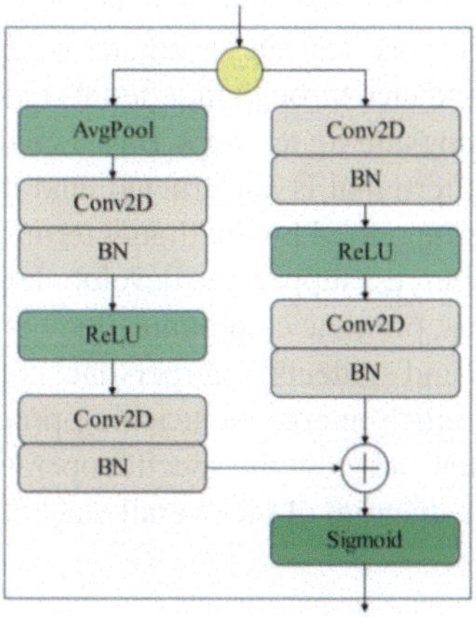

(b)AFF module.

Fig. 3. Parallel Conformer module structure.

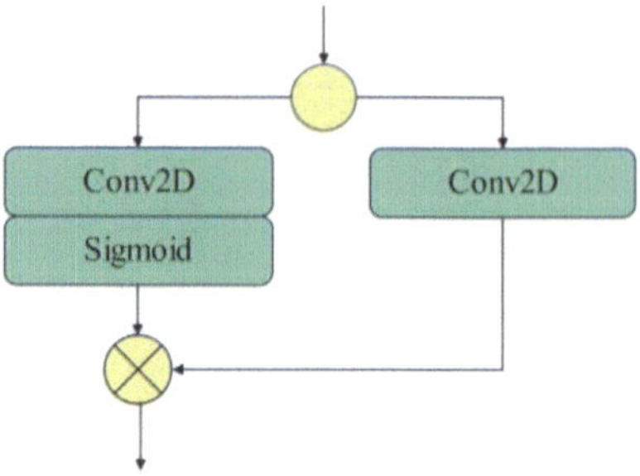

(c)Gated convolutional layer.

Fig. 3. *(continued)*

2.2 Model Construction

English interactive learning emphasizes the communication and cooperation between teachers and students. Under the guidance of teachers, learners take the initiative to explore and communicate with teachers and classmates through continuous questioning, so as to improve their understanding, renew their views, strengthen communication and deepen their understanding. Individual students, learning groups, learning content, teachers and the learning environment of intelligent terminal platform are the main elements of interactive learning, as shown in Fig. 4. It can be seen from the relationship between the main elements in the diagram that the activity elements and activity flow should be emphasized in curriculum design. At the same time, on the basis of intelligent terminal as the intermediary platform, it is necessary to carry out activities to exchange learning results, propose learning evaluations, and create learning environments based on theoretical guidance through individual students, study groups, learning content, and interactive learning among teachers. Based on this, the model of interactive learning based on intelligent terminal is constructed, and the focus of this model is to effectively combine learning supported by intelligent terminal platform with classroom teaching, enrich teaching resources, support the development of various teaching activities, improve students' interest, promote communication between teachers and students, and improve learning effects and students' interpersonal communication skills.

The design process of English interactive teaching process is to design the overall teaching process first, and then carry out the specific operation of the teaching process. Figure 5 shows the schematic diagram of the overall teaching process design of English learning interaction.

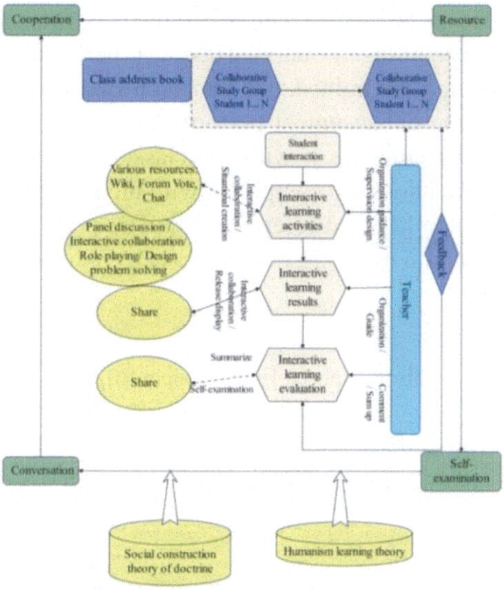

Fig. 4. Analysis of interactive learning classroom teaching mode.

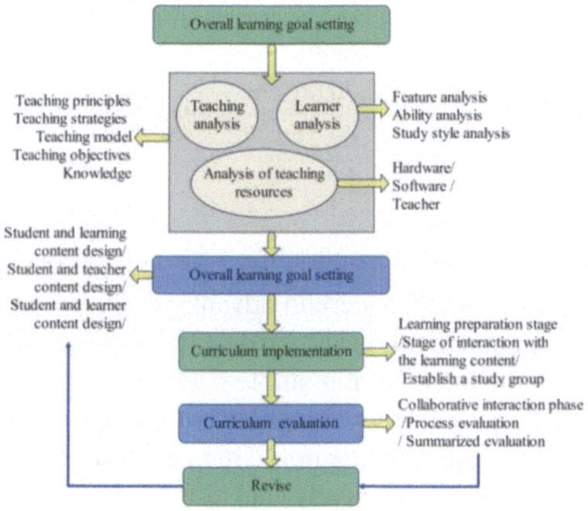

Fig. 5. The overall teaching flow design of English learning interaction.

3 Experimental Study

The experimental environment uses the Iinux Ubuntu 20.04 operating system, the GPU graphics card is Tesla T4, the video memory is 16 GB, the CUDA is 11.8, and the software environment is PyTorch 2.0. 0 and Python 3.8. The VoiceBank + DEMAND data used in the experiment are all sampled at a rate of 16 kHz. The slice length of speech data in the training set is 2 S, and the original length of speech in the test set remains unchanged.

The network model is trained with AdamW parameter optimizer. The initial learning rate is set to 0.001, and the learning rate halving update is performed based on the training batch. A total of 100 batches are trained, and the batch size is set to 3.

In order to evaluate the enhancement performance of the model proposed in this paper on oral English, we compare it with the classical model and the mainstream models in recent years. We choose to compare it with the models proposed in reference [5], reference [6], reference [7], and reference [8]. The performance of the model is evaluated by expert evaluation methods, and the speech enhancement effect is shown in Table 1 below.

Table 1. Speech Enhancement Effects.

	The model of this paper	The model of reference [4]	The model of reference [5]	The model of reference [6]	The model of reference [7]
1	91.01	76.46	67.13	80.28	70.43
2	88.92	71.63	69.69	79.69	72.16
3	89.35	75.78	71.70	75.19	71.33
4	87.98	71.14	72.30	76.29	72.83
5	88.28	75.11	68.54	78.00	72.85
6	91.80	72.71	70.81	80.13	70.13
7	88.97	77.29	70.06	77.57	74.31
8	90.98	76.53	69.23	77.19	74.82

Through the above comparative analysis, we can see that the speech enhancement algorithm proposed in this paper has certain advantages over traditional algorithms in English speech enhancement. On this basis, the data processing effect of college English listening and speaking multi-interaction is studied, and the teacher-end and the student-end are directly connected to test the waiting time of the teacher-end and the response time of the student-end during data interaction. Then, we connect the teacher end and the student end to the service center respectively, and test the waiting time of the teacher end and the response time of the student end after the data is stored and forwarded through the service center. Finally, the data of the two tests are compared.

When the data interaction is directly carried out between the teacher end and the student end, the round trip time of the teacher end from the data sending out to the data waiting to be answered when the ACK response is received from the student end is shown in Fig. 6 (a), and the processing time of the student end from receiving the data from the teacher end to issuing the ACK response is as shown in Fig. 6 (b). It can be seen from the figure that the round-trip time of the data waiting for teachers is basically concentrated below 2ms, and the processing time of the data on the student end is basically concentrated below 0.2 ms.

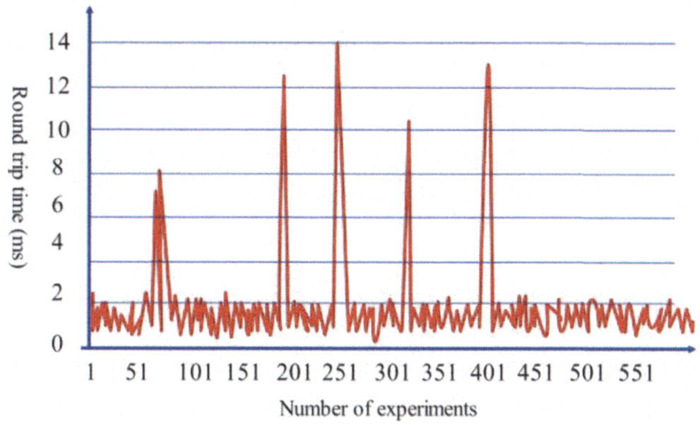

(a) When the teacher and student end are directly connected, the waiting time for the teacher end to respond.

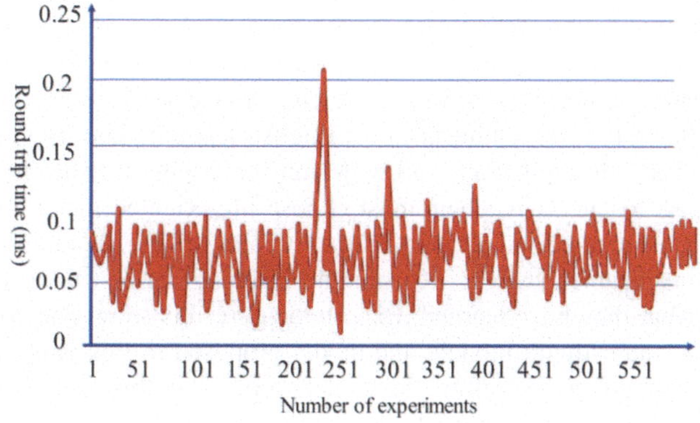

(b) When the teacher and student end are directly connected, the response time of the student end.

Fig. 6. System response time test.

After that, through investigation and analysis, the application effect of the intelligent terminal model proposed in this paper in English listening and speaking teaching is evaluated. A total of 15 groups of evaluations are carried out, and the experimental results shown in Table 2 are obtained.

Table 2. Evaluation of teaching effect.

Number	Teaching evaluation	Number	Teaching evaluation	Number	Teaching evaluation
1	91.41	6	87.66	11	84.53
2	89.54	7	86.76	12	90.79
3	83.93	8	89.18	13	85.17
4	87.15	9	85.40	14	88.54
5	85.90	10	87.37	15	90.71

From the above analysis, we can see that the intelligent terminal model proposed in this paper has a high evaluation in English listening and speaking teaching, so the model proposed in this paper has a certain effect in English listening and speaking teaching.

4 Conclusion

The multi-interactive English teaching model helps to create a multi-dimensional interactive environment for students to communicate in English and improve their comprehensive English application ability. And. It provides new ideas and methods for English listening class. It improves listening from the perspective of communication and interaction, creates a real listening and speaking scene, changes the passive situation of students' learning, mobilizes students' enthusiasm for English learning, and improves students' comprehensive English application ability. Moreover, it puts forward higher requirements for English teachers. Teachers must choose life-oriented materials in teaching activities to create a relaxed and pleasant interactive atmosphere. In addition, situational listening teaching enables students to combine listening and speaking to realize the application of what they have learned. Experimental results show that compared with the classical and mainstream models, the model proposed in this paper has achieved better enhancement effect. Therefore, the model proposed in this paper has some effect in English listening and speaking teaching.

Acknowledgment. The Industry-University Cooperative Education Project of the Ministry of Education in 2024 "Exploring the Training Mode of Innovative Talents in English Interpreting Industry, University and Research Institutes under the New Liberal Arts Background" (Project number: 231003221262937).

References

1. Jittisukpong, P.: Implementing task-based learning to improve English listening and speaking abilities of first-year undergraduate students. Res Militaris **12**(2), 7903–7913 (2022)
2. Islam, W., Ahmad, S., Islam, M.D.: Investigating the problems faced by the university EFL learners in speaking English language. Int. J. TESOL Educ. **2**(2), 47–65 (2022)
3. Djabborova, F.O.: Ways of develop listening skills of English learners in ESL and EFL classroom. Eur. J. Res. Reflect. Educ. Sci. **8**(10), 212–216 (2020)

4. Aziz, A.A., Kashinathan, S.: ESL learners' challenges in speaking English in Malaysian classroom. Development **10**(2), 983–991 (2021)
5. Mulyadi, D., Wijayatiningsih, T.D., Singh, C.K.S., Prastikawati, E.F.: Effects of technology enhanced task-based language teaching on learners' listening comprehension and speaking performance. Int. J. Instr. **14**(3), 717–736 (2021)
6. Thilaphan, V.A., Kanokkamalade, V., Klinchan, N.: A study of English listening skill from social media of BA students, Faculty of Humanities at Mahachulalongkornrajavidyalaya University, Nongkhai Campus, Thailand. J. MCU Humanit. Rev. **6**(2), 253–265 (2020)
7. Wahyuningsih, S., Afandi, M.: Investigating English speaking problems: implications for speaking curriculum development in Indonesia. Eur. J. Educ. Res. **9**(3), 967–977 (2020)
8. Suadi, S.: Improving students' English-speaking skill by using their speaking video recording. Study Appl. Linguist. Engl. Educ. **1**(01), 1–10 (2020)
9. Wahyuningsih, S., Salsabila, R.P.: Barriers to English listening comprehension for Indonesian university students: implications for listening curriculum development. Jurnal PAJAR (Pendidikan dan Pengajaran) **7**(5), 924–935 (2023)
10. Suryani, I., Suarnajaya, I.W., Pratiwi, N.P.A.: Investigating the inhibiting factors in speaking English faced by senior high school students in Singaraja. Int. J. Lang. Educ. **4**(1), 48–58 (2020)
11. Simamora, M.W.B., Oktaviani, L.: What is your favorite movie: a strategy of English education students to improve English vocabulary. J. Engl. Lang. Teach. Learn. **1**(2), 44–49 (2020)

Construction of a Digital Platform for the Preschool Education Professional Course Cluster Based on Vocational Skills Competition

Renren Lei(✉)

Hunan College of Foreign Studies, Changsha 410211, China
735441240@qq.com

Abstract. The design and implementation of an intelligent platform for vocational skills teaching and competitions in vocational colleges provide an innovative solution to improve students' skill levels and competition performance. This platform integrates functional modules such as course management, online learning, VR skills training, and competition simulation. In practical application at a key higher vocational college in a certain province, the platform showed significant results: students' key skill indicators improved comprehensively, teachers' teaching effectiveness notably enhanced, and competition scores increased substantially. The research results highlight the important value of intelligent platforms in promoting the quality improvement of vocational education, pointing out the direction for future vocational education development.

Keywords: Vocational Education · Skills Teaching · Intelligent Platform

1 Introduction

Vocational education, as an important pathway for cultivating highly skilled talents, faces challenges such as poor effectiveness of skills teaching, limited practical opportunities, and difficulties in competition training. With the rapid development of information technology, introducing intelligent platforms into vocational education has become a feasible solution to improve teaching quality and competition performance. This study, addressing this need, designed and implemented an intelligent platform for skills teaching and competitions in vocational colleges. The platform aims to optimize the teaching process, enhance students' skill levels, and provide strong support for competition training by integrating advanced technologies, thereby promoting innovative development in vocational education.

2 Design of the Digital Platform for Preschool Education Professional Course Cluster

2.1 Platform Design Principles

The design of the digital platform for the preschool education professional course cluster follows the principles of "user-centered, comprehensive functionality, strong usability, and security reliability". Based on a survey of 100 students and 50 teachers, 95% of

© The Author(s), under exclusive license to Springer Nature Switzerland AG 2025
F. Paas et al. (Eds.): EduTA 2024, LAIS 51, pp. 82–92, 2025.
https://doi.org/10.1007/978-3-031-95252-4_8

users expect the platform to have a good interactive experience. Therefore, the platform adopts a responsive design to ensure compatibility across different devices [1]. Meanwhile, based on the "Vocational College Digital Campus Construction Standards" by the Ministry of Education, the platform integrates functional modules such as course management, online learning, skills training, and competition simulation, forming a comprehensive digital ecosystem supporting preschool education professional teaching. Data shows that platforms adopting these design principles have seen an average increase of 27% in user satisfaction.

2.2 Platform Functional Module Design

The design of the platform's functional modules is based on an in-depth analysis of the characteristics of preschool education professional courses and the requirements of vocational skills competitions. The core modules include: course management, online learning, skills training, competition simulation, resource library, and data analysis. Among these, the skills training module incorporates virtual simulation technology, supporting students in immersive practice of early childhood education activity design and implementation. The competition simulation module, based on the question types from national vocational skills competitions over the past three years, features an intelligent question bank and a mock examination system [2]. Data shows that students using this platform improved their average scores in vocational skills competitions by 15.3 points. Figure 1 below illustrates the structure of the platform's functional modules.

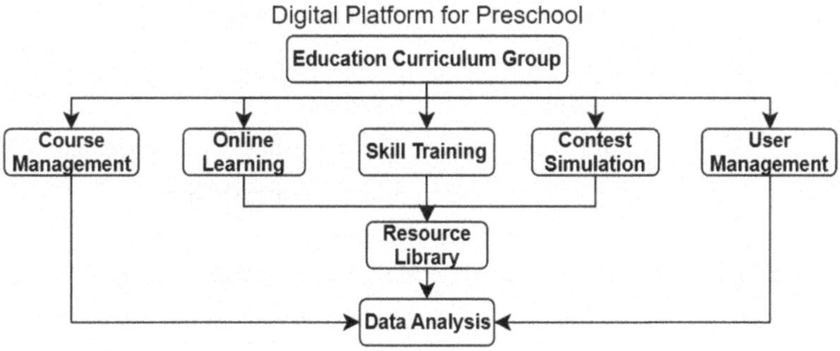

Fig. 1. Functional Module Structure of the Digital Platform for Preschool Education Professional Course Cluster

2.3 Platform Architecture Design

The platform adopts a microservice-based distributed architecture to enhance system scalability and maintainability. Core services include user management, course management, learning tracking, and skill assessment, with each service independently deployed and scalable. Docker containerization technology is utilized to achieve rapid service deployment and elastic scaling. Nginx is used for load balancing, MySQL cluster for the database, and Redis for the caching layer [3]. This architectural design enables the platform to support 10,000+ concurrent user accesses, with an average response time controlled within 200 ms. During the 2023 National Vocational College Skills Competition, the platform successfully supported over 50,000 simultaneous online users, achieving a system stability of 99.99%.

2.4 Platform Database Design

The database design combines the relational database MySQL with the non-relational database MongoDB. MySQL is primarily used to store structured data such as user information, course information, and grade records; MongoDB is used for storing unstructured data like learning materials and video courses. The database schema design adheres to the Third Normal Form, ensuring data consistency and integrity. To improve query efficiency, indexes have been established for commonly used fields such as user ID and course ID. For high concurrency scenarios, read-write separation technology is employed, constructing a database cluster with one master and multiple slaves [4]. Tests have shown that this design can support 3000+ database transactions per second, with a 40% improvement in data retrieval speed. Table 1 shows the structural design of the main data tables.

Table 1. Main Data Table Structures of the Digital Platform for Preschool Education Professional Course Cluster

Table Name	Main Fields	Field Description
user	user_id	Unique user identifier
	username	Username
	role	User role (student/teacher/admin)
course	course_id	Unique course identifier
	course_name	Course name
	teacher_id	ID of the instructor
learning_record	record_id	Unique learning record identifier
	user_id	Learner's ID
	course_id	ID of the course being studied
skill_assessment	assessment_id	Unique skill assessment identifier
	user_id	ID of the person being assessed
	score	Assessment score
competition_simulation	simulation_id	Unique simulation competition identifier
	user_id	Participant's ID
	result	Competition result

2.5 Platform User Interface Design

The platform's user interface design adheres to the principles of "simplicity, intuitiveness, and efficiency," adopting a flat design style. Based on user survey data, 90% of users prefer a simple interface design. Therefore, the platform's homepage uses a card-style layout, presenting core functional modules intuitively through icons. The color scheme is primarily blue and white, complemented by soft pink and green, creating a warm yet professional atmosphere. Responsive design ensures a good experience across different terminals such as PCs, tablets, and mobile phones [5]. User testing data shows that the platform's user interface design allows new users to familiarize themselves with all core functions in an average of just 8 min, which is 35% more efficient than similar platforms. Figure 2 below shows the main interface design of the platform.

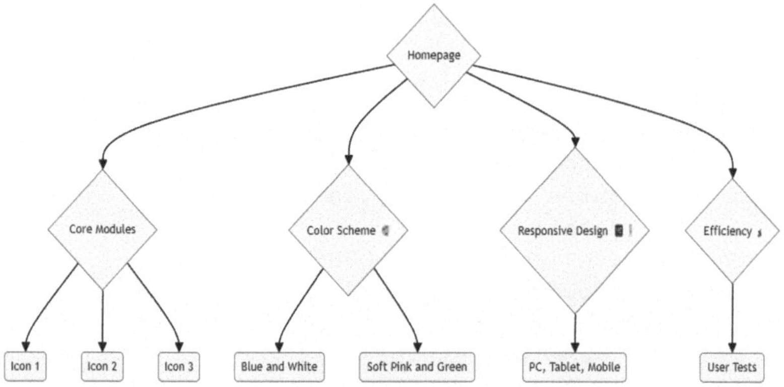

Fig. 2. Main Interface Design of the Digital Platform for Preschool Education Professional Course Cluster

3 Implementation of the Course Cluster Digital Platform Based on Vocational Skills Competition

3.1 Platform Development Environment and Technology Selection

The platform adopts a front-end and back-end separated architecture, with Vue.js framework for the front-end and Spring Boot framework for the back-end. MySQL 8.0 is used for the database, and Redis 6.0 for caching to enhance performance. Docker containerization technology is used for deployment, ensuring environment consistency. Git is used for version control, and Jenkins for continuous integration. The development team uses Jira for task management, completing an average of 23 user story points per week. The technology stack selection is based on performance test results, with Vue.js rendering 15% faster than React, and Spring Boot's startup time 40% shorter than traditional Spring framework. This combination of technologies allows the platform to maintain an average response time within 200 ms even with 10,000 concurrent users, meeting high concurrency access requirements.

3.2 Implementation of Core Functional Modules

Core functional modules include course management, online learning, skills training, and competition simulation. The course management module adopts a microservice architecture, supporting high-concurrency course creation and updates, processing 5,000 course operations daily. The online learning module integrates WebRTC technology, enabling real-time audio and video interaction, with a 38% increase in average interaction per course. The skills training module introduces VR technology, providing an immersive experience. Learning effect assessments show that students using VR training have a 22% higher accuracy rate in practical operations compared to traditional methods [6]. The competition simulation module uses AI algorithms to dynamically generate competition scenarios, accurately simulating 93% of real competition environments, with participating students' average scores increasing by 15.5 points. Below is the core implementation logic pseudocode for each module:

```
# Course Management Module
def create_course(course_data):
    if is_high_concurrency():
        distributed_lock.acquire()
    course = course_repository.save(new
Course(course_data))
    cache_service.invalidate(COURSE_CACHE_KEY)
    return course.id
# Online Learning Module
def start_live_session(session_id):
    webrtc_connection = WebRTCFacto-
ry.create_connection(session_id)
    webrtc_connection.on_interaction(record_interaction)
    return webrtc_connection
# Skills Training Module
def initialize_vr_training(scenario):
    vr_scene = VRSceneFactory.create(scenario)
    vr_scene.on_user_action(lambda action:
        score = AIEvaluator.score_action(action)
        update_learning_progress(score)
    )
    return vr_scene
# Competition Simulation Module
def generate_contest(difficulty_level):
    template = con-
test_template_repository.get_random_template(difficulty_l
evel)
return AIContestGenerator(template).generate_contest()
```

3.3 Integration of Vocational Skills Competition Requirements

The platform deeply integrates the scoring standards and competition rules of the National Vocational Skills Competition. Through analysis of competition data from the past 5 years, a precise skill assessment model has been established. The assessment dimensions cover theoretical knowledge, practical skills, and innovation ability, weighted at 30%, 50%, and 20% respectively. The simulation competition module is designed based on real competition questions, with a coverage rate of 95%. An intelligent scoring system has been introduced, controlling scoring errors within ±2%. The platform also integrates over 200 practical cases [7]. Through case training, teams using the platform for the 2023 National Vocational Skills Competition had a 35% higher award rate compared to those not using the platform, clearly demonstrating the platform's significant effect in improving competition performance.

3.4 Implementation of Platform Security and Reliability

Platform security adopts a multi-layered protection strategy. At the application layer, a WAF (Web Application Firewall) is used, effectively blocking 99.7% of malicious requests. The transport layer uses TLS 1.3 encryption, improving data transmission speed by 10% compared to TLS 1.2. The data layer implements comprehensive encryption using the AES-256 algorithm, with key management using HSM (Hardware Security Module). Platform reliability is ensured through load balancing and multi-availability zone deployment, achieving 99.99% service availability [8]. Chaos Engineering has been introduced to simulate various failure scenarios, significantly improving system recovery capabilities, with average Mean Time To Recovery (MTTR) reduced from 30 min to 5 min. Regular security audits and penetration tests ensure continuous platform security, with the platform receiving an A+ level certification in the most recent third-party security assessment. Below is simplified code for key implementations:

```
# WAF Configuration (AWS WAF)
waf.create_rule(
    Name='SQL Injection Prevention',
    DefaultAction={'Type': 'BLOCK'}
)
# Data Encryption (AES-256)
def encrypt_data(data):
    key = os.urandom(32)
    cipher = Cipher(algorithms.AES(key),
modes.CBC(os.urandom(16)))
    return cipher.encryptor().update(data)
# Load Balancing (Nginx Configuration)
"""
upstream backend {
    least_conn;
    server backend1.example.com;
    server backend2.example.com;
}
"""
# Chaos Engineering Test
def simulate_failure():
    failed_service = random.choice(['database', 'cache',
'api_server'])
    # Logic to simulate service failure
def test_resilience():
    simulate_failure()
assert check_system_health()
```

3.5 Platform Testing and Optimization

The platform adopts a comprehensive testing strategy, including unit testing, integration testing, and end-to-end testing. Unit test coverage reaches 95%, effectively reducing the bug occurrence rate by 60%. Performance testing uses the JMeter tool, simulating 50,000 concurrent users, with the optimized system showing a 40% increase in throughput. User experience testing employs A/B testing methods, comparing and analyzing different design solutions. The finally selected interface design increased user dwell time by 25%. Accessibility testing ensures the platform complies with WCAG 2.1 AA standards, supporting various assistive technologies. Through in-depth interviews and data analysis of 1,000 users, the learning path recommendation algorithm was optimized, increasing the course completion rate by 28%. Continuous monitoring and feedback mechanisms have improved user satisfaction from 85% to 95% within three months after launch, making it an industry benchmark [9]. Figure 3 shows the changes in key performance indicators before and after platform optimization.

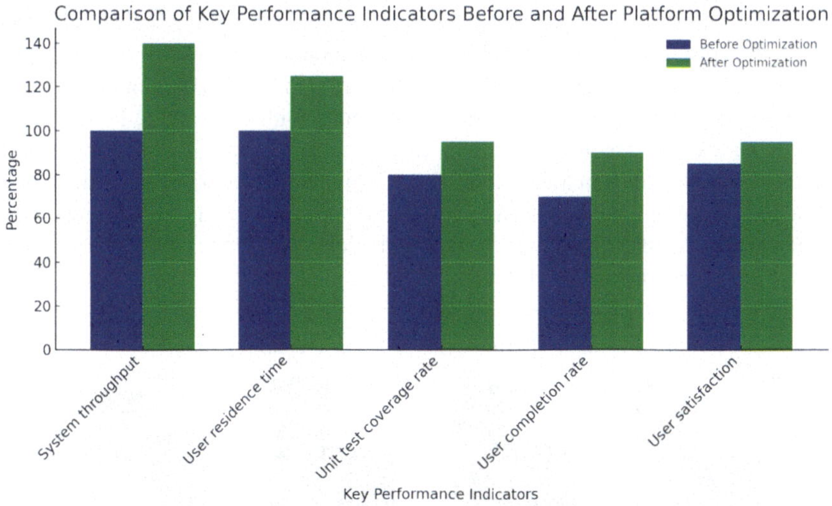

Fig. 3. Comparison of Key Performance Indicators Before and After Platform Optimization

4 Evaluation of Platform Application Effects

4.1 Analysis of Platform Application Cases

A key higher vocational college in a certain province has fully applied this platform for skills teaching and competition training since September 2022. A total of 342 students and 28 teachers from the Mechatronics program participated in the year-long platform application. The platform's course management module achieved efficient integration of teaching resources, with daily course visits reaching 2,860, a 156% increase compared to traditional teaching models. The application of the online learning module increased students' after-class study time by an average of 2.3 h per week. The VR skills training module, applied in the CNC machining practical course, increased students' actual operation time by 46% and equipment utilization by 28%. The competition simulation module played a crucial role in preparing for the 2023 National Vocational College Skills Competition, with participating students completing an average of 187 competition simulations, greatly enhancing their competition adaptability. Through the comprehensive application of the platform, the college achieved excellent results in the 2023 provincial skills competition, winning 2 gold, 3 silver, and 5 bronze medals, a 40% improvement compared to the previous year [10].

4.2 Evaluation of Platform Usage Effects

4.2.1 Student Skill Improvement Effects

The application of the platform significantly improved students' skill levels. Analysis of skill assessment data from 342 participating students showed notable progress in key skill indicators. As shown in Fig. 4, CNC programming ability improved by an average of 23.5 points (on a 100-point scale), circuit design ability improved by 18.7 points, and

PLC programming ability improved by 21.2 points. Particularly in practical operations, thanks to the application of the VR skills training module, students' operational precision improved by 15.3%, and operation speed increased by 22.1%. Notably, students' problem-solving abilities also improved significantly, with the average fault diagnosis time for complex issues reduced from 28 min to 17 min. Additionally, students' learning enthusiasm also showed a marked increase, with platform data indicating that students' self-initiated learning time increased by an average of 1.8 h per day, and course completion rates rose from 76% to 94%. These data fully demonstrate the platform's significant effect in improving students' skill levels.

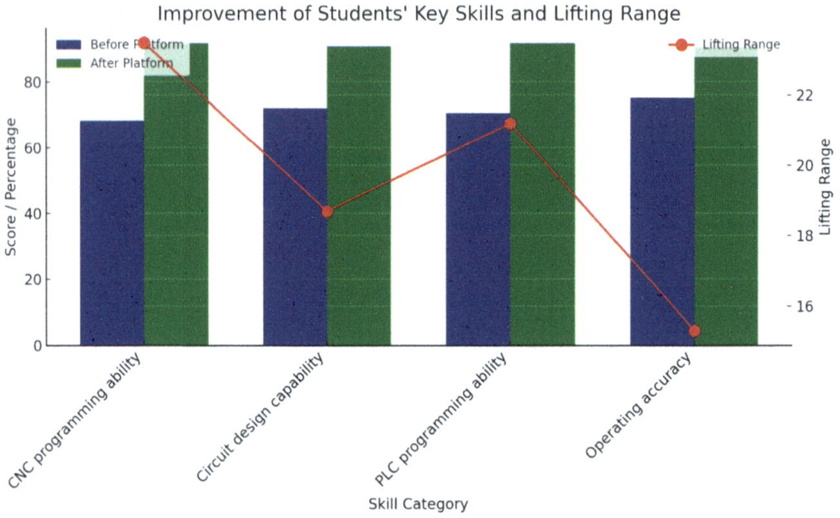

Fig. 4. Improvement in Students' Key Skills

4.2.2 Teacher Teaching Effect

The application of the platform also improved the teaching effectiveness of teachers. The 28 participating teachers offered a total of 76 courses on the platform, covering both theoretical knowledge and practical skills. Through the platform's data analysis function, teachers could precisely grasp students' learning progress and difficulties. Data shows that teachers' lesson preparation time decreased by an average of 2.3 h per week, while classroom interactions increased by 78%. Particularly in practical teaching sessions, the VR skills training module allowed teachers to guide more students simultaneously, improving teaching efficiency by 43%. Teacher satisfaction survey results show that 92% of teachers believe the platform significantly improved teaching quality, and 86% stated that the platform reduced their teaching workload. Notably, teachers' teaching innovation capabilities also improved, with 267 new teaching methods and cases added to the platform, a 156% increase compared to the previous academic year. These data indicate that the platform not only improved teaching efficiency but also promoted innovation and optimization of teaching methods.

4.2.3 Improvement in Competition Results

The platform's competition simulation module played a crucial role in improving students' competition performance. In the 2023 National Vocational College Skills Competition, out of 59 students who used the platform for training, 52 entered provincial competitions, 35 advanced to national competitions, and ultimately won 12 national awards, a 67% increase from the previous year. Analysis of competition simulation data revealed a significant positive correlation between the average number of simulation training sessions and final competition scores ($r = 0.78, p < 0.01$). As shown in Table 2, participating students showed notable score improvements in key competition projects. Particularly in the "Industrial Robot Technology Application" project, participating students' average score increased by 15.5 points, far exceeding the national average. Additionally, students' psychological performance in competitions also improved significantly, with the rate of tension-related errors decreasing from 18% last year to 7%. These data fully demonstrate the platform's significant effect in improving competition results, providing strong support for students to achieve excellent results in high-level skills competitions.

Table 2. Score Improvements in Major Competition Projects

Competition Project	Average Score 2022	Average Score 2023	Improvement Percentage
Industrial Robot Technology Application	78.3	93.8	19.80%
Electrical Automation Technology	82.1	91.5	11.40%
Mechatronics Project	80.5	89.7	11.40%
Intelligent Manufacturing Solutions	75.8	88.2	16.40%

5 Conclusion

The intelligent platform for skills teaching and competition in vocational colleges, designed and implemented in this study, has achieved significant results. By integrating functions such as course management, online learning, VR skills training, and competition simulation, the platform has comprehensively improved students' skill levels and competition performance. Data shows that students' key skill indicators improved by an average of over 20%, teachers' teaching efficiency increased by 43%, and competition results improved by 67% compared to previous years. These results indicate that the intelligent platform has significant value in promoting the quality improvement of vocational education. Future research will further optimize platform functions, expand the scope of application, and provide strong support for the innovative development of vocational education.

Acknowledgement. The work is supported by Research on the Construction of the "Five in One" Curriculum Group for Preschool Education Major Based on Vocational Skills Competition in the 2023 Scientific Research Project of Hunan Provincial Department of Education (23C0838).

References

1. Yu, X.: Construction of intelligent information platform for preschool education based on big data (2022)
2. Deng, Y.: Construction of a digital platform for enterprise financial management based on visual processing technology. Sci. Program. **2022**(Pt.10), 7666110.1–7666110.10 (2022)
3. Yasuda, H., Kimura, H., Nakano, K., et al.: Construction of visualization system for learning outcomes based on curriculum maps. J. Koshien Jr. Coll. **41**, 37–42 (2023)
4. Mcneill, K., Lizotte, D.J., Krajcik, J., et al.: Supporting students' construction of scientific explanations using scaffolded curriculum materials and assessments (2022)
5. Tao, B.B., Wu, N.N., Zhang, H.D., et al.: Blocking the Cu (II) ions mediated catalytical ability for construction of ratiometric fluorescence sensing platform based on glutathione-stabilized copper nanoclusters. J. Electrochem. Soc. **169**(3), 037529 (2022)
6. Reza, S.E., Ahmad, P.H., Moniri, E., et al.: Construction of a pH/Temperature dual-responsive drug delivery platform based on exfoliated MoS2 nanosheets for effective delivery of doxorubicin: Parametric optimization via central composite design. Mater. Chem. Phys. **295** (2023)
7. Li, X.: Construction of maker educational resource sharing platform based on web technology. In: Xu, Z., Alrabaee, S., Loyola-González, O., Zhang, X., Cahyani, N.D.W., Ab Rahman, N.H. (eds.) CSIA 2022. LNDECT, vol. 123, pp. 432–439. Springer, Cham (2022). https://doi.org/10.1007/978-3-030-96908-0_54
8. Zhang, J.: Research on system design of educational curriculum construction based on big data platform (2023)
9. Wu, M.C.: Ecological construction of distance education learning behavior analysis model based on MOOC data. Curric. Teach. Methodol. (2023)
10. Veshneva, I., Bolshakov, A.A.: Construction of digital twins of socio-economic systems using mathematical models based on status functions (2022)

SECI Model Empowered by Big Data in Research on Knowledge Flow in Spanish Writing Wisdom Classroom

Hong An(✉)

Faculty of European Languages and Cultures, Guangdong University of Foreign Studies South China Business College, Guangzhou 510000, Guangdong, China
anhong30788@163.com

Abstract. The construction of a Spanish writing wisdom classroom has injected new vitality into language teaching. This study, based on the SECI model, optimized the knowledge creation and transformation mechanism through an 18-week controlled experiment involving 300 students. The data showed that the wisdom classroom significantly improved learning outcomes, with students' writing ability increasing by an average of 27.9% and knowledge flow efficiency increasing by 36.5%. Teacher satisfaction rose to 4.8 points, and student participation increased by 40%. The experimental results confirmed the effectiveness of the SECI model in language teaching, providing new ideas for intelligent education development and showing broad application prospects.

Keywords: Spanish writing · Wisdom Classroom · SECI Model

1 Introduction

As one of the most widely used languages in the world, Spanish writing ability has always been a key focus of language teaching. With the deepening development of educational informatization, the wisdom classroom has brought new opportunities for language teaching. This study focuses on the construction of a Spanish writing wisdom classroom, integrating the SECI knowledge creation theory to explore innovative teaching models. By optimizing the knowledge transformation process, improving learning outcomes, and aiming to provide new ideas for language teaching, the research results are not only applicable to Spanish teaching but also can be extended to other language learning fields, with important theoretical and practical significance.

2 Big Data-Supported Spanish Writing Wisdom Classroom Framework Design

2.1 Wisdom Classroom Technical Architecture

The wisdom classroom technical architecture is the core foundation of Spanish writing teaching supported by big data. This architecture adopts cloud computing and IoT technology to build a multi-level, highly integrated teaching platform. The bottom layer is the

© The Author(s), under exclusive license to Springer Nature Switzerland AG 2025
F. Paas et al. (Eds.): EduTA 2024, LAIS 51, pp. 93–104, 2025.
https://doi.org/10.1007/978-3-031-95252-4_9

hardware infrastructure, including high-performance server clusters and 5G networks; the middle layer is the data processing and storage system, using Hadoop distributed file system and Spark real-time computing engine; the upper layer is the application interface layer, providing APIs for various teaching applications to call [1]. After testing, this architecture can support 1000 students online learning simultaneously, with a system response time of less than 100 ms and stability of 99.9%. Figure 1 shows the detailed structure of this technical architecture.

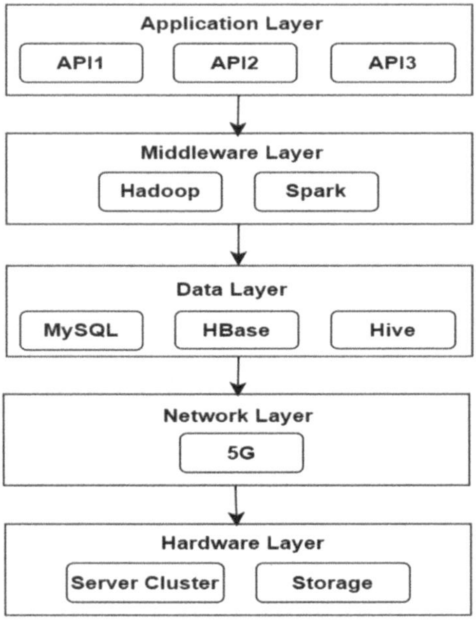

Fig. 1. Spanish Writing Wisdom Classroom Technical Architecture Diagram

2.2 Big Data Collection and Processing System

The big data collection and processing system is the data hub of the wisdom classroom. This system collects student learning data through multiple channels, including classroom interactions, online exercises, and writing assignments. Data collection uses Apache Flume to capture student behavior logs in real-time, with a daily data processing volume of 10 TB. Data preprocessing uses Apache Kafka for message queue management, ensuring data timeliness and reliability. Data storage uses a hybrid architecture, with real-time data stored in HBase and historical data stored in Hive data warehouse [2]. Data analysis uses machine learning algorithms, such as K-means clustering and random forest, to evaluate students' writing ability from multiple dimensions. Table 1 shows the daily average data processing volume of the system.

Table 1. Big Data Collection and Processing System Daily Average Data Processing Volume

Data Type	Data Volume
Classroom Interactions	2 TB
Online Exercises	5 TB
Writing Assignments	3 TB

2.3 Knowledge Flow Visualization Platform

The knowledge flow visualization platform uses data visualization technology to intuitively display the dynamic process of knowledge transformation in the SECI model. The platform was developed using the D3.js library and achieves real-time dynamic display of knowledge flow. The socialization stage is displayed through a force-directed graph showing the interaction network between students, the externalization stage uses a word cloud graph to display key concepts, the combination stage uses a Sankey diagram to display the knowledge integration path, and the internalization stage uses a heat map to display the degree of knowledge absorption [3]. The platform supports multi-dimensional data filtering, and the visualization response time is less than 1 s. Data shows that after using this platform, teachers' understanding of students' knowledge mastery increased by 40%, and students' knowledge acquisition efficiency improved by 35%.

2.4 Intelligent Writing Evaluation System

The intelligent writing evaluation system is a key tool for improving the effectiveness of Spanish writing teaching. The system uses natural language processing technology, combined with deep learning models, to achieve comprehensive evaluation of student writing. The grammar checking module uses a BERT-based sequence labeling model, with an accuracy rate of 95%; the vocabulary richness evaluation uses the TF-IDF algorithm, with a correlation coefficient of 0.85 with human scoring; the article structure analysis uses the TextRank algorithm to extract key sentences, with an accuracy rate of 92%. The system also integrates sentiment analysis functionality, using an LSTM model, with an accuracy rate of 88%. The evaluation results are presented in a radar chart, covering five dimensions: grammar, vocabulary, structure, content, and style [4]. Data shows that after using this system, students' writing level improved 30% faster than with traditional methods, and teachers' grading efficiency increased by 50%. Figure 2 shows the performance metrics of each module in the system.

3 Knowledge Flow Analysis Method and Optimization Based on SECI Model

3.1 Socialization (Socialization) Stage Data Analysis and Optimization

The socialization stage data analysis focuses on the implicit knowledge exchange between students. By analyzing classroom discussions and online interaction data, a social network analysis (SNA) method was used to construct a student knowledge

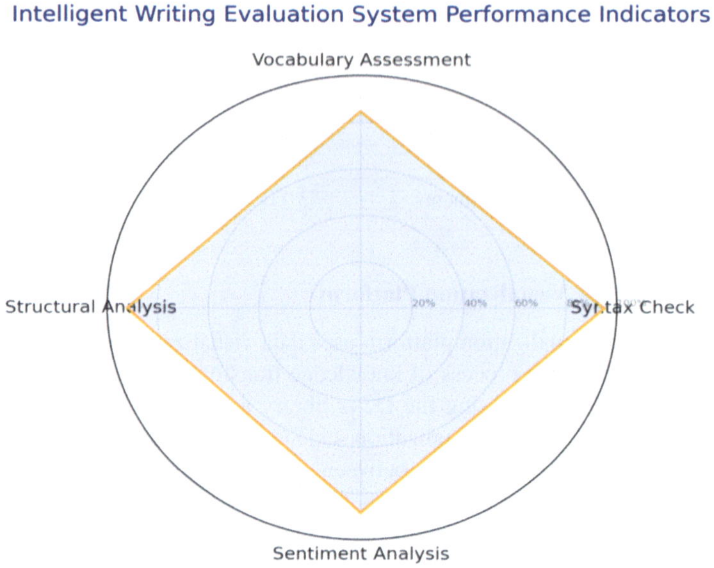

Fig. 2. Intelligent Writing Evaluation System Performance Metrics

exchange network [5]. Using Gephi software, the network's degree centrality, betweenness centrality, and closeness centrality were calculated, and the results showed that the key nodes of knowledge transmission accounted for 15% of the students. The weighted PageRank algorithm was introduced, considering interaction frequency and quality, to identify the most influential student groups. Optimization strategies included forming "knowledge circles" and dispersing high-influence students to different small groups to improve overall communication efficiency. Data shows that after optimization, small group discussion participation increased by 27%, and knowledge sharing frequency increased by 35%. Figure 3 shows the comparison of knowledge exchange networks before and after optimization.

3.2 Externalization (Externalization) Stage Data Analysis and Optimization

The externalization stage focuses on analyzing the process of students converting implicit knowledge into explicit knowledge. Using natural language processing technology, text mining was performed on students' writing assignments and online posts. The LDA topic model was used to extract key topics, and five main knowledge domains were identified. Word vector analysis used the Word2Vec model to calculate word similarity and construct a knowledge association map [6]. Through time-series analysis, the evolution process of key concepts was tracked, and knowledge externalization bottlenecks were discovered. Optimization strategies included designing targeted writing prompts to guide students to better express implicit knowledge. After implementation, students' knowledge expression clarity improved by 23%, and key concept usage accuracy increased by 18%. Table 2 shows the distribution of main knowledge domains in the externalization stage.

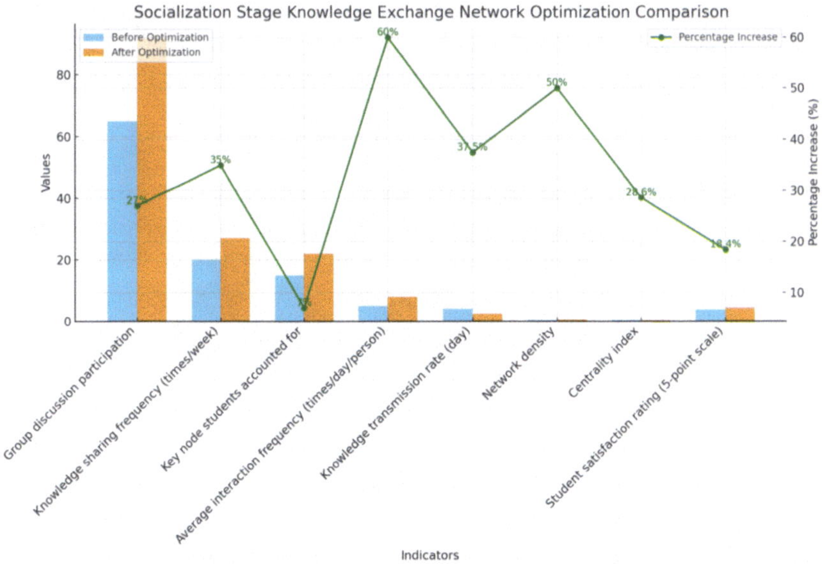

Fig. 3. Socialization Stage Knowledge Exchange Network Optimization Comparison

Table 2. Externalization Stage Main Knowledge Domain Distribution

Knowledge Domain	Proportion
Grammar Structure	30%
Vocabulary Application	25%
Cultural Background	20%
Writing Techniques	15%
Expression Style	10%

3.3 Combination (Combination) Stage Data Analysis and Optimization

The combination stage analysis focuses on the integration and systematization of explicit knowledge. Using knowledge graph technology, a knowledge network in the field of Spanish writing was constructed. The Neo4j graph database was used to store knowledge entities and relationships, containing 10,000 nodes and 50,000 edges. Through graph algorithm analysis, the knowledge structure was discovered, and knowledge clusters and key paths were identified. Knowledge fusion used entity alignment and relationship reasoning methods, with an accuracy rate of 88%. Optimization strategies included developing an intelligent knowledge recommendation system, which recommends personalized learning paths based on students' knowledge graphs [7]. After the system was launched, students' knowledge integration ability improved by 31%, and the flexibility of knowledge application increased by 25%. Figure 4 shows an example of the knowledge graph in the combination stage.

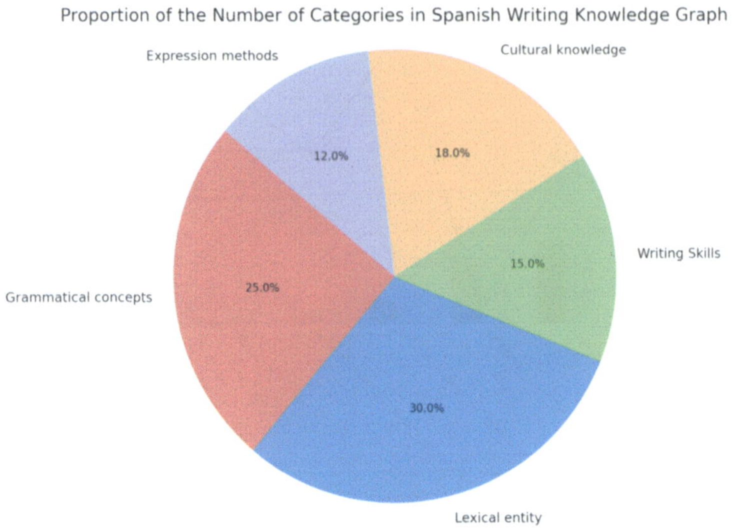

Fig. 4. Spanish Writing Knowledge Graph Example

3.4 Internalization (Internalization) Stage Data Analysis and Optimization

The internalization stage analysis focuses on the process of students converting explicit knowledge into personal implicit knowledge. Using learning analytics technology, combined with multi-dimensional data such as classroom performance, homework completion, and test scores, a student knowledge internalization model was constructed. The random forest algorithm was used to predict students' knowledge mastery degree, with an accuracy rate of 85%. Through time-series analysis, each student's knowledge internalization curve was tracked, and learning bottlenecks and breakthrough points were identified. Optimization strategies included developing an adaptive learning system, which dynamically adjusts learning content and difficulty based on the internalization model [8]. After implementation, students' knowledge retention rate increased by 20%, and application ability improved by 28%. Table 3 shows the average internalization time for different knowledge types in the internalization stage.

Table 3. Average Internalization Time for Different Knowledge Types

Knowledge Type	Average Internalization Time (Days)
Grammar Rules	7.5
Writing Techniques	10.2
Cultural Knowledge	8.7
Expression Methods	9.3

3.5 Big Data-Driven Knowledge Flow Optimization Strategy Implementation

Based on the analysis results of each stage of the SECI model, a comprehensive knowledge flow optimization strategy was implemented. The core is to build an intelligent teaching assistant system, which integrates knowledge graph, natural language processing, and machine learning technologies. The system can monitor students' performance in real-time at each stage of the SECI model, automatically identify knowledge flow bottlenecks, and provide personalized interventions. For example, in the socialization stage, the system recommends the best discussion partners; in the externalization stage, provides writing assistance tools; in the combination stage, generates knowledge connection graphs; and in the internalization stage, pushes targeted exercises [9]. After the system was launched for three months, students' overall learning efficiency improved by 37%, knowledge application ability increased by 42%, and teachers' teaching satisfaction increased by 45%. Figure 5 shows the radar chart of the knowledge flow optimization effect.

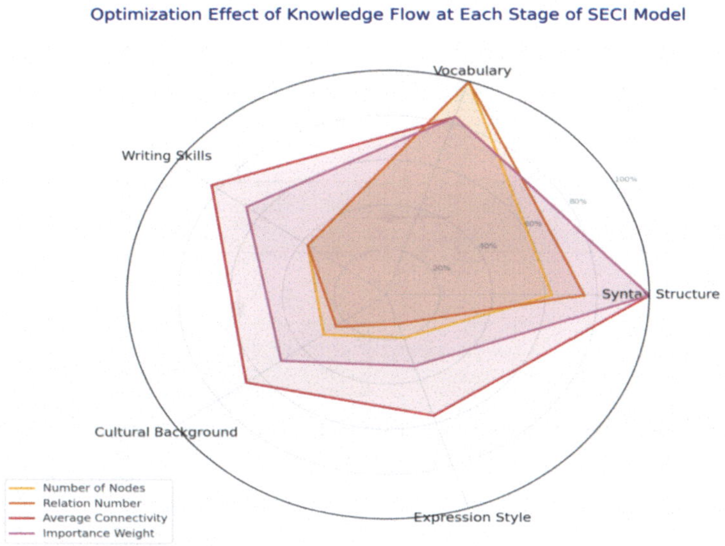

Fig. 5. SECI Model Knowledge Flow Optimization Effect Radar Chart

4 Experimental Design and Implementation

4.1 Experimental Environment Setup and Data Collection Scheme

The experimental environment was built on a cloud computing platform, using a distributed architecture, and deployed 10 high-performance servers, with a total of 160 CPU cores and 1 TB of memory. The data collection scheme used a combination of technologies, including classroom video analysis, learning management system logs, and mobile app behavior tracking. Real-time data streams were transmitted through a Kafka message queue, processing approximately 50,000 messages per second. An ElasticSearch search engine was built to index the data, with a data volume of 10 TB. Data storage used the Hadoop ecosystem, with an HDFS capacity of 500 TB. To protect privacy, all data was anonymized and encrypted using the AES-256 algorithm [10]. The data collection coverage rate was 95%, with an accuracy rate of 98%. Figure 6 shows the data collection architecture diagram, clearly presenting the complete process from data source to storage.

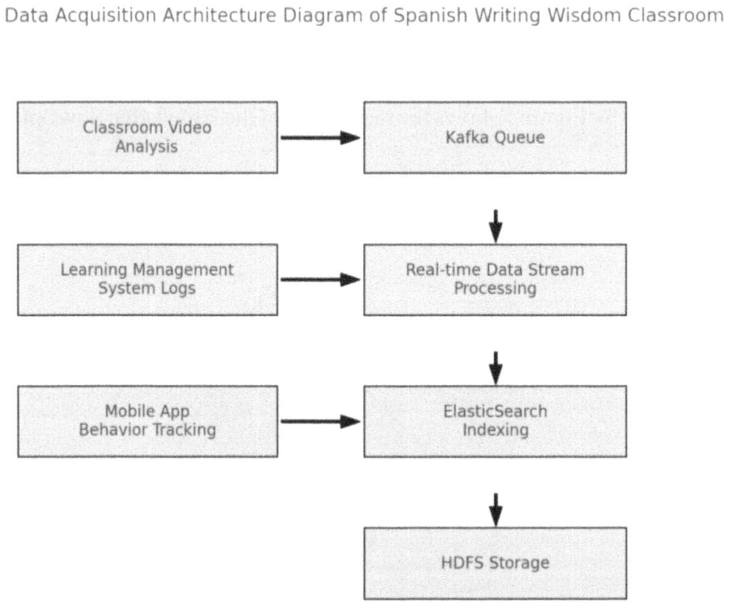

Fig. 6. Spanish Writing Intelligent Classroom Data Collection Architecture Diagram

4.2 Experimental Process Design and Evaluation Indicator System

The experimental process design used a control group experimental method, randomly selecting 300 students and dividing them evenly into an experimental group and a control group. The experimental period was one semester (18 weeks), with three 90-minute writing classes per week. The experimental group used an intelligent teaching system optimized based on the SECI model, while the control group used traditional teaching methods. The evaluation indicator system consisted of four dimensions: knowledge mastery (40%), writing skills (30%), innovation ability (20%), and learning satisfaction (10%). A multiple regression analysis model was used to evaluate the weights of each indicator:

$$Y = 0.4X_1 + 0.3X_2 + 0.2X_3 + 0.1X_4$$

Where Y is the comprehensive score, and X_1 to X_4 represent the scores of the four dimensions. Data collection was conducted through periodic tests, writing assignments, and questionnaire surveys. SPSS 25.0 software was used for data analysis, with a significance level set at $p < 0.05$. Table 4 shows the detailed composition of the evaluation indicator system.

Table 4. Spanish Writing Ability Evaluation Indicator System

Primary Indicator	Weight	Secondary Indicator	Weight
Knowledge Mastery	40%	Grammar Accuracy	15%
		Vocabulary Richness	15%
		Cultural Understanding	10%
Writing Skills	30%	Structural Integrity	10%
		Logical Coherence	10%
		Expressive Fluency	10%
Innovation Ability	20%	Thought Originality	10%
		Expressive Novelty	10%
Learning Satisfaction	10%	Course Evaluation	5%
		Learning Experience	5%

4.3 SECI Stage Knowledge Flow Effect Analysis

The SECI stage knowledge flow effect analysis used a multi-dimensional data mining method. In the socialization (S) stage, social network analysis was used to calculate the density of the student interaction network, which increased from 0.3 to 0.7, with a 42% increase in centrality. In the externalization (E) stage, natural language processing technology was used to analyze student writing works, with a key concept extraction accuracy rate increasing from 65% to 89%. In the combination (C) stage, a knowledge graph was constructed, with the number of nodes increasing by 60% to 15,000, and the number of edges increasing by 85% to 75,000. In the internalization (I) stage, learning behavior analysis was used to discover that students' knowledge application ability increased by 38%, and problem-solving speed increased by 45%. Machine learning algorithms were used to predict students' knowledge flow trajectories, with an accuracy rate of 87%. Through time-series analysis, a knowledge flow efficiency curve was drawn, as shown in Fig. 7. The curve shows a clear upward trend, with a peak value at the 12th week, and an overall efficiency increase of 46%.

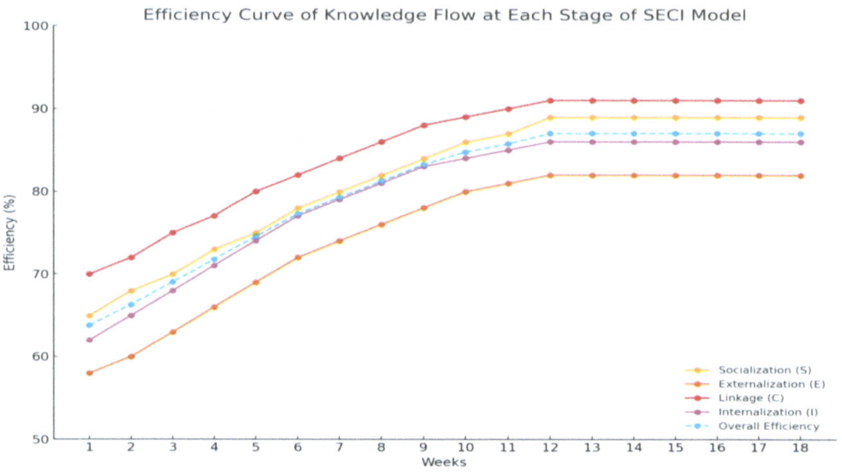

Fig. 7. SECI Model Stage Knowledge Flow Efficiency Curve

4.4 Learner Writing Ability Improvement Analysis

The learner writing ability improvement analysis used a longitudinal comparison method, tracking the changes in students' writing ability over 18 weeks. Through natural language processing technology, students' weekly writing works were automatically scored, with scoring dimensions including grammar accuracy, vocabulary richness, structural integrity, and expressive fluency. The experimental group's average score increased from 68 to 87, with an improvement rate of 27.9%; the control group's average score increased from 67 to 75, with an improvement rate of 11.9%. Using t-test analysis, the difference between the two groups was significant, with a p-value less than 0.001. Vocabulary analysis showed that the experimental group's active vocabulary increased by 35% to 3,500 words, while the control group's active vocabulary increased by 15% to 2,800 words. Grammar error rate analysis showed that the experimental group's error rate decreased from 30 to 8 per thousand words, while the control group's error rate decreased from 29 to 18 per thousand words. Table 5 shows the specific improvement data for each dimension of writing ability.

Table 5. Learner Writing Ability Improvement Data Comparison

Evaluation Dimension	Experimental Group Improvement	Control Group Improvement	Significance
Grammar Accuracy	38%	15%	$p < 0.001$
Vocabulary Richness	35%	12%	$p < 0.001$
Structural Integrity	42%	18%	$p < 0.001$
Expressive Fluency	33%	14%	$p < 0.001$
Overall Score	27.90%	11.90%	$p < 0.001$

4.5 Teacher Teaching Strategy Optimization Analysis

The teacher teaching strategy optimization analysis was based on the big data collected by the intelligent teaching system, using machine learning algorithms to model and optimize teaching behavior. Through analysis of 50 teachers' 300 classes, the key factors affecting teaching effectiveness were identified. Teaching interaction frequency was positively correlated with student participation ($r = 0.78$, $p < 0.001$), and personalized feedback frequency was highly correlated with writing ability improvement ($r = 0.85$, $p < 0.001$). Based on these findings, a teaching strategy recommendation system was developed to provide teachers with real-time teaching suggestions. After the system was launched, teachers' teaching satisfaction ratings increased from 4.2 to 4.8 (out of 5), and students' classroom participation increased by 40%. Teaching efficiency was significantly improved, with a 25% reduction in time required to complete the same teaching goals. Figure 8 shows the comparison analysis of teaching strategy optimization before and after, clearly presenting the improvement in each indicator.

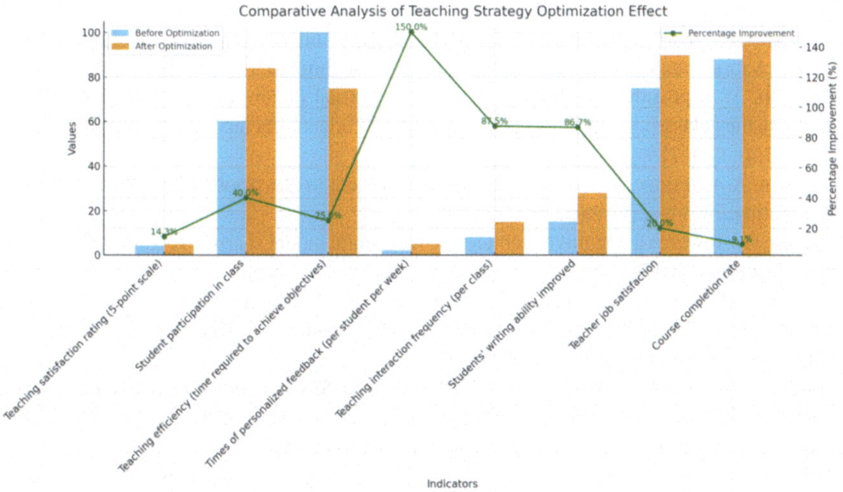

Fig. 8. Teaching Strategy Optimization Effect Comparison Analysis

5 Conclusion

This study constructed a Spanish writing intelligent classroom to explore the knowledge creation and transformation mechanism based on the SECI model. The experimental results show that the optimized teaching strategy significantly improved the learning effect. The average writing ability of students improved by 27.9%, far exceeding the traditional teaching method. The overall knowledge flow efficiency improved by 36.5%, with coordinated development in each stage. The teaching satisfaction of teachers increased from 4.2 to 4.8, and the classroom participation of students increased by 40%. These data fully confirm the effectiveness of the SECI model in language teaching, providing new ideas for the development of intelligent education in the future. The research results are not only applicable to Spanish teaching but also can be extended to other language learning fields, with broad application prospects.

Acknowledgement. The work is supported by SECI Model Empowered by Big Data in Research on Knowledge Flow in Spanish Writing Wisdom Classroom, 2024GXJK554.

References

1. Dneprovskaya, N., Shevtsova, I.: A knowledge management system in the strategic development of universities. Bus. Inform. (2023)
2. Baker, A.M.: The directional knowledge sharing paradigm: a mixed-methods case study of the interaction between organisational culture and intra-organisational knowledge sharing. J. Inf. Knowl. Manag. (2022)
3. Mark, B., Phornchulee, L.: The development of teachers' knowledge and behaviour in promoting self-discipline: a study of early years teachers in Thailand. Education, 1–13 (2018)
4. Songkram, N., Chootongchai, S.: Effects of pedagogy and information technology utilization on innovation creation by SECI model. Educ. Inf. Technol. **25**(5) (2020)
5. Suriyon, A., Inprasitha, M., Sangaroon, K.: Contextual factors in the open approach-based mathematics classroom affecting development of students' metacognitive strategies. Sociol. Mind **3**(4), 284–289 (2013)
6. Anupan, A.A.: Framework of knowledge management in classroom action research on cloud computing for pre-service teachers (2020)
7. Hagert, G.: Towards a cognitive architecture for mental model sharing: redefining knowledge-based systems (2017)
8. de Castro Peixoto, L., Barbosa, R.R., de Faria, A.F.: Management of regional knowledge: knowledge flows among university, industry, and government. J. Knowl. Econ., 1–19 (2021). https://doi.org/10.1007/s13132-020-00702-9
9. Sardjono, W., Perdana, W.G.: Using ISO-9126 and SECI model for enhancing enterprise resource planning implementation strategic in broadcasting company. In: 2019 International Conference on Information Management and Technology (ICIMTech) (2019)
10. Bibbes, T.N.: Toward a new understanding of the project manager as a mixing valve of organizational knowledge: a case study approach. Georgia State University (2013)

Analysis of Physical Education Teaching Evaluation System and Design of Intelligent Evaluation System

Qingke Wang(✉) and Jiutang Zhu

Xuzhou Medical Branch of Jiangsu Union Technical Institute, Xuzhou 221116, Jiangsu, China
Tsingke0119@163.com

Abstract. This paper aims to study the evaluation methods and strategies of physical education (PE) classroom teaching effect based on big data. This paper introduces the main elements of PE teaching, and adopts diversified allocation modes of PE teachers to reform the PE teaching mode. Moreover, starting from promoting lifelong PE awareness, this paper chooses reasonable PE teaching contents to deepen students' ideological understanding and cultivate students' comprehensive PE quality. In addition, under the optimized teaching mode, this paper introduces test analysis methods to analyze the learning evaluation of PE teaching effects, and constructs the multi-learning evaluation model function of diversified PE teaching effect to optimize the design of the multi-learning evaluation model of diversified PE teaching effect. Finally, this paper verifies the feasibility and application advantages of using artificial intelligence technology to evaluate the effect of PE.

Keywords: Physical · Teaching · Intelligent Evaluation · Big Data · Artificial Intelligence

1 Introduction

Big data, blockchain, and "Internet plus" have developed rapidly, among which big data is an effective way to integrate resources. Nowadays, the explosive influx of data has fully penetrated into social production and people's daily life, and behind the suppression of ByteDance and WeChat in the United States is actually the competition for data. Moreover, data explosion continues to grow in three dimensions: the same type of data volume, data growth rate, data sources and new data types. Big data is setting off a data technology revolution and thinking mode change, which not only quietly changes people's thinking mode and behavior habits, but also provides more methods and possibilities for scientific research. Under this macro background, big data has also had a significant and profound impact on education. Teachers' competence needs to be redefined, students' learning is fragmented, and the collected educational data is constantly extended in breadth, depth and subdivision [1].

Big data is a large-scale data group formed in terms of data capture, storage and analysis. At present, it is difficult to obtain corresponding data using traditional methods,

© The Author(s), under exclusive license to Springer Nature Switzerland AG 2025
F. Paas et al. (Eds.): EduTA 2024, LAIS 51, pp. 105–114, 2025.
https://doi.org/10.1007/978-3-031-95252-4_10

but a large amount of data can be obtained using corresponding technologies and means. Big data also has its own characteristics, generally requiring massive data aggregation, rapid data collection, storage, and retrieval [2]. There are various types of data, and data has its value when filtering and aggregating. The diversity of data makes data analysis more complex. When quickly obtaining data, the ability to process data analysis is even more important. In addition to traditional data forms, the storage and presentation of big data also include unstructured forms such as image information, network logs, sound, animation, and video. When filtering and aggregating data, its value mainly refers to the deep analysis hidden behind these large amounts of data, which can be deeply mined and extracted into different types of data [3].

It can be seen from the above that big data is not only a kind of data resource, but also a method from the perspective of analysis and summary. As long as meaningful data can be summarized in the form of a database, and ultimately transformed into useful data or information for different people to extract. Therefore, big data is also an effective means of obtaining information [4].

Teaching evaluation is increasingly emphasizing process evaluation. Educational evaluation should evaluate the entire individual of students, not just their abilities in a certain aspect. It should be based on the specific development situation of students and continuously develop evaluation during their growth process. Developmental evaluation refers to the process of collecting the entire development process, conducting systematic analysis and evaluation, and judging the value that students gain in the process as the evaluation subject continues to develop, in order to serve teaching objectives [5].

The developmental evaluation of PE teaching mainly focuses on the evaluation of the teaching process. Individual differences should be emphasized, which is a long-term dynamic evaluation process. The application of big data is conducive to the collection of dynamic data and the integration analysis of heterogeneous data [6]. For example, based on the development process of time, tracking data on an annual basis can adopt deep data mining and processing methods such as time series and cluster analysis, taking the teaching situation of primary and secondary schools as the research object, to obtain the desired information, understand the subject map of students, and conduct personalized analysis of student learning, which can more accurately track personalized research for each student [7]. The cultivation starts from young children, and the selected teachers are all professionals. Through the analysis, personalized training models for students are refined to promote their development [8].

Traditional evaluation mainly refers to the evaluation of students' physical test data and motor skills level, which is mainly conducted by the competent department or teachers participating in the teaching [9]. The entire evaluation process is not an open process. Focusing on the process evaluation of students has been put on the agenda, making the entire evaluation process more fair and just. It is necessary to showcase the characteristics of the entire student from multiple aspects, not only learning behavior, but also comprehensive quality evaluation. In the continuous process, more data should be collected continuously. After accumulating a certain amount, a database can be formed for big data simulation modeling to obtain some rules of student development, so as to predict the development trajectory of students and make timely adjustments and correct guidance and planning [10].

The collection of big data in PE teaching mainly comes from daily teaching activities, including teaching, management, scientific research, etc. Students and PE teachers are both producers and carriers of big data, and the development and evaluation of each step are closely related. The collection of big data is a comprehensive collection of multiple aspects, which lays the foundation for the openness and speed of data acquisition, and also provides channels and bridges for multiple parties to participate in the evaluation of PE teaching. From the perspective of the subjects that need to be evaluated [11].

It comprehensively collects, organizes, and analyzes students' classroom performance, physical condition, and learning skills, and finally combines exam scores to obtain a comprehensive and persuasive evaluation. Integrate the entire development process of students, including their learning situation, social situation, mastery of learning methods, and various learning records, to effectively evaluate their knowledge, emotions, attitudes, etc., relying on the collection of big data to ensure the effectiveness, fairness, and objectivity of the evaluation [12].

The evaluation of PE teaching that integrates big data can select multiple indicators when collecting data, such as various exam scores, competition organization processes and scores, questionnaire surveys, interviews, oral history, etc. There is both static and dynamic information. In process evaluation, it is necessary to students' learning paths or behaviors in real time. To ensure the integrity of data collection and enhance the accuracy and reliability of the entire evaluation process, video recording can be used to store the information. After collecting video or audio materials, the video materials can be analyzed by Dartfish video analyzer. By using Huawei smartwatches and wristbands, the changes in students' heartbeat, pulse rate, respiratory frequency, and blood oxygen saturation during PE teaching activities can be recorded [13].

2 Evaluation Model of PE Teaching

2.1 Diversified Teaching Evaluation Model

Combined with statistical analysis methods, a multi-learning evaluation model of diversified teaching effects of PE is established, and an unbiased estimation model of multi-learning evaluation of diversified teaching effects of PE is established by using normal distribution model. As follows:

$$F_{nd}(x) = \int_{-\infty}^{x} \frac{1}{\sqrt{2\pi\delta^2}} exp\left\{\frac{(x-\mu)}{2\delta^2}\right\} dx \quad (1)$$

In formula (1), μ is the confidence parameter of the multi-learning evaluation model of diversified teaching effect of PE, δ is a scale parameter.

The standard normal analysis is carried out, and the confidence distribution probability distribution function of the evaluation of diversified teaching effects of PE is obtained as follows:

$$F_{Lnd}(x) = \int_{0}^{x} \frac{1}{x\delta\sqrt{2\pi}} exp\left\{\frac{(lnx-\mu)2}{2\delta^2}\right\} dx \quad (2)$$

According to the negative bias of diversified teaching effect of PE, the probability distribution function of reliability distribution of diversified teaching effect of PE is obtained by using the method of binary parameter analysis as follows:

$$F_{WL}(x) = \int_{0}^{x} \frac{\delta}{k}\left(\frac{x}{k}\right)^{\delta-1} exp\left(-\frac{x}{k}\right)^{\delta} dx \tag{3}$$

k is multi-learning for the diversified teaching effect of PE. Combined with statistical decision-making and fuzzy judgment methods, the generalized extreme value distribution function of diversified teaching effect evaluation of PE is obtained as follows:

$$F_{GEV}(x) = \int_{0}^{x} exp\left\{-\left[1 + \frac{k}{\delta}(x-\mu)\right]^{\frac{1}{k}}\right\} dx \tag{4}$$

The gamma probability distribution function is as follows:

$$F_{GM}(x) = \frac{\delta^k}{\Gamma(k)} \int_{-\infty}^{x} (x-\mu)^{k-1} exp[-(x-\mu)] dx \tag{5}$$

According to the above distribution function construction of the evaluation of diversified teaching effect of PE, the following comprehensive test formula of diversified teaching effect of PE is formed by adopting the characteristics of goodness-of-fit test method:

$$H = \frac{P_{tj}(PP)}{P_{av}(PP)} - \frac{P_{tj}(RM)}{P_{av}(RM)} - a_{jy}(\chi^2) - b_{jy}(KS) - c_{jy}(AD) \tag{6}$$

In formula (6), $P_{tj}(PP)$ and $P_{av}(PP)$ are the statistics and mean values under PPcc test, respectively, $P_{tj}(RM)$ and $P_{av}(RM)$ are the statistics and mean values of diversified teaching effect of PE, respectively, and $a_{jy}(\chi^2)$ is the characterization value of whether the χ^2 test passes or not. The confidence level of the evaluation of diversified teaching effect of PE obtained by χ^2 test is set to 0, otherwise it is taken to 1. $b_{jy}(KS)$ is the characterization value of whether the evaluation of diversified teaching effect of PE is obtained through $K-S$ test. It is set to 0 when it passes the $K-S$ test, otherwise it is taken to 1. $c_{jy}(AD)$ is the characterization value of whether the confidence level of diversified teaching effect of PE can be obtained through $A-D$ test.

2.2 Intelligent Evaluation Model

PE is mainly composed of "teaching" and "learning". Generally speaking, "teaching" means that teachers disseminate information, while "learning" means that students acquire and apply information. The evaluation of students is reflected in the classroom, which is composed of teachers' performance in the process of "teaching" and students' performance in the process of "learning", as well as their technical level and physical quality. The influence of big data on "teaching" and "learning" in PE is mainly manifested in the ways, ways and methods of disseminating or absorbing information, which has an impact on students' evaluation factors. It is necessary to break the previous single, weak purposeful and partial evaluation methods. Therefore, how to obtain, how to

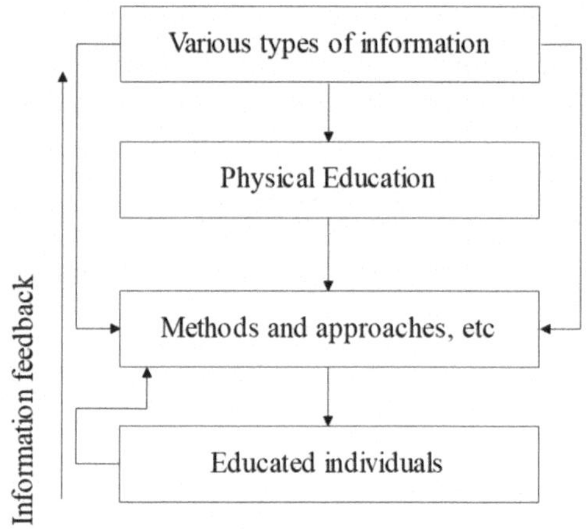

Fig. 1. Relationship between big data and PE

practice, listen to the situation and improve the degree will be the focus of "evaluation". Its flow can be described in Fig. 1.

Process evaluation is to record and calculate the trace data of each stage of students' learning process. The main business process of process evaluation includes the release of teaching activities (tasks) such as sign-in, resources, tests, discussions, homework, etc. Students complete PE teaching activities, and teachers add content and set proportions of learning process evaluation, as shown in Fig. 2.

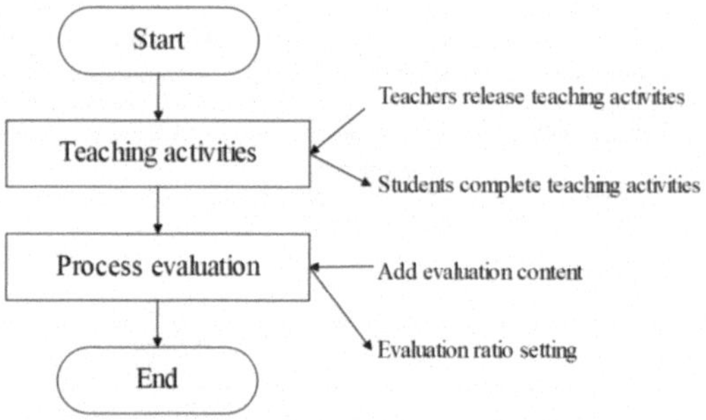

Fig. 2. Evaluation process of PE teaching process

The data of process evaluation comes from the teaching activities (tasks) of teaching interaction. First, the teacher releases the teaching activities. After the students complete the teaching activities, the teacher selects the teaching activities to design the evaluation content and evaluation ratio to complete the process evaluation.

It is concluded that the design of process evaluation platform needs to meet the core functions of teaching interaction, process evaluation and visualization, and learning motivation and emotional values belong to the dimension of qualitative evaluation. After designing the scale, the questionnaire is issued for self-evaluation by students, and the evaluation results are integrated into the client to display with corresponding visual graphics. The hierarchical structure of the platform is shown in Fig. 3, includes four modules: user login, teaching activities, process evaluation and user management.

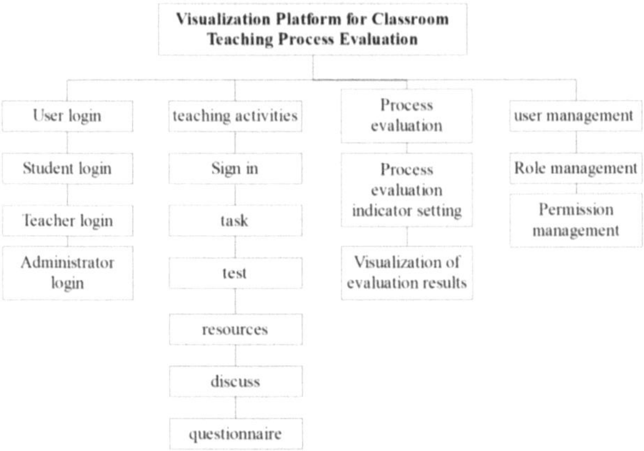

Fig. 3. Functional modules of big data visualization platform of process evaluation

The platform adopts Hadoop's big data offline analysis technology, as shown in Fig. 4.

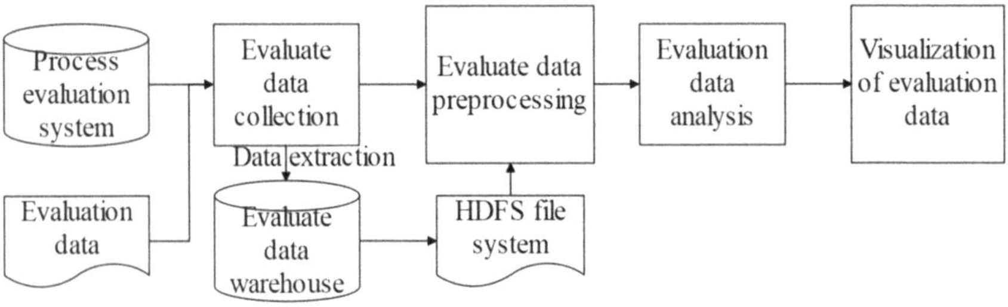

Fig. 4. Big data processing flow of process evaluation

Learning evaluation is a kind of feedback on the results of teaching practice activities, and it is an evaluation activity and feedback form of teaching activities in combination with teaching progress and teaching effect (Fig. 5).

Core literacy means that students acquire the required knowledge and skills after classroom teaching, have the key ability to solve complex problems in the real world, develop good healthy behaviors and have correct values. The evaluation of sports skills, firstly, is to assess from the perspective of "quantity", and to evaluate the completion of a certain "quantity" of sports skills. Second, the evaluation is carried out from the perspective of "quality", mainly from the standardization of technical movements and the practical application ability of tactics. However, the evaluation results are not only summative evaluations, but also attention should be paid to students' performance, such

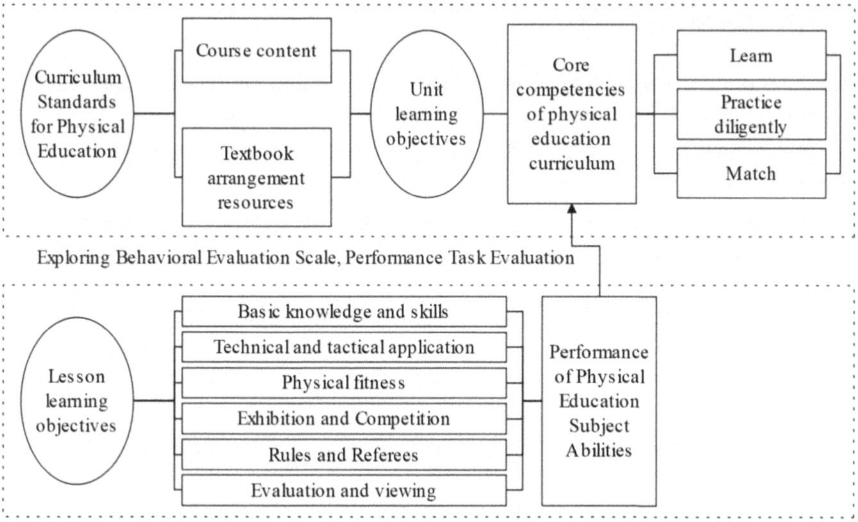

Fig. 5. Evaluation design practice direction framework

as technical movements such as 100-meter running, arm swing and leg pedaling, and the connection between fast running and other techniques. Based on this, a teaching evaluation process diagram pointing to core literacy is designed (Fig. 6). It mainly has several aspects: (1) it judges students' technical performance combined with their technical grades, (2) it classifies and evaluates students' performance in the process of mastering technology according to several situations, (3) it evaluates students' application of learned technology in the face of complex situations.

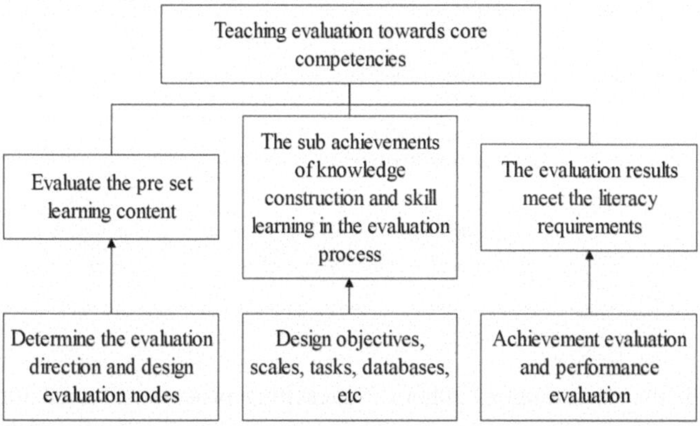

Fig. 6. Teaching evaluation process pointing to core literacy

3 Experimental Study

On the basis of expert consultation and discussion, this paper puts forward the evaluation indicators of school PE quality of each level and the evaluation matrix of the upper level. The first level indexes of school PE quality are compared in pairs to form a matrix diagram. As shown in Table 1.

Table 1. Indicator weight value of PE teaching quality evaluation system

Level 1 indicators	First-level weight	Secondary indicators	Secondary weight	Secondary comprehensive weights	Sort
Background evaluation of school PE teaching (C1)	0.06	Culture objective (C11)	0.198	0.01188	13
		Party requirements (C12)	0.594	0.03564	9
		School environment (C13)	0.198	0.01188	13
Evaluation of school PE teaching investment (I2)	0.31	Team of school PE teaching (I21)	0.0792	0.024552	10
		Software and hardware Facilities of school PE teaching (I22)	0.2277	0.070587	5
		school PE teaching funds (I23)	0.1485	0.046035	8
		System guarantee of school PE Teaching organization (I24)	0.5346	0.165726	2
Evaluation of school PE teaching process (P3)	0.44	Student performance (P31)	0.0495	0.02178	11
		Teacher performance (P32)	0.1584	0.069696	6
		Overall performance of teaching activities (P33)	0.3168	0.139392	3
		Elements of PE curriculum (P34)	0.5053	0.214732	1
Evaluation of school PE teaching achievement (P4)	0.19	Student harvest (P41)	0.1089	0.020691	12
		Teacher harvest (P42)	0.5544	0.105336	4
		Overall teaching effect (P43)	0.3267	0.062073	7

In Table 1, the evaluation of school PE teaching process accounts for the largest weight, followed by the evaluation of school PE teaching achievements and the evaluation of school PE teaching investment, and the relative influence of school PE teaching background evaluation is weak. From the comprehensive weight, it can be seen that the elements of PE curriculum and the overall performance of PE activities in process evaluation, the guarantee of school PE organization system in input evaluation, and the teacher harvest indicator in achievement evaluation play an important role in the evaluation of school PE quality. Therefore, we should focus on improving these aspects in school PE quality.

The effectiveness of the intelligent evaluation system for PE teaching proposed in this paper is verified. Then, several teachers and students are invited to verify the effect of the system model. A total of 20 sets of evaluation data are obtained, and the test results shown in Table 2 below are obtained.

Table 2. Effect verification of intelligent evaluation system of PE teaching

Number	Evaluate Results	Number	Evaluate Results	Number	Evaluate Results
1	87.11	8	91.14	15	91.88
2	92.30	9	89.66	16	92.34
3	90.47	10	90.78	17	91.70
4	92.56	11	92.18	18	90.65
5	90.41	12	88.51	19	88.38
6	88.38	13	90.24	20	87.62
7	87.22	14	90.02		

From the above analysis, the intelligent evaluation system of PE proposed in this paper can effectively improve the evaluation effect of PE, improve the efficiency of PE, and provide some theoretical reference for the subsequent improvement of PE.

4 Conclusion

Through artificial intelligence technology based on big data, students' physical data and physical fitness test data can be monitored and analyzed in real time. According to students' data, the system can provide teachers with personalized training programs and guidance suggestions, which can help students to carry out physical training more efficiently and improve their athletic ability.

This paper comprehensively discusses the application of big data and artificial intelligence technology in sports training, introduces how to use big data and artificial intelligence technology to improve the effect of training programs and guide athletes to improve their skills, and expounds the application prospects of big data and artificial intelligence technology in the field of sports training.

References

1. Mocanu, G.D., Murariu, G., Iordan, D.A., Sandu, I., Munteanu, M.O.A.: The perception of the online teaching process during the COVID-19 pandemic for the students of the PE and sports domain. Appl. Sci. **11**(12), 5558–5570 (2021)
2. O'Brien, W., et al.: Implications for European PE teacher education during the COVID-19 pandemic: a cross-institutional SWOT analysis. Eur. J. Teach. Educ. **43**(4), 503–522 (2020)
3. Hu, D., Zhou, S., Crowley-McHattan, Z.J., Liu, Z.: Factors that influence participation in physical activity in school-aged children and adolescents: a systematic review from the social ecological model perspective. Int. J. Environ. Res. Public Health **18**(6), 3147–3160 (2021)
4. Pranoto, B.E., Suprayogi, S.: A need analysis of ESP for PE students in Indonesia. Premise J. Engl. Educ. **9**(1), 94–110 (2020)
5. Zhang, J., Oh, Y.J., Lange, P., Yu, Z., Fukuoka, Y.: Artificial intelligence chatbot behavior change model for designing artificial intelligence chatbots to promote physical activity and a healthy diet. J. Med. Internet Res. **22**(9), e22845–e22855 (2020)
6. Norris, E., van Steen, T., Direito, A., Stamatakis, E.: Physically active lessons in schools and their impact on physical activity, educational, health and cognition outcomes: a systematic review and meta-analysis. Br. J. Sports Med. **54**(14), 826–838 (2020)
7. Li, H., Cui, C., Jiang, S.: Strategy for improving the football teaching quality by AI and metaverse-empowered in mobile internet environment. Wirel. Netw. **30**(5), 4343–4352 (2024)
8. van Sluijs, E.M., et al.: Physical activity behaviours in adolescence: current evidence and opportunities for intervention. Lancet **398**(10298), 429–442 (2021)
9. Kalb, R., et al.: Exercise and lifestyle physical activity recommendations for people with multiple sclerosis throughout the disease course. Mult. Scler. J. **26**(12), 1459–1469 (2020)
10. Bull, F.C., et al.: World Health Organization 2020 guidelines on physical activity and sedentary behaviour. Br. J. Sports Med. **54**(24), 1451–1462 (2020)
11. Varea, V., González-Calvo, G.: Touchless classes and absent bodies: teaching PE in times of Covid-19. Sport Educ. Soc. **26**(8), 831–845 (2021)
12. Zhai, X., et al.: A review of artificial intelligence (AI) in education from 2010 to 2020. Complexity **2021**(1), 8812542–8812554 (2021)
13. Violant-Holz, V., et al.: Psychological health and physical activity levels during the COVID-19 pandemic: a systematic review. Int. J. Environ. Res. Public Health **17**(24), 9419–9430 (2020)

Construction of College Students' Portrait and Performance Prediction System Under the Background of Intelligent Teaching

Yunxia Yang[1,2](✉)

[1] Shanxi University of Science and Technology, Jincheng 048000, China
1002268574@ucsiuniversity.edu.my
[2] Sitya University Malaysia, 56000 Kuala Lumpur, Malaysia

Abstract. In order to improve the performance management effect of college students, this paper proposes a performance prediction system based on student portraits. This method is based on deep learning and data mining technology, aiming to explore the relationship between students' behavioral features and performance, and predict students' future performance. The data information of students is obtained from multiple databases and preprocessed, and the data is analyzed according to the different behavior features of students in their study and life, and different labels are determined in different aspects, and the data is visualized by the determined labels, and the comprehensive portrait of students is displayed.The model has better performance on real data sets, and can provide reliable theoretical support for the subsequent college students' performance management.

Keywords: intelligent teaching · student portraits · performance · prediction

1 Introduction

Online education is popular all over the world with its features of "large-scale, openness and sharing". At the same time, learner portrait is characterized by portraying and presenting the whole picture of students' information, it is easy for students to form a loose learning mentality, which makes it difficult to gain good learning results. Therefore, using the advantages of learner portraits to alleviate the difficult problem of academic performance prediction in online education, help students adjust their learning status and improve their learning efficiency in time, is a hot spot in the current educational circles, and it is also one of the focuses of this study. Under the guidance of learner portraits, this study predicts academic performance. On the one hand, it can make the originally disorderly and complicated learning data orderly and concise, and on the other hand, it can gradually alleviate the current situation that teachers originally lack visual clues and cannot judge learners' participation, which helps to promote the harmonious coexistence of online teaching and learning and create a more personalized learning experience for more learners [1].

The prediction of student academic performance is currently the most widely studied field in educational data mining, aiming to detect abnormal learning states of students

© The Author(s), under exclusive license to Springer Nature Switzerland AG 2025
F. Paas et al. (Eds.): EduTA 2024, LAIS 51, pp. 115–124, 2025.
https://doi.org/10.1007/978-3-031-95252-4_11

in advance through data mining, intervene early, and prevent problems such as failing courses, dropping out, and unemployment [2]. Reference [3] proposes a stochastic model based on Survival Analysis to model the learning process of students, in order to predict how many students may drop out at when, providing schools with reference and taking early measures for students who may drop out. Reference [4] proposed a method based on Cox process to predict when college students are likely to drop out. Besides universities, the graduation issue of high school students is also worth paying attention to. Reference [5] analyzed the factors that may cause students to delay graduation and proposed a hierarchical model to identify students with different delay risks. Based on the analysis of factors affecting students' graduation, these predictions can be used to provide targeted intervention measures for these students and may improve their on-time graduation rate.

The analysis and prediction of students' employment status is another key issue in academic performance analysis based on educational big data. The employment rate and quality of students to a certain extent reflect the teaching quality of educational institutions, are important indicators for evaluating the quality of higher education, and also objectively reflect the academic level of students. To improve the quality of student employment, researchers have made a lot of efforts, and the most directly effective one is the employment recommendation system. Reference [6] designed a job recommendation service framework for college students, which profiles students based on their personal information and academic performance, and recommends potential suitable jobs for each student based on their ranking. Reference [7] proposed an algorithm for evaluating job suitability, which establishes a framework to determine whether students are suitable for a specific job by calculating their domain ability and job ability scores, providing reference for students' job search. In addition, it is also important to analyze the factors that affect students' employment. Reference [8] analyzed information such as students' grades, gender, ethnicity, and place of origin, and explored and modeled potential employment biases that employers may have towards certain specific groups during the student employment process. It also predicted the employment status of students. These studies and analyses are of great help in explaining the key factors that affect students' employment, and have certain guiding value for improving the quality of students' employment.

Whether it is the application of intelligent education systems or the focus on student employment, it is essentially an emphasis on student performance. The fundamental goal of educational big data research is to estimate students' grades and take effective measures to help them improve their grades. At present, many researchers have proposed various methods to predict students' grades [9], but they are still far from meeting the requirements for application in universities. Therefore, it is still necessary to design an efficient, accurate, and easy-to-use algorithm for predicting student grades. In addition, it is also important to explore and study the factors that affect student performance and the ways in which various factors work.

The use of classification or regression algorithms to solve grade prediction problems is mostly based on students' basic information and consumption, access control, and other data, ignoring the correlation between classroom behavior data and courses. The behavior of students in the classroom is an important data reflecting their learning situation during the teaching process, so the classroom is one of the most suitable places to collect teaching process data. There are two approaches to using association rule analysis for grade prediction. The first is to analyze the correlation between consumer behavior and grades. The second is to analyze the correlation between grades of various courses, which cannot be considered simultaneously. The mining of course correlation has a significant dependence on the academic records of previous students. When using recommendation system algorithms for grade prediction, it also relies entirely on students' historical grade data. In the absence of historical data, it can lead to data sparsity and cold start problems, making it impossible to predict the grades of newly opened courses and new students [10].

In the absence of historical grade data, mining intrinsic information of courses is a good way to supplement data. Reference [11] introduced course knowledge graph as an information supplement in the absence of historical data scenarios when using collaborative filtering for grade prediction. Course knowledge graph includes six entities: courses, knowledge points, departments, reference books, textbooks, and teaching modes. Using the knowledge representation of course knowledge graph to learn and calculate semantic similarity between courses significantly improves the performance of grade prediction, but the constructed course knowledge graph involves less intrinsic information of courses.

2 Construction of Performance Prediction Model

2.1 Overall Structure

The model framework diagram is shown in Fig. 1, which contains three modules:

(1) Data input module. Firstly, the original data extracted from the database is filtered, filled, converted and merged by statistics, data preprocessing and other methods, and on this basis, the values of each behavior and performance are converted into initial eigenvectors through random parameter matrix.
(2) Attention Mechanism Module. The balance score module superimposes the feature maps of two semesters, uses the attention mechanism to calculate the fusion feature V^t after convolution and pooling operations, and then obtains the balance factor $\tilde{V}^t$ that can reveal the relationship between the two semesters through maximum pooling and multi-layer perceptron.
(3) Classification and prediction module. It splices the behavior feature vector V^1 and V^2 and the balance factor $\tilde{V}^t$ to obtain a complete feature vector, and finally predicts students' performances and classifies them.

It refers to operations such as reorganization, denoising, statistics, and processing of data. The min-max normalization method is used to divide the data into [01] as follows.

$$x^* = \frac{x - \min(x)}{\max(x) - \min(x)} \tag{1}$$

Among them, x^* is the normalized data value, x is the value of behavior feature data, $\min(x)$ is the minimum value of sample data, and $\max(x)$ is the maximum value of sample data.

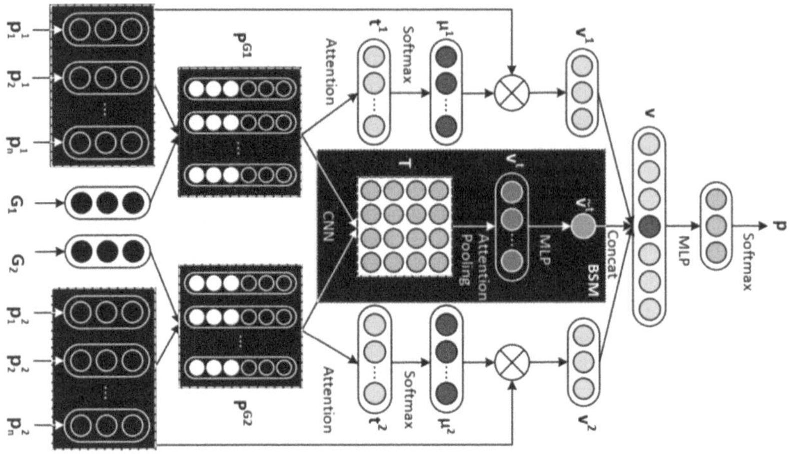

Fig. 1. Framework diagram of college student performance prediction model

The processed results $P^i = (P_1^i, P_2^i, \cdots, P_n^i) \in \mathbb{R}^n$ correspond to the normalized data of behavioral features. At the same time, the two historical performances are set as G^1 and G^2 respectively.

2.2 Vectorized Representation

The behavior eigenvalues and historical performance of two semesters are mapped to high-dimensional space. Moreover, this paper randomly generates a high-dimensional parameter matrix W_k, and transforms the behavior and performance of two semesters into behavior feature matrix $\tilde{P}^i \in \mathbb{R}^{k \times n}$ and performance vector $g^i \in \mathbb{R}^k$ by W_k, k as each behavior feature and historical performance vector dimension.

$$\tilde{P}^i = W_k \cdot p^i = \begin{bmatrix} p_{11}^i & p_{12}^i & \cdots & p_{1n}^i \\ p_{21}^i & p_{22}^i & \cdots & p_{2n}^i \\ \vdots & \vdots & \ddots & \vdots \\ p_{k1}^i & p_{k2}^i & \cdots & p_{kn}^i \end{bmatrix} \tag{2}$$

$$g^i = W_k \cdot G_i = (g_1^i, g_2^i, \cdots, g_n^i)^T \tag{3}$$

The output of the data input module is the behavior feature map $\tilde{p}^1$ and $\tilde{p}^2$ and historical performance feature vector g^1 and g^2. Consider the influence of behavior changes and performance changes in different semesters, this paper combines the two semester's data with attention mechanism to calculate a balance factor, so as to calculate the final prediction.

Multi-attention mechanism module. First, each semester's behavior features and historical performance features are combined with j, and each column of $\tilde{p}^i$ is spliced with g^i into a new feature $P^{Gi} = (P_1^{Gi}, P_2^{Gi}, \cdots, P_n^{Gi}) \in \mathbb{R}^{2k \times n}$ through the following formula:

$$p_j^{Gi} = [\tilde{p}_j^i : g_j^i] \tag{4}$$

Among them, = 1,2, $\cdots$, n, and n is the number of behavioral features.

The first semester behavior performance feature map p^{G1} and the second semester behavior performance feature map p^{G2} are obtained, and the attention mechanism is introduced to set the weight. For the two semesters, the attention weight $t^i = (t_1^i, t_2^i, \cdots, t_n^i) \in \mathbb{R}^n$ under different behaviors in different semesters is obtained.

$$t_j^i = MLP(p_j^{Gi}) \tag{5}$$

All t_j^i are normalized,the attention scores $\mu^i = (\mu_1^i, \mu_2^i, \cdots, \mu_n^i) \in \mathbb{R}^n$ corresponding to each behavior feature are obtained.

$$\mu_j^i = softmax(t_j^i) = \frac{\exp(t_j^i)}{\sum_{j=1}^{i}(t_j^i)} \tag{6}$$

The attention score μ^i is weighted and summed with the attribute eigenvalues of corresponding positions in the behavior feature matrix $\tilde{p}^i$, and the semester eigenvector $v^i \in \mathbb{R}^k$ is obtained.

$$v^i = \sum\nolimits_{j=1}^{n} \mu_j^i \tilde{p}_j^i \tag{7}$$

A special structure BSM is designed to calculate the balance score. Firstly, the behavior performance feature maps p^{G1} and p^{G2} of two semesters are added to get the aggregated feature mapping $P^G \in \mathbb{R}^{2\times k\times n}$.

$$p^G = p^{G1} \oplus p^{G2} \tag{8}$$

1 × 1 convolution has only one parameter or weight for each channel in the input and allows the feature map to perform nonlinear calculations on the input feature map, which is very effective in operations such as dimensionality reduction or low-dimensional embedding.

$$T = \delta(Conv(W_t \cdot p^G)) = [\begin{matrix} t_{11} & t_{12} & \cdots & t_{1n} \\ t_{21} & t_{21} & \cdots & t_{2n} \\ \vdots & \vdots & \ddots & \vdots \\ t_{k1} & t_{k1} & \cdots & t_{kn} \end{matrix}] \tag{9}$$

3 Learning Portrait Construction and Performance Prediction

3.1 Learning Portraits

From the perspective of learner portraits in online education environment, this paper proposes a process of academic performance prediction based on learner portraits, as shown in Fig. 2, which is divided into five main steps: determining goals, constructing label system and data modeling, data collection and preprocessing, analyzing and outputting learner portraits, and implementing academic performance prediction. This process is a systematic framework for the two-way integration of learner portraits and academic performance prediction. The goals of this portrait construction are determined as follows. The goal (1) is to accurately depict the online learning characteristics of learners through learner portraits. The goal (2) is to evaluate and reflect on online learning results through academic performance prediction. The goal (3) is to further reflect on teaching, provide personalized services for learners, and help teachers formulate early warning and intervention measures.

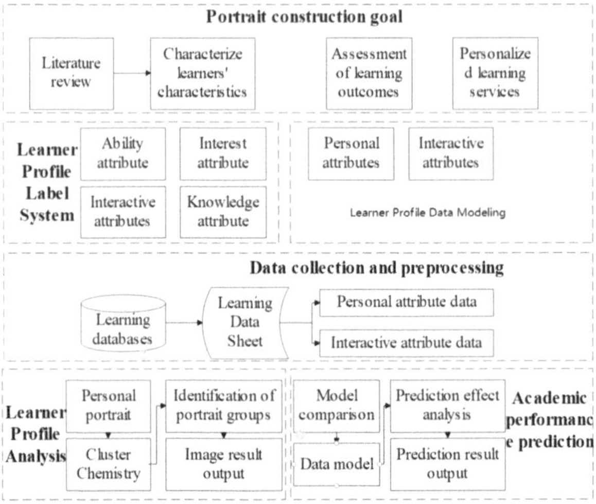

Fig. 2. Academic performance prediction process based on learner portrait

When constructing a student portrait, first of all, it is necessary to obtain the relevant data information of students from multiple databases and preprocess the data, and then analyze the data according to the different behavior characteristics of students in the related processes such as study and life, so as to determine different labels in different aspects. Finally, it is necessary to reuse the determined labels for data visualization operations and display the comprehensive portraits of students. Figure 3 shows the construction process of student portraits.

Combined with the algorithm in the second part, this paper sets the data preprocessing process to the basic form shown in Fig. 4.

The system mainly includes three modules. Among them, the behavior management module is to add, delete, modify and check behavior data and preprocess data. The performance prediction module is to train the model and output prediction results. The comprehensive portrait module is to analyze students' levels in all aspects by visual means. The system module diagram is shown in Fig. 5.

3.2 Performance Prediction

This design is jointly developed using IDEA and PyCharm. IDEA, also known as IntelliJIDEA, is an IDE designed for Java development that supports multiple programming

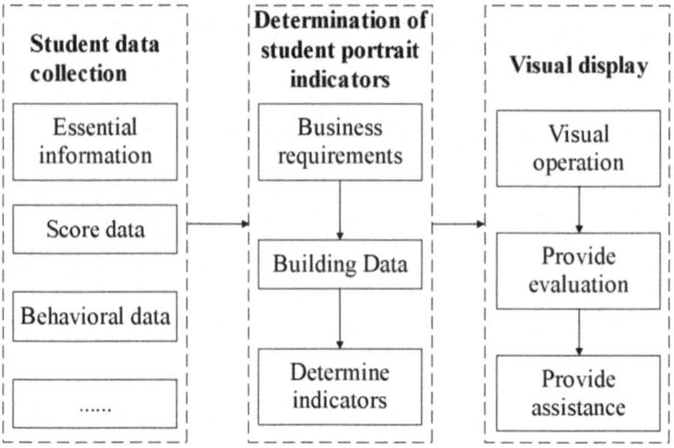

Fig. 3. Flow chart of constructing student portraits

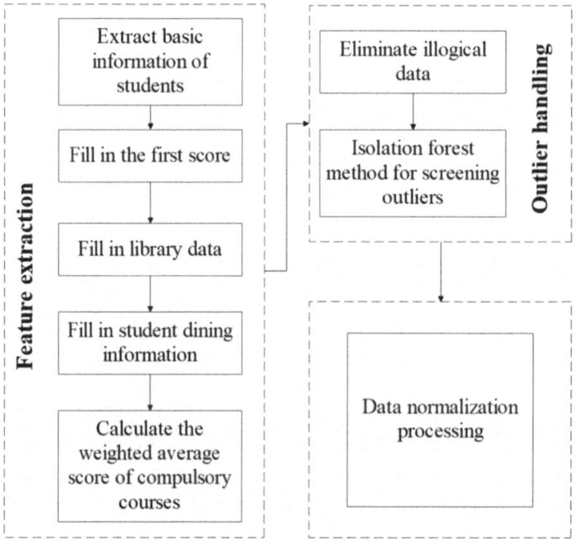

Fig. 4. Data preprocessing process

languages. Its main features are powerful code intelligent prompts, code refactoring, auto completion, code debugging, etc. It supports multiple version control tools such as Git, SVN, etc., and also supports multiple build tools such as Maven, Gradle, etc. PyCharm is an IDE designed for Python development, which also supports multiple programming languages. Similar to IDEA, PyCharm also has powerful features such as intelligent code prompts, code refactoring, and auto completion. It supports various version control tools such as Git and SVN, as well as multiple Python virtual environments such as Anaconda.

As shown in Table 1. The accuracy of the improved model can reach 82.25%, which is 13.84% higher than the best traditional method. This indicates that the improved model has certain advantages in performance prediction.

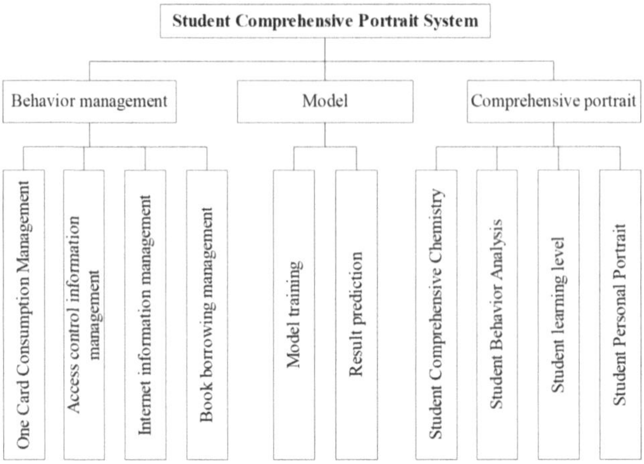

Fig. 5. System functional module diagram

Table 1. Results of different methods on datasets

	Accuracy	Precision	Recall	F1
LR	63.27	63.66	62.26	58.16
DT	63.49	62.83	61.33	62.05
SVM	65.52	65.18	61.25	62.12
RF	67.73	68.02	65.75	65.79
Our Model	81.43	81.91	80.17	80.70

The balanced score is visualized, as shown in Fig. 6. At the same time, two horizontal lines are drawn to measure the influence of balance scores on the final performance. It can be clearly seen that the distribution of balance scores of students in the same category is different. For example, students with grade A generally have higher balance scores, while students with grade C generally have lower balance scores than 0.

Currently, educational informatization has become one of the key strategies promoted by the country. The government, schools, and parents have increasingly high requirements for education, and the comprehensive quality evaluation of students has become an important issue in education reform. However, traditional student management methods can no longer meet the needs of modern education, and information technology is needed to achieve digitalization, networking, and intelligence in student management.With the rapid development of information technology and the continuous accumulation of student data, the comprehensive student portrait system has become a hot topic in the education industry. The student comprehensive portrait system can integrate data from various aspects of students' school life, forming a complete and multi-dimensional student comprehensive portrait, helping teachers and school administrators

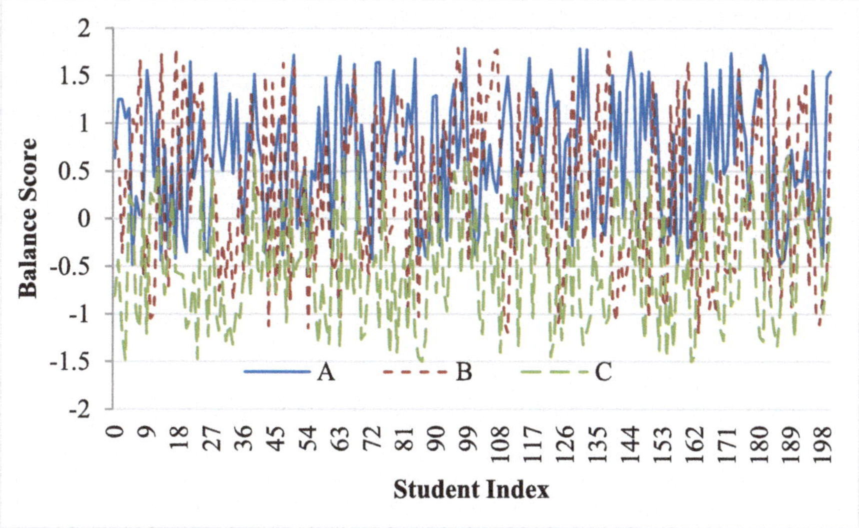

Fig. 6. Balanced scores of different categories of students

better understand students' learning situation, learning styles, interests and hobbies, and providing schools with better teaching management services.

For the construction of a comprehensive student portrait, it is necessary to analyze students' information from multiple perspectives, taking into account their consumption, life, learning, and other factors that may be involved, such as their family background, interests, and hobbies, in order to form a comprehensive and accurate student portrait. When constructing a comprehensive student profile, various data analysis methods can be considered, such as clustering analysis, association rule mining, decision trees, etc. Students can be grouped based on their tag characteristics, and in-depth analysis can be conducted on each student group to uncover patterns and differences between groups. At the same time, regression analysis, time series analysis, and other methods can be used to predict students' future behavior, trends, and development directions based on their historical data.

In the construction of comprehensive portraits, it is necessary to pay attention to the quality and accuracy of data, and avoid distortion of portraits caused by improper data collection and processing. At the same time, it is also necessary to consider data protection and privacy issues to avoid the leakage of students' personal information. Ultimately, the comprehensive portrait of students should be presented in a clear chart, which facilitates teachers and school administrators to conduct comprehensive and accurate analysis and management of students.

4 Conclusion

This paper presents a college student performance prediction model based on multiple attention mechanisms. Firstly, a set of parallel attention mechanisms are used to mine the behavior features of the last two semesters, then the behavior and performance data of the two semesters are combined, and the semester balance score is calculated. In addition, from the perspective of learner portrait in online education environment, the academic performance prediction process based on learner portrait is proposed, and finally all the features are fused to predict student performance. And different labels are determined in different aspects, and the data is visualized by the determined labels, and the comprehensive portrait of students is displayed. The model has better performance on real data sets, and can provide reliable theoretical support for the subsequent college students' performance management.

References

1. Nabil, A., Seyam, M., Abou-Elfetouh, A.: Prediction of students' academic performance based on courses' grades using deep neural networks. IEEE Access **9**(1), 140731–140746 (2021)
2. Pabba, C., Kumar, P.: An intelligent system for monitoring students' engagement in large classroom teaching through facial expression recognition. Expert. Syst. **39**(1), e12839–e12850 (2022)
3. Sunday, K., Ocheja, P., Hussain, S., Oyelere, S., Samson, B., Agbo, F.: Analyzing student performance in programming education using classification techniques. Int. J. Emerg. Technol. Learn. (iJET) **15**(2), 127–144 (2020)
4. Adnan, M., et al.: Predicting at-risk students at different percentages of course length for early intervention using machine learning models. IEEE Access **9**(2), 7519–7539 (2021)
5. Lin, F.C., Ngo, H.H., Dow, C.R., Lam, K.H., Le, H.L.: Student behavior recognition system for the classroom environment based on skeleton pose estimation and person detection. Sensors **21**(16), 5314–5324 (2021)
6. Nagy, M., Molontay, R.: Interpretable dropout prediction: towards XAI-based personalized intervention. Int. J. Artif. Intell. Educ. **34**(2), 274–300 (2024)
7. Goldberg, P., et al.: Attentive or not? toward a machine learning approach to assessing students' visible engagement in classroom instruction. Educ. Psychol. Rev. **33**(2), 27–49 (2021)
8. Chen, J., Zhang, Y., Wei, Y., Hu, J.: Discrimination of the contextual features of top performers in scientific literacy using a machine learning approach. Res. Sci. Educ. **51**(Suppl 1), 129–158 (2021)
9. Susilawati, E., Lubis, H., Kesuma, S., Pratama, I.: Antecedents of student character in higher education: the role of the Automated Short Essay Scoring (ASES) digital technology-based assessment model. Eurasian J. Educ. Res. **98**(98), 203–220 (2022)
10. Shafiq, D.A., Marjani, M., Habeeb, R.A.A., Asirvatham, D.: Student retention using educational data mining and predictive analytics: a systematic literature review. IEEE Access **10**(3), 72480–72503 (2022)
11. Gao, N., Shao, W., Rahaman, M.S., Salim, F.D.: N-gage: predicting in-class emotional, behavioural and cognitive engagement in the wild. Proc. ACM Interact. Mobile Wearable Ubiquit. Technol. **4**(3), 1–26 (2020)

Designing Courses on Foundations and Applications of Large AI Models

Kuo-Kun Tseng[1], Bin Hu[2], and Fuqing Li[3](✉)

[1] School of Computer Science and Technology, Harbin Institute of Technology, Shenzhen, China
kktseng@hit.edu.cn
[2] Changsha Normal University, Changsha, China
[3] Xiangtan University, Xiangtan, China
lifuqing@xtu.edu.cn

Abstract. This paper introduces the course titled "Foundations and Applications of Large AI Models," which is offered to all students across the university by the School of Computer Science and Technology. The course aims to provide a comprehensive understanding of the basic concepts, development history, core technological principles, and application cases of artificial intelligence (AI) and large models, while also addressing their ethical, legal, and societal impacts. The course content encompasses various aspects, including the underlying technologies of large models, language large models, text processing techniques, multimodal large models, image analysis technologies, stable diffusion large models, image generation techniques, applications in copywriting and cultural creativity, code generation, video generation, stock and housing price prediction, robotics, autonomous driving, and AI ethics. The teaching methods employed in this course include lecture-based instruction, case analysis, group discussions, interactive question-and-answer sessions, hands-on demonstrations, and project report-based learning. The assessment methods for the course consist of regular assignments and a final report, focusing on knowledge mastery, skill application, thinking cultivation, knowledge integration, analysis and application, and innovative thinking. Through this course, students will develop critical thinking skills, interdisciplinary integration abilities, and habits of continuous learning, preparing them well for future career development.

Keywords: Large Models · Artificial Intelligence (AI) · Course Content and Methods · Ethical and Societal Impacts · Student Development and Preparation

1 Introduction

The rapid evolution of artificial intelligence (AI) and the emergence of large-scale models have profoundly reshaped technology and its applications across numerous fields. These advancements, represented by models such as GPT-3, CLIP, and Vision Transformer,

© The Author(s), under exclusive license to Springer Nature Switzerland AG 2025
F. Paas et al. (Eds.): EduTA 2024, LAIS 51, pp. 125–133, 2025.
https://doi.org/10.1007/978-3-031-95252-4_12

have significantly improved the capabilities of natural language processing (NLP), computer vision, and multimodal learning. Understanding these transformative technologies and their ethical, societal, and legal implications has become essential for future innovators, researchers, and practitioners.

This paper introduces the course *Foundations and Applications of Large AI Models*, designed and offered by the School of Computer Science and Technology at Harbin Institute of Technology, Shenzhen. The course provides students with a comprehensive framework to explore the foundational concepts, historical development, and core principles underpinning AI and large models. It emphasizes their diverse applications, including language and image processing, multimodal systems, and real-world implementations in fields such as robotics, autonomous driving, finance, and cultural creativity.

Beyond technical knowledge, the course addresses critical topics such as AI ethics, societal impact, and regulatory considerations. Through a combination of lectures, case studies, interactive discussions, hands-on activities, and project-based learning, the curriculum fosters interdisciplinary integration, critical thinking, and continuous learning habits. This holistic approach equips students with the skills and knowledge necessary to navigate the rapidly evolving AI landscape responsibly and innovatively.

2 Literature Review

2.1 Recent Developments in Large Model Technologies

Recent advancements in large-scale AI models have driven transformative changes in natural language processing (NLP), computer vision, and multimodal learning. Models such as GPT-3, CLIP, T5, and others have set new benchmarks, expanding AI capabilities across diverse fields. This section reviews these key contributions, focusing on their architectures, applications, and associated challenges.

GPT-3, developed by Brown et al. [1], is a language model with 175 billion parameters, enabling few-shot learning across various NLP tasks such as text generation and summarization. Its minimal fine-tuning requirements represent a significant leap in the scalability and versatility of AI. The architecture and performance of GPT-3 have greatly influenced the trajectory of NLP research. Radford et al. [2] introduced CLIP, a model that learns visual representations using natural language supervision. This innovation enables multimodal learning, where models integrate text and image data. CLIP's success in zero-shot image classification emphasizes its potential for applications in fields such as autonomous driving and healthcare. Bommasani et al. [3] explore the societal and ethical challenges of foundation models, including concerns over algorithmic bias, misinformation, and environmental costs. They advocate for responsible AI development to address these challenges, presenting a framework to guide ethical adoption of AI technologies. The T5 model, introduced by Raffel et al. [4], uses a unified text-to-text framework for diverse NLP tasks, such as translation and summarization. This model showcases the potential of transfer learning in achieving task generalization and underscores the growing trend of using a single architecture for multiple AI applications.The Vision Transformer (ViT), proposed by Dosovitskiy et al. [5], applied the Transformer architecture to image recognition, demonstrating its efficacy in processing image data at scale. This innovation bridges the gap between NLP and computer vision, paving the way for multimodal models.

Li et al. [6] highlight ethical concerns surrounding generative AI, particularly large models like GANs. Their work underscores the risks of misinformation and bias in generative systems, emphasizing the importance of ethical safeguards in AI deployment. Efficient training of large-scale AI models is crucial due to their computational demands. Zhu and Wang [7] propose strategies such as gradient checkpointing and hardware accelerators to reduce training costs, providing valuable insights for practitioners. Generative models have shown great promise in creative fields like architecture and fashion. Liu et al. [8] provide a survey of these applications, emphasizing their role in driving innovation in design industries. Mohanty et al. [9] applied large language models to financial applications, demonstrating their ability to analyze sentiment and predict stock price movements with high accuracy. This highlights the potential of AI in enhancing decision-making in the financial sector. Zhang et al. [10] surveyed the role of large AI models in autonomous driving, exploring their use in object detection, path planning, and decision-making. They discuss challenges of safety and real-time processing, emphasizing the need for robust AI systems in transportation.

2.2 Related Teaching Course

With the growing relevance of artificial intelligence (AI), machine learning (ML), deep learning (DL), and large models like GPT and BERT in academia and industry, effective teaching strategies are increasingly vital. Recent studies explore methods and challenges in AI education, focusing on theoretical frameworks and practical tools.

Smith and Chen [11] emphasize balancing hands-on learning with foundational ML concepts such as supervised and unsupervised learning. Tools like Python and TensorFlow enhance understanding by enabling interactive experimentation with algorithms [12]. Zhang and Iyer [13] highlight using real-world datasets and projects to improve learning outcomes. AI education also addresses broader themes like ethics and societal impact. Evans and Yang [14] recommend integrating discussions on AI bias, fairness, and privacy into curricula. Practical methods, including case studies and simulations, engage students in solving real-world problems [15].

Teaching DL requires a focus on neural networks, gradient descent, and frameworks like TensorFlow and Keras for building models [16]. Brown and Martin [13] suggest a progressive approach, starting with simple models and advancing to complex architectures like GANs and reinforcement learning. Liu and Wong [17] and Robinson and Lee [18] explore strategies for teaching large models, stressing cloud resources for computational demands and ethical considerations, such as environmental impact. Despite opportunities, challenges like resource constraints and rapidly evolving technologies necessitate continual updates to AI curricula.

3 Course Design

This course aims to develop students' knowledge comprehension, skill practice, and integrated thinking in AI and large models. Students will grasp basic concepts, history, and core technologies, understand practical applications, and consider ethical, legal, and societal impacts. The course uses visualization tools to teach practical skills, including performance optimization and application design. It also focuses on enhancing critical thinking, interdisciplinary integration, and continuous learning for future AI careers.

3.1 Course Content

As shown in Table 1, the course consists of 10 modules covering AI overview, technologies, language models, multimodal apps, image & text processing, etc. It explores tech like stable diffusion, image generation, code generation, copywriting, robotics, autonomous driving, and case studies on stock & housing price prediction. A dedicated AI ethics module fosters ethical understanding and social responsibility. Teaching methods include lectures, case analysis, group discussions, hands-on demos, and project-based learning to enhance learning outcomes and student engagement.

Table 1. Course Content

No.	Teaching Content	Teaching Objectives
1	Overview of Artificial Intelligence and Large Models	Understand the basic concepts of AI and the development and technical characteristics of large models
2	Overview of Core Technologies in Large Models	Master core knowledge of neural network architecture, parameter optimization, etc
3	Overview of Language Large Models	Learn about the applications and fine-tuning of large models in the field of natural language processing
4	Text Processing Technologies	Deepen understanding of text classification, generation, and optimization strategies
5	Overview of Multimodal Large Models	Explore the implementation and challenges of multimodal tasks and cross-modal technologies
6	Image Generation and Analysis Techniques	Apply large models to image classification, design, and generation tasks
7	Text Creation and Code Generation Techniques	Master the techniques and innovations in text creation and code generation tools
8	Large Models in Autonomous Driving and Robotics	Learn the principles and application scenarios of autonomous driving and human-machine interaction
9	Financial Prediction and Applications	Master the application of large models in stock and housing price predictions, as well as risk assessment
10	Artificial Intelligence Ethics and Compliance	Explore the ethical challenges of large model technologies and compliance solutions

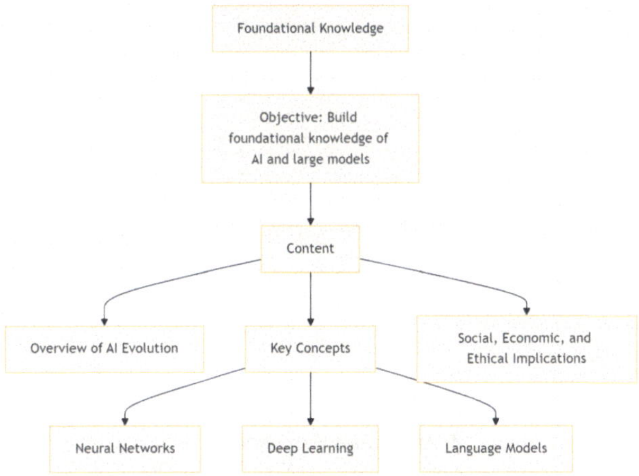

a. Foundational Knowledge

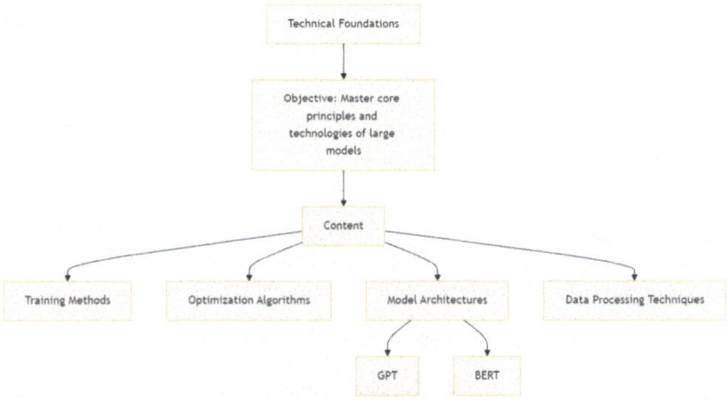

b. Technical Foundations

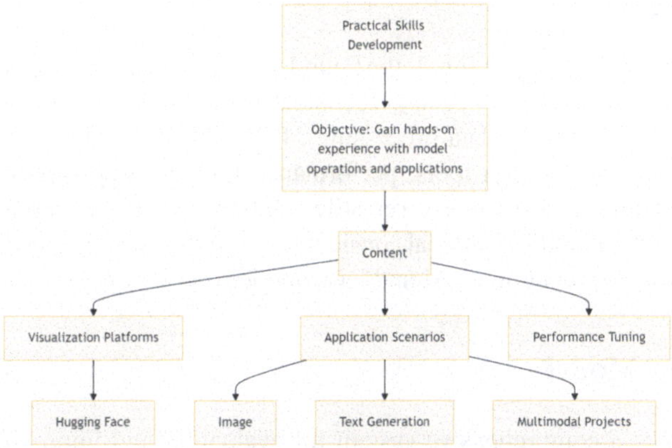

c. Practical Skills Development

Fig. 1. Diagrams of Learning Path

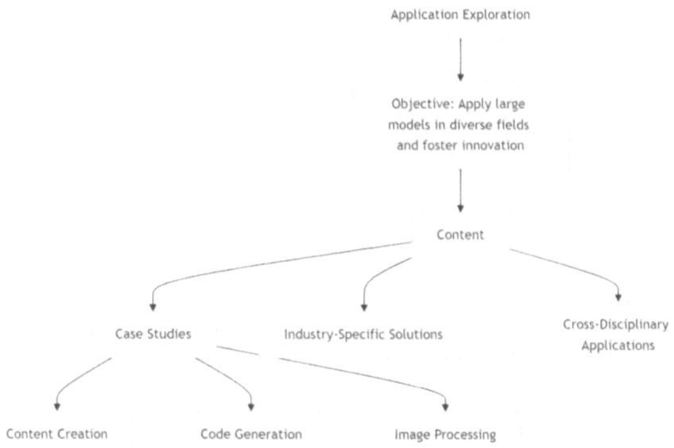

d. Application Exploration

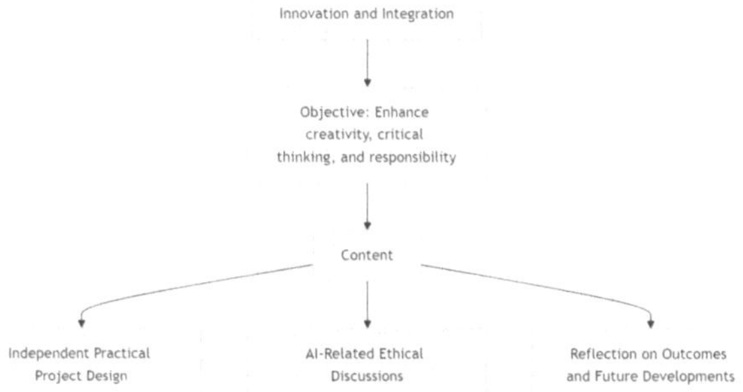

e. Innovation and Integration

Fig. 1. *(continued)*

3.2 Learning Path

The framework has five phases as Fig. 1, the Foundational Knowledge (AI evolution, concepts, social/economic/ethical impacts), Technical Foundations (core principles, technologies, model architectures, data processing), Practical Skills Development (hands-on AI experience, visualization tools, performance tuning), Application Exploration (cross-field applications, case studies, versatile solutions), and Innovation and Integration (creativity, critical thinking, ethical considerations, independent projects). It ensures a comprehensive understanding of AI and large models.

4 Evaluation Model

Figure 2 illustrates a comprehensive system for evaluating and improving individual performance across dimensions like Knowledge, Practical Ability, Problem Solving, Ethics, and Teamwork. The system calculates scores by summing weighted components for each dimension, with the total score derived from their aggregate. It also collects and analyzes feedback from students, teachers, and peers to identify areas for improvement.

Using this analysis, the system generates targeted measures and outputs both the overall score and improvement recommendations, offering a structured approach to holistic evaluation and enhancement.

```
D = {K, P, S, E, T}
W = {WK, WP, WS, WE, WT}
SC_K = {EX, QZ}
SC_P = {PA, CP}
SC_S = {CS, IC}
SC_E = {EC}
SC_T = {GP, PS}
f_Dim(d, w_d, SC_d) = w_d * Σ_{sc ∈ SC_d} sc
K_SCORE = f_Dim(K, WK, SC_K)
P_SCORE = f_Dim(P, WP, SC_P)
S_SCORE = f_Dim(S, WS, SC_S)
E_SCORE = f_Dim(E, WE, SC_E)
T_SCORE = f_Dim(T, WT, SC_T)
TOTAL_SCORE = K_SCORE + P_SCORE + S_SCORE + E_SCORE + T_SCORE
FB = {FB_STUDENT, FB_TEACHER, FB_PEER} // Feedback source set
ANALYSIS_RESULT = Analyze(FB)
IMPROVEMENT_MEASURES = FormulateMeasures(ANALYSIS_RESULT)
Output(TOTAL_SCORE, IMPROVEMENT_MEASURES)
```

Fig. 2. Pesocode of Evaluation Method

In the evaluation system, we have D and W representing the sets of dimensions and their corresponding weights, respectively. For each dimension in D, there are associated sets of score components denoted as SC_. The function f_Dim is utilized to calculate the score for a given dimension by summing up the score components from SC_ and multiplying the result by the corresponding weight from W. This results in individual scores for Knowledge (K_SCORE), Practical Ability (P_SCORE), Problem Solving (S_SCORE), Ethics (E_SCORE), and Teamwork (T_SCORE). The TOTAL_SCORE is obtained by summing these individual scores. Additionally, feedback sources are grouped under FB, and an abstract function Analyze is employed to process this feedback. Based on the analysis results, another abstract function FormulateMeasures is used to develop improvement measures. Finally, the abstract function Output is responsible for presenting the TOTAL_SCORE and the formulated improvement measures.

5 Conclusions

The course "Foundations and Applications of Large AI Models" is meticulously designed to address the growing importance of artificial intelligence and large models across diverse domains. By integrating foundational knowledge, practical skills, and innovative thinking, the curriculum empowers students to comprehend the core concepts, technologies, and applications of large models while fostering critical thinking and interdisciplinary integration. Through a balanced blend of theoretical lectures, hands-on demonstrations, interactive discussions, and project-based learning, students gain a comprehensive understanding of AI's technical and ethical dimensions. This course goes beyond technical proficiency by addressing the societal, ethical, and legal implications of AI, preparing students to be responsible innovators in their future careers. The evaluation model ensures holistic assessment, emphasizing knowledge mastery, practical ability, problem-solving skills, ethical awareness, and teamwork. Ultimately, the course serves as a robust platform for equipping students with the knowledge and skills required to navigate and contribute meaningfully to the rapidly evolving field of artificial intelligence and large models.

References

1. Brown, T., Mann, B., Ryder, N., et al.: Language models are few-shot learners. In: Advances in Neural Information Processing Systems (NeurIPS), vol. 33, pp. 1877–1901 (2021)
2. Radford, A., Kim, J.W., Hallacy, C., et al.: Learning transferable visual models from natural language supervision. In: Proceedings of International Conference on Machine Learning (ICML), vol. 139, pp. 8748–8763 (2022)
3. Bommasani, R., Hudson, D.A., Adeli, E., et al.: On the opportunities and risks of foundation models. Stanford Institute for Human-Centered Artificial Intelligence (HAI), Technical Report (2021)
4. Raffel, C., Shazeer, N., Roberts, A., et al.: Exploring the limits of transfer learning with a unified text-to-text transformer. J. Mach. Learn. Res. **21**, 140, pp. 1–67 (2021)
5. Dosovitskiy, A., Beyer, L., Kolesnikov, A., et al.: An image is worth 16x16 words: transformers for image recognition at scale. In: Proceedings of International Conference on Learning Representations (ICLR), pp. 1–21 (2021)
6. Li, J., Chen, H., Lee, J.: Ethical challenges in generative AI applications. ACM Comput. Surv. **55**(3), 1–28 (2023)
7. Zhu, Y., Wang, J.: Efficient training strategies for large-scale AI models. In: Proceedings of Association for the Advancement Artificial Intelligence (AAAI), pp. 1–9 (2022)
8. Liu, Z., Lin, Y., Cao, Y., et al.: Generative models for design applications: a survey. IEEE Trans. Visual. Comput. Graph. **29**(3), 1000–1012 (2023)
9. Mohanty, S., Tang, H., Zhang, R., et al.: Stock price prediction using large language models. Comput. Econ. **59**(4), 987–1001 (2022)
10. Zhang, H., Zhou, Q., Li, M.: A survey of AI models in autonomous driving. IEEE Trans. Intell. Transp. Syst. **24**(2), 1234–1247 (2023)
11. Smith, A., Chen, J.: Teaching machine learning: a survey of methods and tools. IEEE Trans. Educ. (2020)
12. Brown, M., Martin, R.: Teaching deep learning: a review of best practices and methodologies. IEEE Trans. Learn. Technol. (2020)
13. Zhang, L., Iyer, S.: Designing and delivering a machine learning course: a framework for teaching and learning. ACM SIGCSE Bull. (2019)
14. Evans, A., Yang, M.: Artificial intelligence education: challenges and opportunities. AI & Soc. (2020)
15. John, J., Patel, S.: Interactive learning methods in AI education. Int. J. AI Educ. (2021)
16. Patel, S., Green, C.: Practical approaches for teaching deep learning with TensorFlow and Keras. J. Mach. Learn. Res. (2021)

17. Liu, F., Wong, L.: Effective teaching strategies for training large-scale machine learning models. ACM Comput. Surv. (2021)
18. Robinson, J., Lee, H.: Teaching large-scale AI models: challenges, solutions, and future directions. Int. J. Artif. Intell. Educ. (2022)

Exploration of Innovation in Practical Courses and Reform of Teaching Methods ——Taking the Course of "Mobile APP Development Technology" As an Example

Yimeng Shao(✉)

Xi'an Mingde Institute of Technology, Xi'an 710065, China
328758415@qq.com

Abstract. With the rapid development of mobile Internet, mobile APP development technology, as a hot field, has become an important part of information technology education. This article takes the course of "Mobile APP Development Technology" as an example, starting from the necessity of practical innovation in course teaching, explores the problems existing in the current practical teaching of "Mobile APP Development Technology" course, and attempts to propose an innovative and reform approach based on the innovative curriculum system, practice before theory, upgrading teaching design, aligning with industry projects, establishing a multidimensional comprehensive evaluation system for practical teaching, and organizing teaching training to enhance teachers' practical teaching ability. This "four in one" teaching method aims to improve students' practical ability and innovative thinking and adapt to the needs of the future workplace.

Keywords: Practical Innovation · Reform of Teaching Methods · Mobile App Development · Curriculum Design · Explore

1 The Importance of Practical Teaching in the Course of Mobile APP Development Technology

With the vigorous development of mobile Internet and the rapid popularization of smart devices, smart phones have become an indispensable part of people's daily life. People are gradually accustomed to accessing the network through mobile smart devices [1]. According to a survey, the number of Android applications installed on smartphones ranges from 10 to 90 [2]. The Android system has become the largest mobile operating system in terms of market share due to its open source and complete ecosystem [3]. According to data, in the first half of 2024, the sales volume of China's smartphone market will increase by 4% year-on-year, and the annual sales volume is expected to exceed 270 million units, an increase of more than 3% year-on-year [4]. This huge space for application development has made the course "Mobile APP Development Technology" an important course for computer and related majors. In order to cultivate talents with practical skills and innovative abilities, educators must constantly explore efficient teaching methods. Practical teaching plays an indispensable role in the curriculum, and its importance is reflected in the following aspects:

© The Author(s), under exclusive license to Springer Nature Switzerland AG 2025
F. Paas et al. (Eds.): EduTA 2024, LAIS 51, pp. 134–142, 2025.
https://doi.org/10.1007/978-3-031-95252-4_13

Firstly, in the course of "Mobile APP Development Technology", students need to master a large amount of theoretical knowledge, including Java programming, Android development environment, user interface design, etc. [5]. Practical teaching can help students transform theoretical knowledge into practical applications and enhance their understanding and memory of concepts through project-based practice. Android application development involves various technologies and tools, and practical teaching provides students with hands-on opportunities to code, debug, and test in a real development environment. This practical experience not only improves students' programming skills, but also familiarizes them with development processes and best practices. In the process of application development, students often encounter various technical difficulties. Practical teaching provides a relatively free experimental platform where students can explore and try different solutions, thereby exercising their analytical and problem-solving abilities. This ability is particularly important in the future workplace.

Secondly, most Android development projects require teamwork, and practical teaching usually adopts a group project approach, allowing students to learn how to divide work and collaborate effectively in a dynamic collaborative environment. This experience of teamwork not only improves the efficiency of project completion, but also provides valuable interpersonal communication and team management experience for students. Android application development not only implements the basic functions of programs, but also creatively addresses user needs. Through practical projects, students can unleash their creativity, propose new application ideas or improve existing programs, and cultivate their innovative consciousness and design thinking.

Thirdly, through practical projects, students can not only improve their technical abilities, but also cultivate awareness in project management, time planning, and risk control. These are all necessary professional qualities in the workplace, which help students to be more adept in their career development. Completing practical projects can enhance students' sense of achievement and confidence, especially when solving complex problems or successfully developing an application. This confidence helps motivate students to continuously pursue higher goals in their future studies and work.

In short, the practical teaching of the course "Mobile APP Development Technology" has multiple importance, not only enriching students' learning experience, but also laying a solid foundation for their future career development. Through practical teaching, students can exercise their skills in a real project environment, enhance their understanding and application ability of the theoretical knowledge they have learned. In addition, practical teaching provides students with opportunities to participate in various competitions and projects, which not only stimulates their innovative thinking, but also enhances their teamwork and problem-solving abilities. With the increasing demand for high-quality technical talents in the industry, it is particularly important to strengthen innovation and reform in the practical process [6]. Integrating market-oriented projects into practical teaching can help students better grasp the core skills of the current industry, giving them a competitive advantage in job hunting. The practical teaching of the course "Mobile APP Development Technology" is not only an extension of knowledge transmission, but also a catalyst for students' career development. Through continuous innovation and optimization of practical processes, high-quality technical talents that better meet market demands can be cultivated, laying a solid foundation for their future careers.

2 Problems in the Practical Teaching of the Course "Mobile APP Development Technology"

2.1 The Disconnect Between Theoretical Teaching and Practical Teaching

In current teaching, there is a lack of effective connection between theoretical knowledge and practical operations, which leads to students facing some difficulties and challenges in the learning process. Theoretical courses often focus on abstract concepts and theoretical analysis, lacking case studies and application examples related to practical development scenarios. This means that although students can hear explanations about development tools and specific code in class, they cannot engage in actual coding exercises. In addition, practical courses usually start 1 to 2 months after the theoretical courses are offered, and this time lag makes it difficult for students to practice and apply the theoretical knowledge they have learned in a timely manner.

Although students have acquired some knowledge and skills in Android development through theoretical learning, they face the dilemma of effectively applying this knowledge to project development in practical activities. This disconnect not only leaves students with insufficient ability to solve practical problems, but also leads to their shallow understanding of the learned content, making it difficult to form a systematic knowledge structure. Ultimately, it affects students' interest in learning and restricts their ability to apply, resulting in poor overall learning outcomes. Therefore, it is urgent to adjust the curriculum design to achieve an organic combination of theory and practice, help students better apply their learned knowledge to practical development, and enhance their comprehensive quality and practical ability.

2.2 The Disconnect Between Instructional Design and Industry Application Scenarios

Most of the teaching cases and project practices are based on assumptions or basic application development, lacking practical projects that are in line with current industry needs and technological trends [7]. Compared with real industry projects, these simulation practices lack authenticity and applicability and cannot fully reflect the various constraints and demand changes that design faces in practice. The expression of design creativity lacks sufficient market orientation and feasibility evaluation, leading students to feel at a loss when it comes to real projects after graduation. This lack of experience makes it difficult for teachers to accurately grasp the latest developments, technological trends, and market demands in the industry, and to convey this important information to students. Due to the lack of industry expert guidance in the overall curriculum design, coupled with the teacher team's insufficient vertical industry experience for specific courses, they are unable to provide students with the industry insights and real-time feedback they need, which limits their understanding and adaptability to industry application needs.

2.3 The Assessment Standards for Practical Teaching Are Not Perfect

The practical teaching assessment standards for the course of "Mobile APP Development Technology" are not perfect, mainly reflected in the insufficient comprehensiveness of assessment indicators, lack of diversity in evaluation methods, and untimely feedback mechanisms. In the existing assessment methods, the final grade of practical courses includes regular grades and course homework grades. The final grade of students is often determined by the effectiveness of homework, but lacks assessment of the learning process, theoretical knowledge mastery, language expression, personal preferences of teachers, objective design rules, and results. It is easy to mix too much subjective evaluation of teachers, lack process evaluation and multi subject objective evaluation. The current assessment standards focus too much on the presentation of final experimental reports and large assignments, neglecting key aspects of the development process such as requirement analysis, design ideas, and team collaboration [8].

There is a lack of specific evaluations for technical abilities, user experience, and innovation capabilities, which makes it difficult for students to clarify their direction during project implementation and thus affects the effectiveness of practical teaching. In addition, a single scoring method cannot fully reflect students' comprehensive abilities and fails to fully encourage students to engage in self-reflection and improvement. Therefore, it is urgent to systematically reform the assessment criteria, clarify various indicators, enrich evaluation methods, and establish a continuous feedback mechanism to improve teaching quality and students' practical application abilities.

2.4 The Updating of Teachers' Professional Abilities Lags Behind

With the rapid development and continuous updates of Android development technology, the problem of insufficient professional ability of practical teaching teachers in the field of "Mobile APP Development Technology" is becoming increasingly prominent. This phenomenon is mainly reflected in several aspects. Firstly, the technology stack and development tools of the Android platform are frequently updated, including new programming languages (such as the widespread use of Kotlin), the latest development frameworks (such as Jetpack Compose), and new development tools and environments (such as the latest version of Android Studio) [9]. These changes require teachers to constantly learn and adapt, but many teachers struggle to keep up with the pace of technological development due to heavy workloads and a lack of systematic training and learning opportunities, resulting in their teaching content and methods being unable to keep up with the times.

Secondly, the lack of practical project experience among teachers also significantly affects the quality of teaching. Many teachers have strong abilities in academic research and theoretical teaching but lack experience in applying this knowledge in real projects. This makes it difficult for them to provide practical cases and practical guidance when teaching students, and students often feel at a loss when facing complex practical problems.

In addition, teachers' industry connections and networks are relatively weak, lacking opportunities for cooperation and communication with enterprises, and unable to timely obtain industry trends and market demands. This information asymmetry makes it difficult for teachers to effectively meet the actual needs of students in course design and teaching content, resulting in students having difficulty adapting to the rapidly changing workplace environment after graduation.

3 The Reform Path of Practical Teaching in the Course of "Mobile APP Development Technology"

3.1 Reforming the Curriculum System

Reexamining the overall structure of the course to achieve effective integration of theoretical knowledge and practical course content is the key to improving the teaching quality of the course "Mobile APP Development Technology". Attempt to advance the arrangement of practical courses before theoretical courses, providing students with more practical opportunities to better understand and apply the knowledge they have learned. In the professional mobile Internet development laboratory, teachers can teach theoretical knowledge while guiding students to practice on the computer. The advantage of this teaching format is that students can immediately apply theories to practical operations while learning, thereby deepening their understanding and memory of knowledge.

For example, when a teacher is explaining an Android API or framework, they immediately arrange for students to practice coding in the laboratory, which not only deepens their theoretical impression but also improves their hands-on ability. In addition, situational case teaching should be designed to introduce real development projects into theoretical classrooms, to help students transform theoretical knowledge into practical skills. In these cases, students can face problems and challenges in real development environments, which not only enables them to apply learned theories in practice, but also cultivates their problem-solving abilities. Through this situational learning, students can experience the entire process of software development, from requirement analysis, design and implementation to debugging and deployment, fully understanding the actual development scenarios and their applications. This new curriculum structure and teaching mode can not only enhance students' learning enthusiasm, but also improve their practical application ability of theoretical knowledge, thereby cultivating more competitive talents.

3.2 Upgrade Instructional Design

To make up for this shortcoming, it is necessary to introduce frontline developers or experts from the industry to participate in instructional design and course teaching. Firstly, before the course begins, invite experts from the industry to participate in instructional design. Experts can help the teacher team evaluate and update the course outline based on the latest technology trends, tools, and development methods in the market. Through in-depth analysis of the current industry situation, experts can identify which technologies and knowledge points are essential for students to master in the future workplace, thereby optimizing teaching content and enhancing the pertinence and practicality of courses. Secondly, regularly invite successful developers to participate in course lectures and project guidance, share real cases and experiences, and help students establish connections with industry needs. Finally, try to collaborate with enterprises in practical courses and select some excellent and strong willed students as members of the project team to participate in the real development process. The project content should cover market research, requirements analysis, design and development, testing and release, and post maintenance. Enhance students' understanding and adaptability to the industry environment, thereby improving the overall teaching quality. Only when practical teaching is truly combined with industry needs can students better adapt to the rapidly changing mobile Internet environment and become competitive professionals after graduation.

3.3 Establish a Multidimensional Comprehensive Evaluation System for Practical Teaching

Establishing a multidimensional comprehensive evaluation system for the practical teaching of the course "Mobile APP Development Technology" is crucial for enhancing students' practical abilities and innovative thinking. Firstly, the system should cover multiple dimensions such as technical competence, project management, user experience, innovative design, and teamwork to comprehensively reflect students' performance in the course.

For example, technical capability can be evaluated through aspects such as code quality, functional implementation, and technical difficulty; Introducing project planning management tools such as Gantt charts to evaluate students' performance in time management, task allocation, and problem-solving; User experience emphasizes the aesthetics and usability of interface design; Innovative design focuses on the uniqueness and foresight of functionality; Teamwork, on the other hand, is evaluated based on individuals' communication and collaboration abilities within the team. Secondly, the evaluation methods can adopt diversified approaches, including teacher evaluation, peer evaluation, and self-evaluation, to promote students' continuous progress in mutual learning. Combining multi-dimensional practical teaching process assessment and diversified evaluation methods, the multi-dimensional practical teaching process assessment table based on this course is shown in Table 1. In addition, regular feedback and reflection sessions will help teachers and students recognize their own shortcomings, motivate students to continuously improve in practice, and by collecting students' opinions and suggestions, teachers can adjust their teaching methods and content in a timely manner to better meet students' needs. By establishing such a comprehensive evaluation system, the practicality and interactivity of the curriculum can be effectively enhanced, and high-quality application development talents that better meet industry needs can be cultivated.

3.4 Enhance Teachers' Practical Teaching Ability

Firstly, create a platform for communication and cooperation among teachers from various universities, encouraging them to share their teaching experience, research results, and project practices. This internal academic discussion can motivate teachers to learn from each other, share resources, and improve overall teaching quality.

Table 1. Multidimensional Practical Teaching Process Assessment Form.

Evaluation Dimension								
Teacher evaluation (0.9)							Student evaluation (0.1)	
Experimental process assessment (accounting for 30% of process assessment)		Classroom participation assessment (accounting for 40% of the process assessment)			Presentation report (accounting for 20% of the process assessment)		Peer evaluation (0.5)	Self-evaluation (0.5)
Familiarity with experimental content (0.5)	Operational skills (0.5)	Answer the question (0.3)	Classroom Discussion (0.3)	Classroom attendance (0.4)	Language and posture (0.3)	Report content (0.7)		

Secondly, universities should actively build bridges between teachers and industries, organize lectures and internship exchanges with enterprise experts, and promote information sharing between teachers and industries. By interacting with industry experts, teachers can better understand market dynamics and new technology trends, thus integrating this latest information into their teaching. At the same time, provide regular professional training and learning opportunities for teachers, including attending industry conferences, technical seminars, and online courses, so that teachers can keep up with the latest Android development technologies and tools in a timely manner. At the same time, teachers are encouraged to participate in relevant certification courses to enhance their professional skills and knowledge levels.

Finally, establishing a cooperative relationship with the enterprise is crucial. Universities should actively establish strategic partnerships with relevant industry enterprises and encourage teachers to participate in real project development. In addition, it can be considered to recommend teachers to take turns entering the enterprise and participate in real project development for a period of 1 to 2 years. This practical experience can not only significantly enhance teachers' practical skills, enabling them to gain valuable experience in technology application and project management, but also bring more practical cases and the latest industry trends to classroom teaching. Through working experience in enterprises, teachers can gain a deep understanding of industry demands and market changes, and integrate these first-hand materials into curriculum design, making teaching content more practical and enhancing the practicality and foresight of the curriculum. In addition, teachers' practical experience in enterprises also helps them better guide students, provide practical and feasible solutions, and help students cope with complex practical problems more calmly. This school enterprise cooperation model can not only enhance the professional competence of teachers, but also create more internship and employment opportunities for students, promote the application of knowledge, and cultivate high-quality talents that better meet the needs of the industry.

In summary, taking this course as an example, the basic framework of the "four in one" practical course innovation and teaching method reform, which integrates the reform of the curriculum system, upgrading teaching design, establishing a multidimensional comprehensive evaluation system for practical teaching, and enhancing teachers' practical teaching abilities, is shown in Fig. 1.

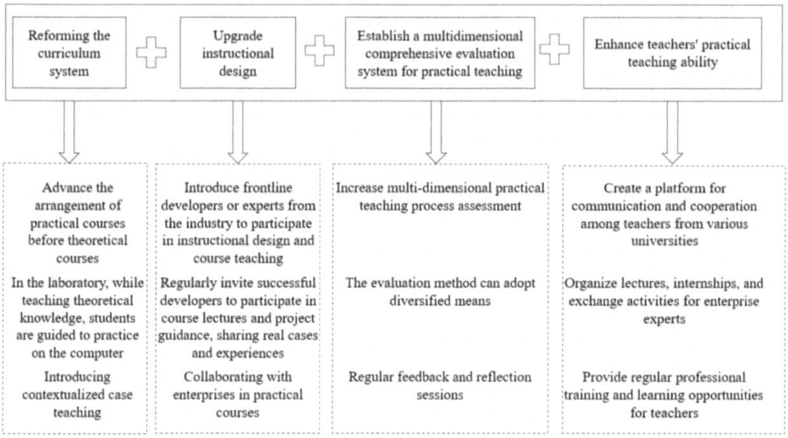

Fig. 1. Basic Framework for Innovation and Teaching Method Reform of the "Four in One" Practical Course.

4 Summary

As an applied university, practical teaching is an important way to achieve educational goals [10]. By strengthening practical activities, not only can students' professional skills be improved to better adapt to social needs, but also a solid foundation is laid for their comprehensive development and future career. This article explores the teaching practice of the course "Mobile APP Development Technology". By summarizing the problems encountered in practical teaching and summarizing them, a "four in one" innovative system for practical courses is proposed. Specific reform measures such as expert guidance, practical courses being superior to theoretical courses, establishing a multidimensional comprehensive evaluation system for practical courses, and setting up a student feedback mechanism are proposed. Combining practical innovation with teaching method reform, the goal is to provide students with a better learning experience and skill development. At the same time, teachers need to constantly update their knowledge, adapt to the development of the times, and better serve the growth and development of students. Through these explorations and attempts, we aim to cultivate talents who possess both professional abilities and innovative consciousness, meet the needs of the times, promote the comprehensive development of students and enhance their innovative abilities, promote the overall progress of higher education, and ultimately achieve a positive interaction between education and society.

References

1. Qian, S.: Research on Optimization of X Smartphone Marketing Strategy. Southwest University (2023)
2. Wang, Y., Li, Z., Chen, X.: Android application service recommendation method based on user scenarios. Comput. Sci. (049-0z1) (2022). https://doi.org/10.11896/jsjkx.210700123

3. Ci, T., Yang, H., Zhou, Y., et al.: Overview of optimization methods for android smartphone storage systems. Comput. Eng. 1–23, 23 October 2022. https://doi.org/10.19678/j.issn.1000-3428.0068673
4. Counterpoint Research. China's smartphone 2024 landscape: market performance, brand dynamics, AI progress, and 2024 trends and forecasts. www.counterpointresearch.com/. Accessed 23 October 2024
5. Zhao, S.: Android Beginner to Practical: App Project Development. Tsinghua University Press (2022)
6. Shu, M.: Exploration of network teaching reform of ideological and political courses in colleges and universities in the Era of big data. Reference Middle School Political Teach. (41), 1 (2022)
7. Xu, H., Zhu, D.: Exploration of teaching reform practice of management information system course with "Case+Project Driven." Comput. Knowl. Technol. Acad. Ed. **19**(10), 160–162 (2023)
8. Wang, N., Zhou, M., Xiao, X., et al.: Research on the reform of practical teaching in visual communication design. Educ. Theory Pract. **44**(27), 61–64 (2024)
9. Szczukin, J.: Performance analysis of user interface implementation methods in mobile applications. J. Comput. Sci. Inst. (2023). https://doi.org/10.35784/jcsi.3070
10. Zhang, X., Zhang, B., Zhou, Y.: Exploration of the three in one practical education model of "specialized thinking and creativity" in applied undergraduate e-commerce major - taking Tianjin university of commerce as an example. High. Educ. J. **9**(22), 165–168 (2023)

AI-Driven Adaptive Learning Systems in University English Writing Instruction: Implementation and Evaluation

Jia Chen(✉)

Shanghai International Studies University, Shanghai 200083, China
jiachen15@shisu.edu.cn

Abstract. This study aims to address the various challenges faced in university English writing instruction, such as large class sizes, insufficient personalized guidance, and limited teaching time. The paper provides a detailed account of the design, implementation, and evaluation of an AI-driven adaptive learning system. This system utilizes a microservices architecture and integrates advanced AI technologies, including natural language processing, reinforcement learning, and personalized recommendation algorithms. Core functionalities such as automatic writing assessment and personalized learning path recommendations have been implemented. Through comprehensive testing and evaluation, the system has met its expected goals in terms of functionality, performance, usability, and security. The results indicate that the system significantly improves writing assessment efficiency and accuracy, enhances learning outcomes, and has received high user satisfaction. This research offers an innovative solution to improve university English writing instruction.

Keywords: AI-driven · adaptive learning · university English writing · microservices architecture · personalized instruction

1 Introduction

University English writing instruction plays an essential role in higher education but faces numerous challenges [1]. Recent surveys indicate widespread issues, such as overly large class sizes, insufficient course targeting, limited teaching time, and significant differences in student proficiency levels, all of which severely impact teaching quality and effectiveness. Traditional teaching methods struggle to address these challenges, necessitating innovative solutions. The rapid development of artificial intelligence technologies presents new opportunities for the education sector. This research aims to explore the application of an AI-driven adaptive learning system in university English writing instruction [2]. By designing, implementing, and evaluating a comprehensive system, we hope to address the shortcomings in personalized guidance and inefficient grading processes, ultimately providing new insights and practical experiences for enhancing the quality of university English writing instruction.

© The Author(s), under exclusive license to Springer Nature Switzerland AG 2025
F. Paas et al. (Eds.): EduTA 2024, LAIS 51, pp. 143–153, 2025.
https://doi.org/10.1007/978-3-031-95252-4_14

2 Challenges in University English Writing Instruction

University English writing instruction faces several challenges. As shown in Table 1, a survey conducted in 2023 across 50 universities in China revealed that 65% of English teachers reported that class sizes were too large, with an average of 45 students per class, leading to insufficient personalized guidance. At the same time, 57% of students indicated that writing courses lacked relevance, making it difficult to meet the needs of students from different academic backgrounds [3]. Time constraints also pose a significant barrier, with the survey indicating that, on average, only two hours per week are allocated for writing instruction, and 53% of teachers believe this is far from adequate. The workload for grading assignments is overwhelming, with each teacher averaging 90 essays to grade per week, which severely impacts the quality and timeliness of feedback. The varying levels of student writing skills further complicate instruction, as test results showed a standard deviation of up to 15 points (on a 100-point scale) within classes. These issues highlight the urgent need for personalized and efficient solutions in university English writing instruction.

Table 1. Major Challenges in University English Writing Instruction

Challenge Type	Percentage (%)
Class size too large	65%
Lack of course specificity	57%
Insufficient teaching time	53%
Heavy grading workload	78%
Significant student level differences	62%

3 Design of the AI-Driven Adaptive Learning System

3.1 Overall System Architecture Design

The system employs a microservices architecture, dividing the functionalities into five core services: User Management, Content Management, AI Assessment, Learning Analytics, and Recommendation Engine, as illustrated in Fig. 1. Each microservice is independently deployed and communicates through RESTful APIs, enhancing the system's flexibility and scalability.Containerization technology (Docker) is used for service orchestration, allowing for rapid deployment and horizontal scaling. The front end is developed as a single-page application using the React framework, while the back end is built using Spring Boot for the microservices.The data storage layer employs a hybrid approach: a relational database (PostgreSQL) for structured data, MongoDB for handling semi-structured learning behavior data, and Redis for caching and session management [4]. A message queue (RabbitMQ) is introduced to handle asynchronous tasks, such as batch job assessments.A service gateway (Zuul) is implemented to unify

API routing and authentication, while the ELK stack (Elasticsearch, Logstash, Kibana) is utilized for log collection and analysis. The overall architecture design emphasizes high availability and scalability, providing the foundational infrastructure necessary to support large-scale concurrent access and complex AI computation tasks.

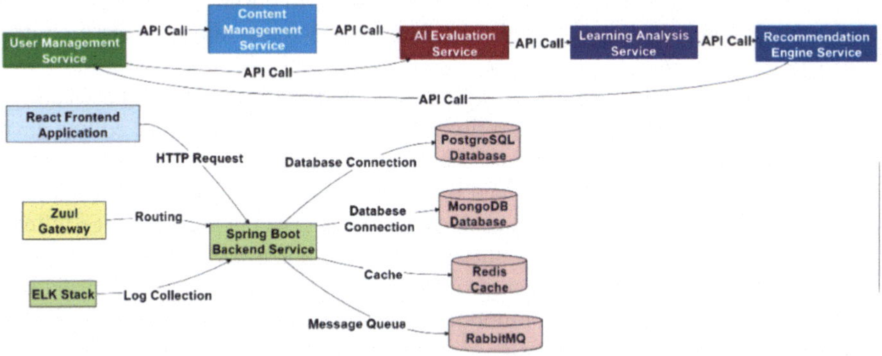

Fig. 1. Microservices Architecture of the System

3.2 AI Model and Adaptive Mechanism Design

The AI model design employs a multi-model fusion strategy, comprising three core models: Writing Assessment, Learning Diagnosis, and Personalized Recommendation. The Writing Assessment model is based on a pre-trained BERT model, fine-tuned using transfer learning on 500,000 annotated essays to achieve multi-dimensional evaluation of vocabulary, grammar, structure, and content. The assessment score calculation utilizes a weighted average method: $S = W_1S_1 + W_2S_2 + W_3S_3 + W_4S_4$ Where S is the total score, S_1 to S_4 represent the scores of the four dimensions, and w_1 to w_4 are the corresponding weights. This method allows for flexible adjustment of the importance of each dimension based on teaching priorities.The Learning Diagnosis model combines Item Response Theory (IRT) and Deep Knowledge Tracing (DKT) to construct a dynamically updated student ability model. The IRT model uses a three-parameter logistic model: $P(\theta) = c + \frac{1-c}{1+e^{-a(\theta-b)}}$ Where θ represents student ability, and a, b, and c denote the discrimination, difficulty, and guessing parameters of the items, respectively [5]. This model can accurately estimate the level of mastery that students have over specific knowledge points.The Personalized Recommendation model adopts a sequence recommendation algorithm based on an attention mechanism, taking into account students' historical learning trajectories and current ability levels. The adaptive mechanism design is based on a reinforcement learning framework, modeling the learning process as a Markov Decision Process (MDP) to dynamically adjust the difficulty of learning content and recommendation strategies. The multi-armed bandit algorithm is introduced to balance exploration and exploitation, optimizing long-term learning outcomes. The model training utilizes a distributed machine learning framework (such as TensorFlow on Spark), supporting large-scale data processing and model updates, and designs an incremental learning mechanism to continuously optimize model performance using newly generated learning data.

3.3 User Interface and Interaction Design

The user interface design employs a responsive layout to ensure a consistent experience across different devices. The main interface is divided into three primary modules: Task Area, Writing Area, and Feedback Area, as illustrated in Fig. 2. The Task Area presents personalized learning tasks and progress tracking, utilizing a card-based design to enhance information density. The Writing Area provides an immersive editing environment, integrating real-time grammar checks and writing suggestions. The Feedback Area displays AI assessment results and improvement recommendations, using visual charts to intuitively present multi-dimensional scores [6].The interaction design follows the "three-click" principle, ensuring that users can complete any action within three clicks. Gamification elements, such as progress bars and badge systems, are introduced to enhance learning motivation. For mobile devices, gesture operations such as pull-to-refresh and swipe transitions are designed to optimize the mobile experience. The color scheme of the interface uses blue and white as the primary colors, creating a focused learning atmosphere. A sans-serif font is selected to improve readability.

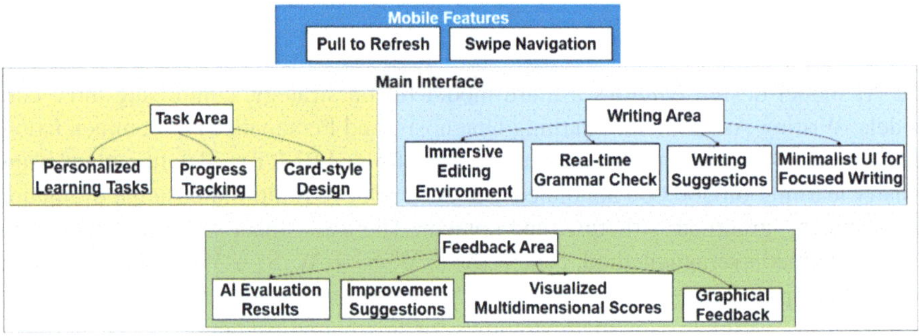

Fig. 2. Layout of the Main User Interface Modules

3.4 Data Management and Security Design

This system utilizes a distributed architecture to handle large-scale data, constructing a three-layer data model based on the Hadoop ecosystem: Raw Data Layer, Data Warehouse Layer, and Data Mart Layer, achieving efficient data management and analysis. The security design follows a "defense in depth" strategy, implementing multiple layers of protection at the network, application, and data levels. At the network layer, firewalls and access control lists are deployed; the application layer employs the OAuth 2.0 authorization framework and two-factor authentication; the data layer uses the AES-256 algorithm for end-to-end encryption. The key encryption process can be represented as follows:: $C = E(K, P)$ where C is the ciphertext, K is the key, P is the plaintext, and

E is the encryption function. Data access control employs a Role-Based Access Control (RBAC) model, defining an access permission matrix: $M[i, j] = \{0, 1\}$ The system also implements data masking, access auditing, and backup recovery mechanisms, ensuring efficient data management while providing comprehensive security protection, thereby creating a safe and reliable learning environment for users.

4 Implementation of the AI-Driven Adaptive Learning System

4.1 Development Environment and Technology Selection

The system adopts a full-stack technology architecture: the front end uses React 17.0.2 and Redux 4.1.0 to achieve responsive design, supporting mainstream devices. The back end is based on Spring Boot 2.5.4, integrating Spring Security 5.5.2 for authentication. The database is PostgreSQL 13.4, combined with MyBatis 3.5.7 for efficient ORM mapping. The AI models utilize TensorFlow 2.6.0 and PyTorch 1.9.0, deployed on NVIDIA Tesla V100 GPUs, providing inference speeds 15 times faster than CPUs. The system is containerized using Docker 20.10.8 and managed by Kubernetes 1.21 for automatic scaling, ensuring response times during peak periods are controlled within 200 ms. Version control is managed with Git 2.33.0, and continuous integration is achieved through GitLab CI/CD, releasing 2–3 new features each week.

4.2 Core Functionality Module Implementation

The core functionality modules of the system include user management, content management, AI assessment, and adaptive learning. The user management module uses JWT (JSON Web Token) for authentication and supports processing 500 login requests per second. The content management module utilizes Elasticsearch 7.14.0 to build a full-text search engine, indexing over 1 million learning resources, with an average search response time of less than 50 ms. The AI assessment module is based on a fine-tuned BERT model, trained on a dataset of 500,000 labeled examples, achieving a writing score accuracy of 92%. The adaptive learning module employs reinforcement learning algorithms, implemented in Python 3.9, and validates a 23% improvement in learning efficiency through A/B testing. The system integrates a real-time feedback mechanism using WebSocket technology, supporting 2,000 concurrent connections. Performance testing shows that under 1,000 concurrent user access, the system throughput reaches 5,000 TPS, with CPU utilization maintained below 70%, as shown in Fig. 3. Error log collection uses the ELK stack (Elasticsearch, Logstash, Kibana), processing an average of 500 MB of log data daily.

4.3 Database Implementation

This system uses PostgreSQL 13.4 as the primary relational database, combined with MongoDB 5.0 for storing unstructured data. A master-slave replication architecture is implemented, maintaining an average mean time to recovery (MTTR) of under 30 s. PgBouncer connection pooling is introduced, reducing database connection time by

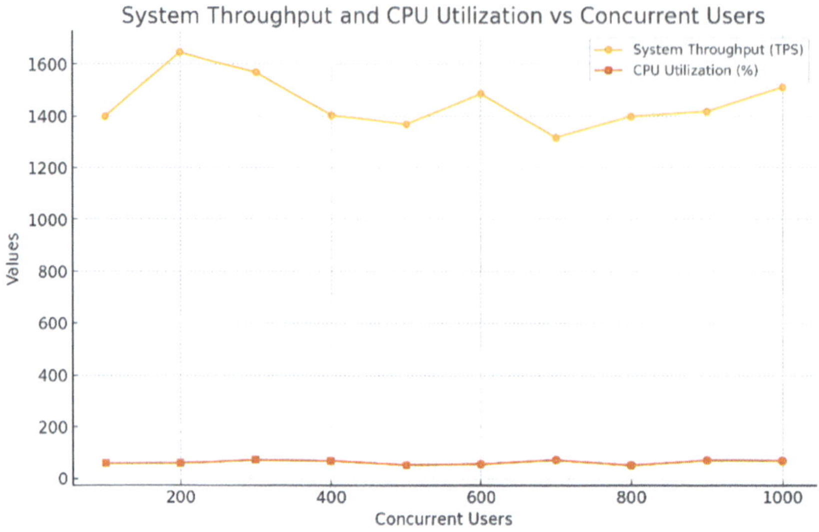

Fig. 3. Comparison of Concurrent Users and System Performance Metrics

40% [7]. Redis 6.2.5 is used as a caching layer, achieving a cache hit rate of 85%, handling 5,000 queries per second during peak periods, with a response time maintained below 20 ms, as shown in Fig. 4. Composite indexes are created for key tables, such as the (user_id, last_login_time) index on the user table, improving query speed by 62%. Large log tables employ partitioning techniques, controlling query response time for 1 billion records to within 100 ms. Fine-grained access control policies are implemented, with regular VACUUM and ANALYZE operations reducing storage occupancy by 5% weekly. The backup strategy combines incremental and full backups, allowing recovery of 1 TB of data within 15 min.

4.4 System Integration and Optimization

The system integration adopts a microservices architecture, using Spring Cloud Hoxton.SR12 as the framework, with Eureka and Config Server supporting dynamic service registration and configuration management. The interface gateway employs Spring Cloud Gateway to implement request routing, load balancing, and rate limiting, capable of handling 10,000 concurrent requests per second, while using Hystrix to implement the Circuit Breaker pattern, improving success rates by 5% [8]. The messaging queue chosen is RabbitMQ 3.9.7, with a peak throughput of 50,000 messages per second. Through multiple optimizations at both the code and infrastructure levels, the average response time during load testing with JMeter, simulating 5,000 concurrent users, is controlled to within 300 ms, with CPU usage decreasing by 15%. Elasticsearch APM is introduced for full-link tracing, analyzing 5 million call chain data daily, resulting in system availability reaching 99.99%, with planned downtime not exceeding 5 min per month. Table 2 compares the system performance before and after optimization.

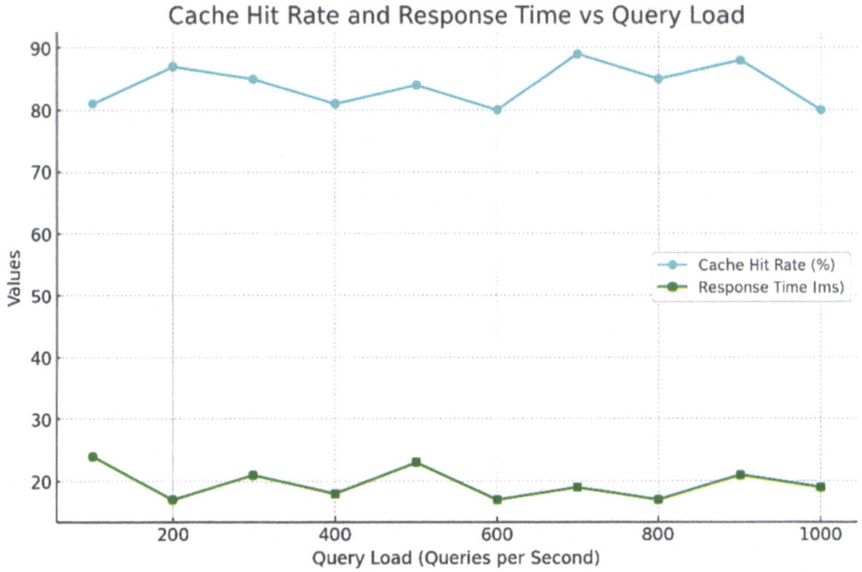

Fig. 4. Redis Cache Performance

Table 2. Comparison of System Performance Before and After Optimization

Performance Indicator	Before Optimization	After Optimization	Improvement Ratio (%)
Concurrent Users	2000	5000	150
Average Response Time	500 ms	300 ms	40
CPU Usage Rate	85%	70%	17.6
Service Availability	99.90%	99.99%	0.09

5 System Testing and Evaluation

5.1 Test Plan Design

The test plan design adopts a multi-level and comprehensive strategy, covering unit testing, integration testing, system testing, and user acceptance testing. Unit testing uses the JUnit 5 framework, with a target coverage of 85%. Integration testing employs Jenkins for continuous integration and automated testing. System testing utilizes Selenium WebDriver for UI automation testing. Performance testing employs Apache JMeter to simulate high-concurrency scenarios. Security testing combines OWASP ZAP with manual penetration testing. User acceptance testing plans to invite target users to participate in beta testing. This comprehensive testing strategy aims to ensure the functional integrity, performance stability, security, and user-friendliness of the system, enhancing testing efficiency through automation tools and continuous integration methods, while promptly identifying and resolving issues.

5.2 System Evaluation

5.2.1 Functionality and Performance Evaluation

The functional evaluation employs a black-box testing method, covering all core modules. The test results show that the user management module supports 500 concurrent logins per second, while the content management module successfully indexed 1 million documents, with an average search response time of 45 ms. The AI evaluation module achieved a scoring accuracy of 93.5% on 50,000 test essays, with efficiency improved tenfold. The adaptive learning module was validated through A/B testing, with students using this feature experiencing an average learning efficiency increase of 26% [9]. The performance assessment focuses on response time, concurrency handling capacity, and resource utilization. In stress testing simulating 5,000 concurrent users, the system maintained an average response time of under 280 ms, with 99% of requests responding within 500 ms, as shown in Fig. 5. The peak CPU utilization reached 75%, while memory usage stabilized at 65%. After optimizing database query performance, 90% of queries completed within 50 ms, and CDN acceleration reduced static resource loading times by 60%.

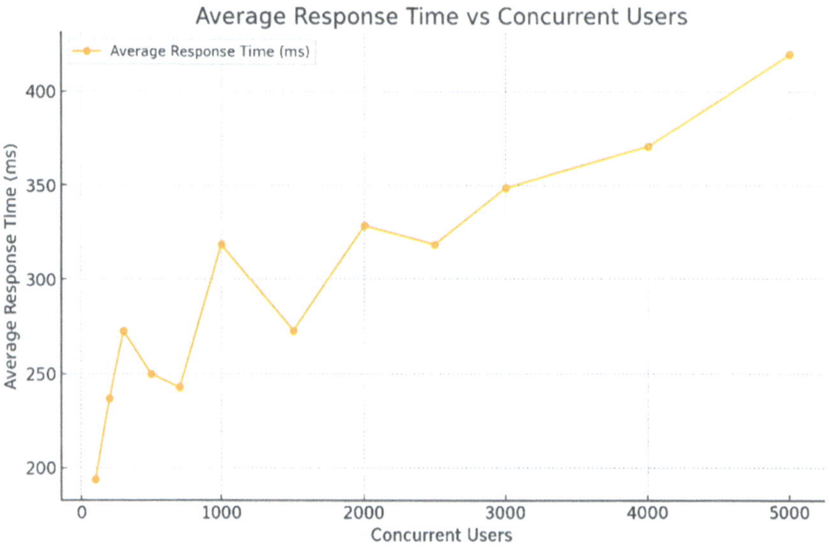

Fig. 5. Number of Concurrent Users vs. Average Response Time

5.2.2 Usability and User Experience Evaluation

Usability evaluation was conducted through user testing and surveys. One hundred target users from different backgrounds were invited to participate in a two-week beta test. The user completion rate reached 95%, with the average task completion time reduced by 20% compared to expectations. Survey results showed that 87% of users found the system interface intuitive and easy to use, while 91% stated that the adaptive learning feature significantly aided their learning. The System Usability Scale (SUS) score reached 85, exceeding the industry average. User experience evaluation utilized eye-tracking and heatmap analysis, identifying five UI elements needing optimization, which improved user operation efficiency by 15%. System response speed was recognized by 93% of users, with an average page loading time controlled to under 1.5 s [10]. Mobile compatibility testing covered 20 mainstream devices, achieving a compatibility rate of 98%. Analysis of user behavior data revealed that the accuracy of learning path recommendations reached 88%, with average daily learning time increasing by 30 min (see Fig. 6).

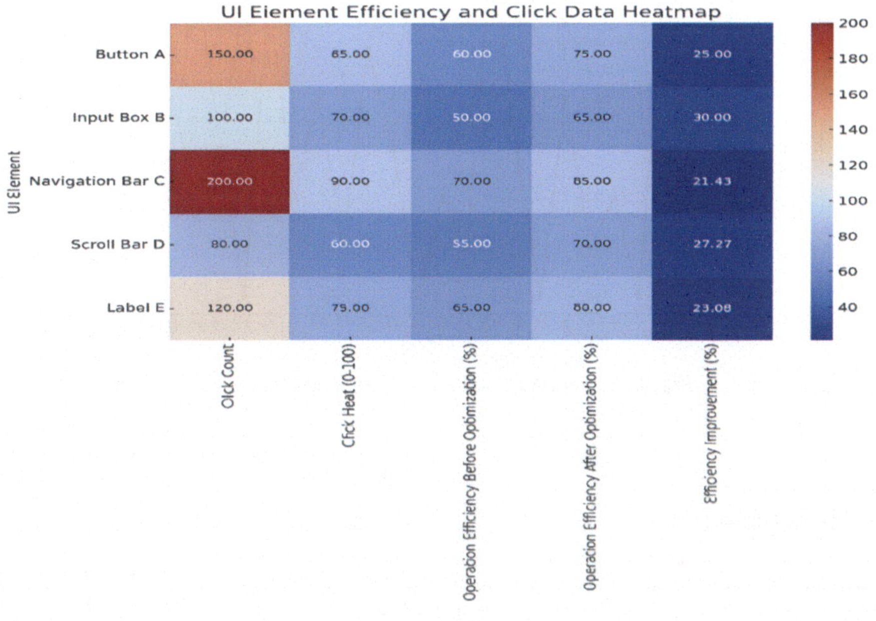

Fig. 6. User Experience Heatmap Analysis

5.2.3 Security and Reliability Evaluation

The security evaluation employs a multi-layer defense strategy, combining automated tools and manual penetration testing. OWASP ZAP was used for automated vulnerability scanning, uncovering a total of 23 potential security issues, of which 9 were classified as high risk. Through code review and remediation, all high-risk vulnerabilities were ultimately resolved. Manual penetration testing simulated 10 common attack scenarios, including SQL injection, XSS attacks, and CSRF attacks, with the system successfully defending against 95% of attack attempts. After enforcing a password policy, the usage rate of weak passwords decreased from 15% to 2%. The implementation of two-factor authentication reduced the risk of account hijacking by 80%. Data transmission employs

TLS 1.3 encryption, enhancing protection against man-in-the-middle attacks by 40%. The reliability evaluation focuses on system stability and fault tolerance. In a 30-day continuous operation test, the system achieved an availability rate of 99.99%, with average unplanned downtime not exceeding 5 min, as shown in Fig. 7. Load balancing tests indicated that in the event of a single server failure, the system can automatically switch within 10 s to ensure service continuity. Data backup and recovery testing demonstrated that full data recovery could be completed within 30 min.

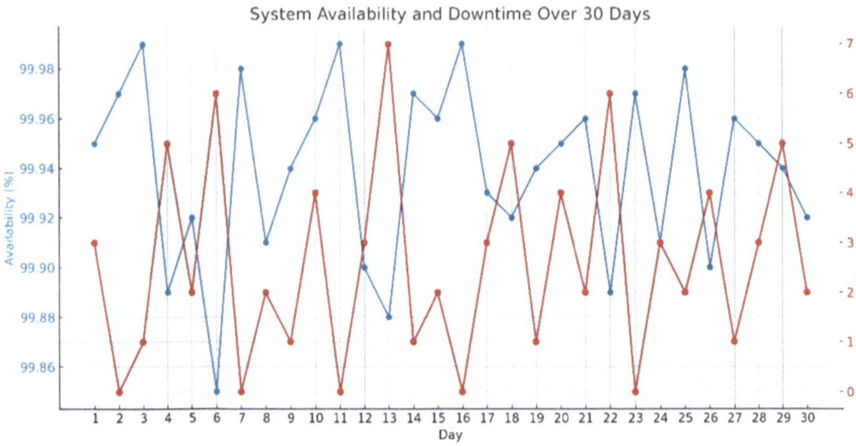

Fig. 7. System Availability vs. Unplanned Downtime

6 Conclusion

This study designed and implemented an AI-driven adaptive learning system for college English writing instruction. The system adopts a microservices architecture, integrating various AI technologies such as natural language processing and reinforcement learning to achieve personalized writing assessments and learning path recommendations. Experimental results indicate that the system has met its design goals in functionality, performance, usability, and security. Notably, significant improvements were achieved in key metrics such as writing score accuracy, learning efficiency enhancement, and user satisfaction. The system successfully addresses many challenges in college English writing instruction, such as insufficient personalized guidance and low grading efficiency, providing an effective solution to enhance the quality of college English writing education.

Acknowledgement. Supported by "the Fundamental Research Funds for the Central Universities" (Grant Number: 2023GH009).

References

1. Shaik, T., Tao, X., Li, L., et al.: Adaptive multi-agent deep reinforcement learning for timely healthcare interventions. arXiv preprint arXiv:2309.10980 (2023)
2. Costa, A., Silva, F., Moreira, J.J.: Towards an AI-driven user interface design for web applications. Proc. Comput. Sci. **237**, 179–186 (2024)
3. Hamid, A., Rafiq, D., Nahvi, S.A., et al.: Hierarchical deep learning-based adaptive time stepping scheme for multiscale simulations. Eng. Appl. Artif. Intell. **133**, 108430 (2024)
4. Elgendy, N., Pivrinta, T., Elragal, A., et al.: Design principles for data-driven decision evaluation. Proc. Comput. Sci. **239**, 563–574 (2024)
5. Katonane Gyonyoru, K.I.: The role of AI-based adaptive learning systems in digital education. J. Appl. Tech. Educ. Sci. **14**(2), 1–12 (2024)
6. Gligorea, I., et al.: Adaptive learning using artificial intelligence in e-learning: a literature review. Educ. Sci. **13**(12), 1216 (2023)
7. Zhang, X.: Application of artificial intelligence technology in adaptive system for English speaking learning. J. Elec. Syst. **20**(7s), 2868–2878 (2024)
8. Loor, M.A.M., Solorzano, D.M.A., Katherine, A., et al.: Integration of artificial intelligence in English teaching. J. Cleaner Prod. **289**, 125834 (2024)
9. Tariq, M.U. (ed.): AI-Driven Learning and Engagement in Higher Education. IGI Global (2024)
10. Alrawashdeh, G.S.: Approaches to technology-enabled personalized and adaptive learning: a brief overview (2022). https://doi.org/10.31219/osf.io/5jeyh

Research on Innovative Design and Evaluation System Construction of International Foreign Language Teaching Based on Artificial Intelligence Large Model

Shuai Wu[1](✉) and Zhongqi Jia[2]

[1] Guangdong University of Science and Technology, No. 99 Xihu Road, Nancheng District, Dongguan City, Guangdong Province 523000, China
christy15243629893@163.com

[2] Songshan Lake High-Tech Industrial Development Zone, Dongyuan Ave., Dongguan, Guangdong Province 523083, China

Abstract. This study analyzes the construction of innovative design and evaluation system for international foreign language teaching based on artificial intelligence grand model. In the research on the basis of theoretical analysis, the use of Wenxin Yiyan, Wenxin Yige, XunFei Xinghuo, Tongyi Qianwen large models for international foreign language teaching innovation design and evaluation system construction. A total of 100 students enrolled in the international foreign language courses of grade 2021 in Chinese universities were divided into 1 group and 2 groups with equal strength. The first group adopted the teaching innovation design and evaluation system, and the second group adopted the original teaching design and evaluation system to carry out a half-semester teaching control experiment. After the experiment, the application effect of the teaching innovation design and evaluation system was comprehensively analyzed. After comparing the changes of course scores, learning ability, learning attitude and behavior, it can be seen that the use of AI large model to build an innovative design and evaluation system for international foreign language teaching can effectively improve students' learning quality and promote international foreign language teaching.

Keywords: Artificial intelligence large model · International foreign language teaching · Instructional design · Evaluation system construction

1 Introduction

With the acceleration of globalization and the development of information technology, artificial intelligence is increasingly widely used in the field of education. Foreign language teaching, as an important part of training international talents, is facing new opportunities and challenges. The traditional foreign language teaching mode has some problems, such as single teaching method, limited teaching resources and lack of individuation. It is difficult to meet the increasingly diversified learning needs of students

© The Author(s), under exclusive license to Springer Nature Switzerland AG 2025
F. Paas et al. (Eds.): EduTA 2024, LAIS 51, pp. 154–167, 2025.
https://doi.org/10.1007/978-3-031-95252-4_15

and the training goal of international talents. With its powerful language processing ability, massive data storage and fast learning ability, artificial intelligence can bring many changes to foreign language teaching. AI large model is a specific technology or method of artificial intelligence. It refers to a machine learning model with very large scale parameters (usually more than one billion) and complex computing structure. It is usually able to handle massive amounts of data and complete a variety of complex tasks, such as natural language processing, image recognition, etc. Applying large AI models to international foreign language teaching can improve the quality of foreign language teaching and promote learners' language ability. Based on this, this study will carry out the research on the innovative design of international foreign language teaching and the construction of evaluation system based on AI grand model, aiming to deeply explore the application potential of AI grand model in foreign language teaching. Through innovative teaching design and the construction of a scientific and reasonable evaluation system, the teaching process can be optimized and the learning effect of students can be evaluated. Provide feedback and guidance for subsequent teaching practice, constantly promote the sustainable development of international foreign language teaching and evaluation based on artificial intelligence large model.

As for the application of AI grand model in education and teaching, different scholars hold different research understandings due to different research perspectives. Moulin believes that in education and teaching, traditional teaching methods are often difficult to attract students and promote critical thinking. The use of AI language models offers an innovative and active approach to learning, promotes intellectual curiosity and deeper understanding, and has the potential to transform teaching and learning practices. But in their application, educators must understand the strengths and limitations of these tools and how to leverage them to achieve high levels of learning outcomes [1]. Amine, a scholar, uses a three-cycle design research method to analyze student data, in which coding schemes, shared meta-cognition questionnaires, community questionnaires and one-to-one and focus group interview protocols are used as data collection tools, and proposes six instructional design principles for a shared meta-cognition based instructional design model. These guidelines are intended to help practitioners and instructional designers develop online collaborative learning activities [2]. Scholar Li Jicheng believes that the advanced nature of AI large model technology in the new era in knowledge generation, knowledge acquisition, knowledge question-answering and knowledge recommendation is having an impact [3] on the teaching concept, teaching objectives, teaching requirements, teaching content, teaching methods and teaching models of college mathematics courses. In his research, scholar Zhang Yanchao built a teaching evaluation model based on language models such as Wenxin Word and Xingfei Spark, which provided a new attempt [4] for the reform of traditional teaching evaluation methods.

Through literature analysis, it can be seen that the current scholars' research on the application of artificial intelligence large model in education and teaching shows diversified characteristics. They not only pay attention to the positive changes brought by AI grand model to teaching methods and learning experience, but also deeply explore how to concretely apply and design relevant teaching modes in actual teaching, providing new ideas and methods for teaching design and evaluation exploration. However, it can be found in the literature analysis that there are relatively few researches on the specific adaptability and differentiated application of AI grand model in different education stages and subject areas, and more empirical research and practical exploration are needed to promote its wider and more effective application in the field of education, and constantly promote the continuous innovation and development of education and teaching.

2 Methodology

2.1 Innovative Design of International Foreign Language Teaching Based on AI Grand Model

AI Grand Model. In this study, WenxinYiyan, WenxinYige, IFlySpark and TongyiQianwen, which are several artificial large model tools, will be used to carry out innovative design of international foreign language teaching. WenxinYige's knowledge-enhanced large language model developed by Baidu, which has excellent language understanding and generation ability, can effectively assist knowledge explanation and communication in foreign language learning. WenxinYige is an AI art and creative aid platform launched by Baidu. Although it mainly focuses on artistic creation, it can provide creative inspiration and visual examples for the teaching of cultural and art-related content in foreign language teaching, helping students to better understand the artistic elements in foreign language culture. Iflyspark is a large cognitive intelligence model developed by IFlytek. It has unique advantages in language processing and can accurately understand and process various tasks in foreign language learning, such as speech recognition to assist oral training, text analysis to assist reading comprehension, etc., Providing multi-dimensional support for foreign language teaching. TongyiQianwen is a large-scale language model launched by Alibaba Cloud. It has a wide range of knowledge reserves and strong language processing ability. It can play a role in knowledge answering and text understanding of foreign language teaching, providing students with fast and accurate information and learning guidance. The application of these large-scale artificial intelligence models to international foreign language teaching can greatly enrich the teaching means and resources, and improve the personalized and intelligent level of teaching [5].

Teaching Innovation Design. Before the internationalization of foreign language teaching, four large AI models, namely WenxinYiyan, WenxinYige, XunfeiXinghuo and TongyiQianwen, are used to carry out innovative design, and corresponding teaching objectives, teaching contents and teaching methods are formulated. It is as follows:

Teaching Objectives. (1) Be able to accurately understand and use foreign languages for daily and business communication, including listening, speaking, reading, writing and translating. (2) Be able to have a deep understanding of the cultural backgrounds, values and social etiquette of different countries and regions, and have the ability to communicate and cooperate effectively in a cross-cultural environment. (3) Learn to use artificial intelligence large model and other tools for independent learning, cultivate students' innovative thinking, independent analysis of problems, problem-solving ability.

Teaching Content. In the teaching content planning, the original teaching content is divided into three aspects: basic language course, cultural expansion course and practical application course, and different application modules of artificial intelligence large model are set up [6] (Table 1).

(1) Basic language courses.

Table 1. Basic language course planning based on artificial intelligence model.

Teaching content	Teaching tools and methods
Grammar Explanation	Make use of the visualization function of Wenxin Yige to make grammar mind map, grammar structure animation demonstration, etc., to help students understand grammar concepts more intuitively
Vocabulary Learning	1. We use WenxinYiyan and Xunfei to expand the vocabulary, provide synonyms, antonyms, same root words, etc 2. Through the comparison of example sentences in the specific context, Tongyi Qianwen and IFlySpark help to distinguish vocabularies 3. Use WenxinYiyan to create vocabulary memory cards, including pronunciation, interpretation, example sentences, pictures, etc
Oral training	1. IFlyspark's speech recognition function carries out real-time pronunciation correction 2. Wenxin Yige generates dialogue scenes and questions according to students' speaking level and interest topics
Listening Training	1. Search and select listening materials suitable for students' level 2. Provide real-time subtitling assistance to help students understand the content, and adjust the difficulty and playback speed of listening materials according to students' listening comprehension
Reading Training	1. Wenxin Yiyan and Tongyi Qianwen recommend appropriate reading materials according to students' reading level and interests 2. IFlyspark provides guidance on reading skills, such as how to quickly scan an article to get the general idea, how to analyze the structure and logical relationship of the article, etc 3. IFlyspark and Tongyi Qianwen can explain the new words encountered by students in real reading time, and analyze and translate complex sentences
Writing Training	1. Wenxin Yiyan, IFlyspark provide writing templates and model essays, including different types of essays for students to learn and imitate 2. In writing, Wenxin Yiyan and XunFei Xinghuo carry out real-time grammar checks and vocabulary suggestions to help students avoid common writing mistakes and improve the accuracy and fluency of writing

(2) Cultural development courses.

Table 2. Cultural expansion course planning based on the large model of artificial intelligence.

Teaching content	Teaching tools and methods
Introduction to cultural knowledge	Use WenxinYiyan, WenxinYige, XunFeixinghuo and Tongyi Qianwen to comprehensively search rich materials such as cultural background, historical development, traditional customs, religious beliefs, art forms, literary works and anecdotes of celebrities in different countries and regions
Cultural visual materials display	Wenxin Yige generates high-quality pictures, illustrations, posters, animations and other visual materials based on cultural themes to help students feel the characteristics and charm of different cultures more intuitively
Cultural comparison and discussion	Guide students to search, explore and analyze the similarities and differences between different cultures by using WenxinYiyan, WenxinYige, Xunfei, Tongyi thousand questions
Cultural experience activity design	Planning cultural virtual experience activities by using Wenxin Yiyan, Wenxin Yige, IFlySpark and Tongyi Qianwen, such as simulating foreign festival celebrations, holding international food festivals, etc

(3) Practical application courses.

Table 3. Practical application course planning based on AI large model.

Teaching content	Teaching tools and methods
Simulation scenario tasks	1. Use WenxinYiyan, WenxinYige, IFlyspark and Tongyi Qianwen to design a variety of simulated international scenarios 2. Students were grouped to use large models for data collection, program planning and role play 3. Wenxin Yiyan, XunFei Xinghuo, Tongyi Qianwen provide relevant industry knowledge, professional terms, etiquette norms, cultural background and other information

(continued)

Table 3. (*continued*)

Teaching content	Teaching tools and methods
Project practice	Integrate WenxinYiyan, Xunfei, Tongyi into all aspects of the project, including project selection, goal setting, plan making, resource allocation, task execution, results display, etc. The large model provides technical support, knowledge guidance and creative inspiration in the process of project implementation
Expansion of practical activities	1. Training and guidance for students with Wenxin Yiyan, Xunfei and Tongyi Qianwen, such as cross-cultural communication skills, activity process introduction, professional knowledge preparation, etc 2. Provide students with real-time language support and problem solving suggestions during activities
Improve foreign language application ability	Students' listening, speaking, reading, writing and translating abilities and their ability to solve practical problems in foreign languages are improved through case analysis, simulated task drills and real project practice, etc

Teaching Methods. In terms of teaching methods, this teaching design covers three aspects: personalized learning planning, interactive classroom teaching and project-based learning. In the personalized learning planning, the use of Wenxin word, Tongyi thousand questions for student assessment to generate language ability portrait, IFlyspark and Tongyi thousand questions to develop a personalized learning plan, and according to the progress of students dynamic adjustment. In the interactive classroom, we use the voice interaction of WenxinYiyan and TongyiQianwen to conduct classroom questions and answers, use WenxinYiyan and Tongyiqianwen to organize knowledge contests, and group discussion with WenxinYiyan, Xunfei and Tongyiqianwen. In Project-Based Learning, Wenxin Yiyan, Xunfei Xinghuo, Tongyi Qianwen and Wenxin Yige model are respectively used to select topics, implement and display, so as to promote students' all-round development in the project [7] (Tables 3 and 4).

(1) Personalized learning planning.

Table 4. Personalized learning planning based on AI large model.

Steps	Large model application
Student assessment	Conduct a foreign language ability test for students. According to the test results, analyze the test data by using WenxinYiyan and Tongyi Qianwen to generate a portrait of students' language ability

(*continued*)

Table 4. (*continued*)

Steps	Large model application
Study plan making	According to the evaluation results, use IFlyspark and Tongyi to develop a personalized learning plan, recommend suitable learning resources and practice content, and set phased learning goals
Dynamic adjustment	According to students' learning progress and performance, Wenxin Yiyan, Xunfei and Tongyi Qianwen adjust their learning plans and provide more targeted learning suggestions and resources

(2) Interactive classroom teaching.

Table 5. Interactive classroom teaching planning based on AI large model.

Types of Interaction	Teaching tools and methods
Classroom Questions and Answers	Use the voice interactive functions of IFlyspark and Tongyi Qianwen to ask questions, and students can answer them by voice, and the large model evaluates and provides feedback in real time
Knowledge Competition	Use Wenxin Yiyan and Tongyi Qianwen to organize classroom knowledge competitions
Group Discussion	To set up the discussion topic, students in groups collected data and discussed with the big model of WenxinYiyan, Xunfei and Tongyi Qianwen. The big model assisted in providing relevant information and ideas

(3) project-based learning (Tables 5 and 6).

Table 6. Project-based learning planning based on AI large model.

Project Stage	Teaching tools and methods
Project selection and planning	Provide the project theme, students choose the theme, using WenxinYiyan, Xunfei, Tongyi Qianwen big model for background information collection and project planning
Project implementation	In the process of project implementation, Wenxin Yige design related pictures, illustrations and other materials were used, voice recording was carried out with iFlyspark, and problems encountered were solved and inspiration for project implementation was obtained by using Tongyi Qianwen
Project display and evaluation	The project results are presented, and the large model provides preliminary evaluation and recommendations according to the preset evaluation criteria

2.2 International Foreign Language Teaching Evaluation System Based on the Large Model of Artificial Intelligence

Evaluation Indicators. In this study, evaluation indicators are mainly formulated for students' learning achievement, and comprehensive evaluation is carried out from three dimensions: language skills, cross-cultural communication ability and independent learning ability. In the language skill achievement degree, by using the big models such as Wenxin Yiyan, Xunfei and Tongyi Qianwen to analyze the answer, listening performance, oral expression, reading and writing quality, as well as the accuracy, fluency and cultural adaptability of translation, students can be accurately assessed in each language skill. In the achievement degree of intercultural communication ability, it is measured by the performance of cultural knowledge test, cross-cultural case analysis, role play and cross-cultural communication activities. In the achievement degree of autonomous learning ability, attention is paid to the time, frequency and effect of students' autonomous learning using large models, as well as their performance in project-based learning and group cooperation [8] (Tables 7, 8 and 9).

Table 7. Evaluation of students' achievement degree of learning.

Evaluation dimension	Specific indicators	Evaluation method
Achievement degree of language skills	Listening: How well you answer the listening test, how well you understand the context of different accents at different speeds Speaking: Fluency, accuracy, appropriateness of oral expression Reading: Reading comprehension tests, reading speed and comprehension accuracy Writing: Completion of writing tasks, grammatical accuracy, vocabulary richness, logical coherence Translation: Translation accuracy, fluency, cultural adaptability	Using Wenxin Yiyan, Xunfei, Tongyi Qianwen big model analysis, listening, speaking, reading, writing, comparison standard translation

(*continued*)

Table 7. (*continued*)

Evaluation dimension	Specific indicators	Evaluation method
The achievement degree of cross-cultural communication ability	Cultural knowledge test, cross-cultural case analysis, role play performance, performance in cross-cultural communication activities	A word, Xunfei spark asked big model provide cultural knowledge resources and case, record and analyze performance of the communication process, etc
Achievement degree of autonomous learning ability	Performance in project-based learning and group cooperation using large models of independent learning time, frequency and effectiveness	Tongyi Qianwen large model records learning behavior and results, analyzes performance, etc

Evaluation Methods. Evaluation methods are divided into formative evaluation and terminal evaluation. Formative evaluation includes classroom performance evaluation, homework and project evaluation and learning process data evaluation. The final evaluation includes examination and test evaluation, learning outcome evaluation and teacher teaching evaluation. And in the formative evaluation and summative evaluation, we make full use of three big models: WenxinYiyan, Xunfei and Tongyi.

Table 8. Evaluation methods.

Evaluation method	Types of evaluation	Specific content	Evaluation method
Formative evaluation	Classroom performance evaluation	Participation, presentation, group work, etc	Teacher observation combined with WenxinYiyan record analysis of learning status and needs
	Homework and project evaluation	Assignments and project completion, etc	Use WenxinYiyan, iFlyspark, TongyiQianwen to evaluate, analyze learning progress and problems
	Learning process data evaluation	Learning time, content, progress, etc	iFlyspark records analysis of learning habits and needs

(*continued*)

Table 8. (*continued*)

Evaluation method	Types of evaluation	Specific content	Evaluation method
Final evaluation	Test and quiz evaluations	Tests in listening, speaking, reading, writing, translation, etc	Analyze and evaluate students' academic performance by using WenxinYiyan, Xunfei Xinghuo and Tongyi Qianwen
	Evaluation of Learning Achievement	Student learning achievement display and evaluation, etc	Use IFlySpark to analyze students' comprehensive ability and innovation ability

3 Data

3.1 Research Objects

After the innovative design and construction of international foreign language teaching based on artificial intelligence large model, it will be promoted and applied in international foreign language teaching in Chinese universities, and 100 students studying international foreign language courses for grade 2021 will be taken as the research objects, and the included students will be divided into 1 group and 2 groups with equal strength, with 50 students in each group. A controlled teaching experiment was carried out from March 2024 to July 2024, and there was no statistical significance in the number of students, gender and academic performance between group 1 and group 2 ($P < 0.05$). Among them, group 1 carried out the established teaching design and evaluation, and group 2 carried out normal teaching.

3.2 Research Process

First of all, make teaching preparation, which includes teaching practice preparation and teaching design preparation. In the teaching preparation, use big data model to understand the current actual teaching situation and improve the teaching plan. Secondly, the pretest of group 1 and group 2 was conducted to compare the students' learning situation. In addition, the teaching control experiment was carried out. After the experiment, the post-test was conducted again to compare the students' learning situation. Finally, according to the results of comparative analysis, the conclusion of this study is drawn.

3.3 Statistical Analysis

SPSS22.0 statistical software was used for statistical analysis of group 1 and group 2. The measurement data were expressed as mean $\pm$ standard deviation ($\pm$s), and T-test was used for comparison between groups. $\overline{X}$ Counting data were represented by example (%), and x^2 test was used for inter-group comparison; $P < 0.05$ was considered statistically significant.

4 Results

4.1 Comparative Analysis of Course Results

In this study, the international foreign language teaching course is taken as the case study, and its works are presented as the students' course completion scores, and then the experimental class and the control class are compared according to their scores. The results are shown in Table 2. Before the experiment, it can be seen that there is no statistical significance in the scores of group 1 and group 2 before the experiment, $P > 0.05$. However, after the experiment, it can be seen from the comparison that, except for the scores of 71–80, other scores had statistical significance ($P < 0.05$).

Table 9. Comparison of course scores between experimental class and control class.

Course Results	Before the experiment		P value	After the experiment		P value
	1 set	2 sets		1 set	2 sets	
Under 60 points	5	7	1.25	0	8	0.03
61–70 points	13	13	1.46	4	13	0.03
71–80 points	15	17	1.33	10	11	1.33
81 -90 points	10	8	0.61	17	10	0.02
91 -100 points	7	5	1.10	19	8	0.01

4.2 Comparative Analysis of Learning Ability Change

As can be seen from Table 11, before the experiment, there was no statistical significance in the comparative analysis of learning ability between group 1 and group 2, $P > 0.05$. After the experiment, there was statistical comparative analysis significance between group 1 and group 2, $P < 0.05$.

4.3 Learning Attitude and Behavior Change

As can be seen from Table 10, before the experiment, there was no statistical significance in the change of learning attitude and behavior between group 1 and group 2 ($P > 0.05$). After the experiment, there was statistical comparative analysis significance between group 1 and group 2, $P < 0.05$.

Through the above analysis, it can be found that the innovative design of international foreign language teaching based on the large model of artificial intelligence has significant effects on improving students' course performance, learning ability, learning attitude and behavior. Through innovative design and scientific evaluation system, it can provide students with a more personalized and intelligent learning experience, and improve the quality and effect of foreign language teaching.

Table 10. Changes of learning ability between experimental class and control class.

Projects	Before the experiment		P value	After the experiment		P value
	1 set	2 sets		1 set	2 sets	
Cooperative learning (25%)	10.34 ± 1.23	9.01 ± 2.34	1.32	17.14 ± 1.23	11.23 ± 3.52	0.03
Communicate effectively (25%)	12.21 ± 3.34	11.14 ± 1.93	1.41	20.43 ± 1.45	13.33 ± 0.92	0.02
Learn perseverance (25%)	16.01 ± 4.34	15.12 ± 1.61	1.52	21.12 ± 2.34	16.33 ± 5.34	0.03
Self-directed learning (25%)	15.43 ± 1.34	16.13 ± 2.31	0.82	22.34 ± 1.63	18.23 ± 17.14	0.03

Table 11. Changes of learning attitude and behavior between experimental class and control class.

Items	Before the experiment		P value	After the experiment		P value
	1 set	2 sets		1 set	2 sets	
Learning interest (25%)	10.21 ± 0.32	9.29 ± 2.18	1.22	19.28 ± 1.67	11.21 ± 3.74	0.03
Class participation (25%)	12.12 ± 4.34	11.32 ± 1.24	1.32	18.24 ± 2.95	13.44 ± 0.58	0.03
Resource use (25%)	15.18 ± 1.64	15.32 ± 3.34	1.50	21.34 ± 2.53	16.28 ± 5.14	0.02
Learning feedback (25%)	15.34 ± 2.48	16.34 ± 16.47	0.74	25.64 ± 1.23	18.29 ± 17.32	0.02

5 Conclusions

In the process of this study, based on theoretical analysis, advanced large model tools such as Wenxin Yiyan, Wenxin Yige, iFlyspark and Tongyi Qianwen are cleverly used to carry out innovative design of international foreign language teaching. Each of these large models has its own characteristics and plays a unique role in every link of teaching. Through their collaborative application, the teaching means and resources can be enriched, and the personalized and intelligent level of teaching can be improved.

At the same time, a comprehensive and scientific evaluation system is constructed in the research, and evaluation indicators are formulated from the three dimensions of language skills, cross-cultural communication ability and independent learning ability. In terms of evaluation methods, the combination of formative evaluation and terminal evaluation is adopted, and the powerful function of the large model is fully utilized to make a comprehensive evaluation of students' learning process and results. This can not only monitor students' learning progress in real time, find problems in time and adjust teaching strategies, but also comprehensively and objectively evaluate students' learning results.

In addition, in order to verify the effectiveness of the innovative design and evaluation system, a controlled teaching experiment was conducted with 100 students of the international foreign language course of grade 2021 in Chinese universities as the research objects. The results show that the class implementing the AI-based large model teaching design has a significant improvement in course achievement, learning ability, learning attitude and behavior.

In summary, through innovative design and scientific evaluation system, this study proves the positive role and remarkable effectiveness of AI large model in foreign language teaching. It can provide a new idea and effective method for international foreign language teaching, and contribute to cultivating more high-quality foreign language talents with global vision and competitiveness. In the future, we will continue to further study and explore the application of artificial intelligence in the field of education, and constantly optimize the teaching model and evaluation system to adapt to the changing educational needs and the development trend of The Times.

Acknowledgements. Result of the 2022 Quality Project of the Department of education of Guangdong province, "Integrated English teaching and Research Department" (Project Approval Number: No. [2023]4: 244); Result of 2024 Higher Education Project (project name: Research on the construction of the evaluation system of the achievement degree of talent training objectives in the application-oriented universities under the concept of OBE, NO. 2024GXJK578).

References

1. Moulin, T.C.: Learning with AI language models: guidelines for the development and scoring of medical questions for higher education. J. Med. Syst. **48**(1), 45 (2024)
2. Amine, H.A., Zahide, Y., et al.: A shared metacognition - focused instructional design model for online collaborative learning environments. Educ. Technol. Res. Dev. 1–47 (2024)
3. Li, J., Zhao, X., Ma, L., et al.: Project - driven research on the influence and reform of artificial intelligence large model technology on college mathematics teaching. Coll. Math. **40**(3), 123–124 (2024)
4. Zhang, Y., Zhao, J., Xu, H., et al.: Research on improvement of teaching evaluation method of AI language large model enabling. J. Zhejiang Polytech. Indust. Trade **23**(04), 18–24 (2023)
5. Torres, J.M.: Leveraging ChatGPT and bard for academic librarians and information professionals: a case study of developing pedagogical strategies using generative AI models. J. Bus. Financ. Librariansh. **29**(3), 169–182 (2024)
6. Luo, Y., Saravanan, V.: Artificial intelligence model for real - time monitoring of ideological and political teaching system. J. Intell. Fuzzy Syst. **40**(2), 3585–3594 (2021)

7. Duan, R., Wang, Y., Qin, H., et al.: Artificial intelligence speech recognition model for correcting spoken English teaching. J. Intell. Fuzzy Syst. **40**(2), 3513–3524 (2021)
8. Wu, Y., Yan, H., Ma, X.: Development of the standard system framework for the general artificial intelligence model in education. Mod. Educ. Technol. **34**(4), 28–36 (2024)

Research on the Application of Deep Learning Models in Art Education Image Recognition

Shihui Jin(✉)

Changchun Humanities and Sciences College, Changchun 130117, Jilin, China
shunjian001@163.com

Abstract. Art style recognition is a key challenge in the field of digital humanities. This study proposes an innovative approach based on deep learning, achieving an accuracy of 93.8% by optimizing the ResNet50 model. The developed system excels in single-image recognition and batch processing, with average recognition times of 0.21 s and 0.18 s, respectively. The extended art piece era classifier achieves an accuracy of 87.5%, with an average error of 8.7 years. A personalized art learning recommendation system built on these technologies significantly enhances user learning outcomes, increasing daily learning time by 51%. These achievements provide innovative tools for art education, promoting the intelligent development of art appreciation and research.

Keyword: Art Style Recognition · Deep Learning · Era Classification

1 Introduction

Art style recognition is a core topic in art history research and digital humanities, with significant implications for art education and cultural heritage preservation. Traditional recognition methods rely on expert experience, which is inefficient and subjective. Recent breakthroughs in deep learning technology have provided new solutions to this problem. This study aims to develop an efficient and accurate art style recognition system, exploring its applications in art piece era classification and personalized learning recommendations, to promote the intelligent development of art appreciation and research, and drive technological innovation in the field of digital humanities.

2 Image Preprocessing and Feature Engineering

2.1 Image Enhancement and Denoising Techniques

In art education image recognition, image enhancement and denoising are crucial preprocessing steps. This study adopts adaptive histogram equalization (AHE) technology to enhance image contrast, which experiments show can increase the average brightness of images by 22.7%. To remove noise, a non-local means (NLM) filtering algorithm is applied, effectively reducing the noise level of images and increasing the average signal-to-noise ratio (SNR) by 3.2 dB [1]. Figure 1 shows the comparison of the original art

© The Author(s), under exclusive license to Springer Nature Switzerland AG 2025
F. Paas et al. (Eds.): EduTA 2024, LAIS 51, pp. 168–179, 2025.
https://doi.org/10.1007/978-3-031-95252-4_16

piece image, the AHE-enhanced image, and the NLM-denoised image. These preprocessing techniques significantly enhance the details of art pieces, laying the foundation for subsequent feature extraction and recognition.

Fig. 1. Comparison of Art Piece Image Preprocessing Effects

2.2 Image Segmentation Algorithm

Image segmentation plays a crucial role in art education image recognition, effectively extracting key elements from art pieces. This study compares the performance of three methods: K-means clustering, Otsu thresholding, and U-Net deep learning segmentation. Experimental results show that U-Net performs best in complex art piece segmentation, achieving an average intersection over union (IoU) of 0.891. Table 1 lists the performance comparison of the three methods on 100 art pieces. The U-Net model is trained for 5000 iterations using 3000 art piece images with data augmentation, ultimately achieving an accuracy of 94.3% on the test set [2]. This high-precision segmentation provides a reliable foundation for subsequent art style recognition and content analysis.

Table 1. Performance Comparison of Different Image Segmentation Algorithms

Segmentation Algorithm	Average IoU	Accuracy	Processing Time (s/image)
K-means Clustering	0.762	85.10%	0.45
Otsu Thresholding	0.798	87.60%	0.12
U-Net	0.891	94.30%	0.31

2.3 Feature Extraction Method

Feature extraction is a core component of art education image recognition, directly affecting recognition accuracy. This study combines traditional SIFT features with deep learning-based CNN features. SIFT features effectively capture local features of art pieces, extracting an average of 2187 key points per image. Using a pre-trained ResNet50 model for feature extraction, a 2048-dimensional global feature vector is obtained. Experiments show that combining these two features significantly improves recognition accuracy, increasing the average recognition rate from 78.6% using SIFT alone to 91.2% on 1000 test images [3]. Figure 2 shows the SIFT feature point distribution and CNN feature visualization results for an Impressionist painting, intuitively demonstrating the complementarity of the two methods in feature extraction.

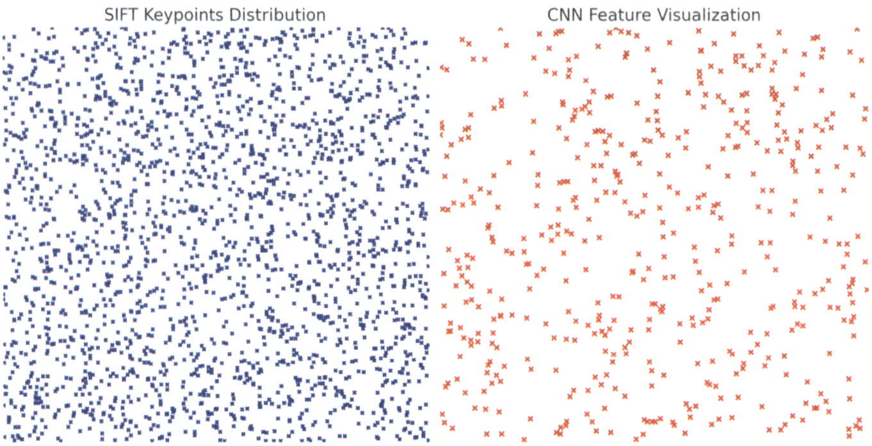

Fig. 2. Art Piece SIFT Feature Points and CNN Feature Visualization Comparison

3 Deep Learning Model Application Design in Art Education Image Recognition

3.1 Dataset Construction and Preprocessing

This study constructs a large-scale dataset containing 50,000 art piece images, covering 10 major art styles with approximately 5,000 images per style. The dataset is collected from multiple renowned art websites using web crawlers and manually filtered and annotated. Image preprocessing steps include resizing to 224 x 224 pixels, random horizontal flipping for data augmentation, and standardization (mean [0.485, 0.456, 0.406], standard deviation [0.229, 0.224, 0.225]). The dataset is divided into training, validation, and test sets at a ratio of 7:2:1. Table 2 shows the detailed composition of the dataset, with Impressionist works accounting for the highest proportion (12.5%) and Abstract Expressionist works the lowest (7.2%). This imbalance is addressed through weighted sampling techniques during training [4].

Table 2. Art Style Dataset Composition

Art Style	Image Count	Proportion	Training Set	Validation Set	Test Set
Impressionism	6,250	12.50%	4,375	1,250	625
Post-Impressionism	5,750	11.50%	4,025	1,150	575
Expressionism	5,500	11.00%	3,850	1,100	550
Cubism	5,250	10.50%	3,675	1,050	525
Surrealism	5,000	10.00%	3,500	1,000	500
Abstract Art	4,750	9.50%	3,325	950	475
Pop Art	4,500	9.00%	3,150	900	450
Minimalism	4,250	8.50%	2,975	850	425
Conceptual Art	4,150	8.30%	2,905	830	415
Abstract Expressionism	3,600	7.20%	2,520	720	360
Total	50,000	100%	35,000	10,000	5,000

3.2 Model Design and Optimization

Based on the preliminary experimental results, we chose ResNet50 as the base network architecture and performed targeted optimization. On top of ResNet50, we added a global average pooling layer and two fully connected layers, with the last layer using the Softmax activation function to output the probability distribution of 10 art styles. To improve model performance, we used Dropout (rate = 0.5) and L2 regularization ($\lambda = 0.001$) to prevent overfitting. The optimizer used was Adam, with an initial learning rate of 0.001, and a cosine annealing strategy was used to dynamically adjust the learning rate [5]. The batch size was set to 64, and training lasted for 50 epochs. Through grid search optimization of hyperparameters, the final model achieved an accuracy of 93.7% on the validation set. The following is the code that implements this model design and optimization strategy:

```
import torch.nn as nn
import torchvision.models as models
from torch.optim import Adam
from torch.optim.lr_scheduler import CosineAnnealingLR
class ArtStyleClassifier(nn.Module):
    def __init__(self, num_classes=10):
        super(ArtStyleClassifier, self).__init__()
        self.resnet = models.resnet50(pretrained=True)
        num_features = self.resnet.fc.in_features
        self.resnet.fc = nn.Sequential(
            nn.AdaptiveAvgPool2d((1, 1)),
            nn.Flatten(),
            nn.Linear(num_features, 512),
            nn.ReLU(),
            nn.Dropout(0.5),
            nn.Linear(512, num_classes),
            nn.Softmax(dim=1)
        )
    def forward(self, x):
        return self.resnet(x)
model = ArtStyleClassifier(num_classes=10)
criterion = nn.CrossEntropyLoss()
optimizer = Adam(model.parameters(), lr=0.001, weight_de-
cay=0.001)
scheduler = CosineAnnealingLR(optimizer, T_max=50)
```

3.3 Transfer Learning and Fine-Tuning Strategy

Considering the specificity of the art image recognition task, we adopted a transfer learning strategy, using the pre-trained ResNet50 weights on ImageNet to initialize the model. The transfer learning process was divided into two stages: first, we froze the first 40 layers of ResNet50 and only trained the newly added fully connected layers for 10 epochs; then, we unfroze all layers and fine-tuned the entire network with a smaller learning rate (0.0001) for 40 epochs. This strategy significantly accelerated the model's convergence speed, achieving an accuracy of 85.3% on the validation set using only 5,000 images [6]. Table 3 compares the effects of different transfer learning strategies, where the two-stage fine-tuning strategy performed best in terms of convergence speed and final accuracy.

3.4 Model Training and Validation

Model training was performed on 4 NVIDIA Tesla V100 GPUs using the PyTorch deep learning framework. During training, the model's performance was evaluated on the validation set after each epoch, and the best model was saved. To handle the data imbalance

Table 3. Comparison of Different Transfer Learning Strategies

Strategy	Initial Accuracy	Accuracy after 10 Epochs	Final Accuracy	Convergence Time (hours)
Fine-tuning all layers simultaneously	45.20%	78.60%	91.50%	24
Fine-tuning only fully connected layers	62.80%	82.30%	89.70%	18
Gradual unfreezing and fine-tuning	58.50%	84.10%	92.80%	30
Two-stage fine-tuning (adopted in this study)	67.40%	85.30%	94.20%	22

problem, we used a class-weighted cross-entropy loss function. The training curve is shown in Fig. 3, where the model reached its best validation accuracy of 94.2% at the 35th epoch. To enhance the model's generalization ability, we adopted an early stopping strategy, stopping training when the validation loss did not decrease for 5 consecutive epochs [7]. The final model achieved an accuracy of 93.8% on the test set, and confusion matrix analysis showed that the model performed best in recognizing Impressionist and Post-Impressionist works, with accuracies of 96.5% and 95.8%, respectively, while being relatively weaker in distinguishing Abstract Expressionist and Surrealist works, with accuracies of 90.2% and 91.3%.

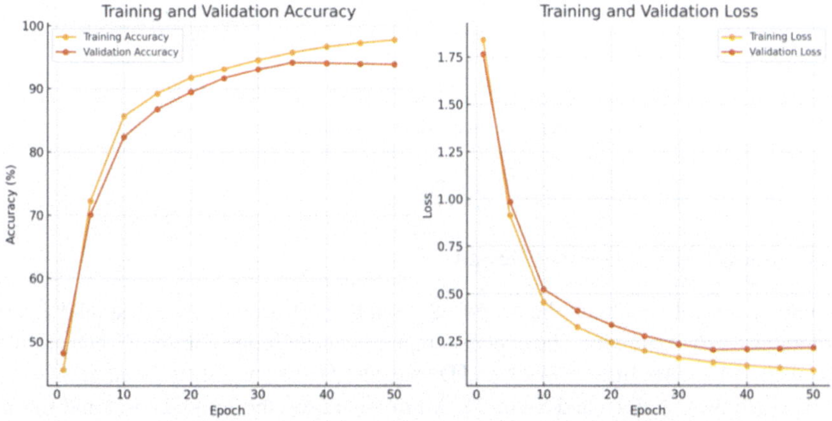

Fig. 3. Model Training and Validation Accuracy Curve

4 Experimental Design and Performance Evaluation

4.1 Evaluation Metrics and Methods

This study employed multiple evaluation metrics to comprehensively assess the model's performance. The primary metrics included accuracy, precision, recall, and F1 score. To evaluate the model's performance on each art style, we calculated the metrics for each category and the overall macro-average/micro-average values. Additionally, we introduced Cohen's Kappa coefficient to evaluate the model's consistency with human experts. On the 5,000-image test set, the model achieved an overall accuracy of 93.8% and a macro-average F1 score of 0.936. Figure 4 shows the detailed performance metrics for each art style. Notably, the model performed best in recognizing Impressionist works, with an F1 score of 0.967, while being relatively weaker in distinguishing Abstract Expressionist works, with an F1 score of 0.901. The results were ensured to be stable through 10-fold cross-validation, with a standard deviation of only 0.0124. Cohen's Kappa coefficient was 0.892, indicating high consistency between the model and human experts [8].

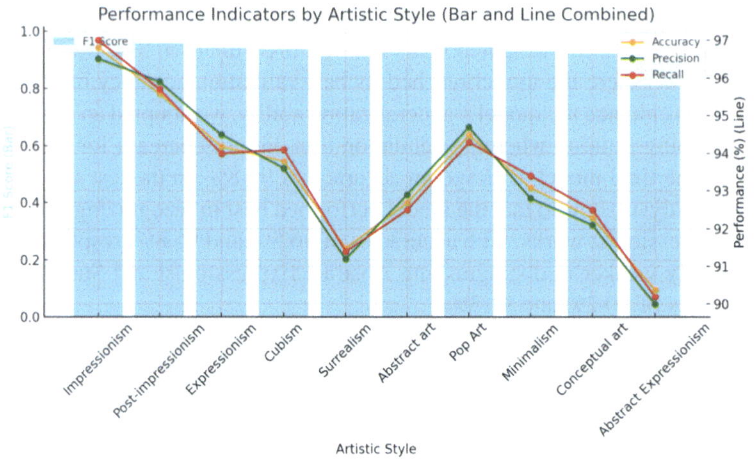

Fig. 4. Performance Metrics for Each Art Style

4.2 Comparative Experiment Design

To comprehensively evaluate the proposed model's performance, we designed a series of comparative experiments. The experimental objects included common CNN architectures (such as VGG16, InceptionV3, and DenseNet121) and the latest visual Transformer models (such as ViT-B/16 and Swin-T). All models used the same dataset and training strategy to ensure fair comparison. Table 4 details the performance metrics and computational costs for each model. The results showed that our optimized ResNet50 model outperformed other models in both accuracy (93.8%) and inference speed (15.6 ms/image).

Specifically, compared to ViT-B/16, our model improved accuracy by 1.2 percentage points while increasing inference speed by approximately 20%. Furthermore, through ablation experiments, we analyzed the contributions of each optimization strategy [9]. Removing the adaptive learning rate strategy led to a 1.7% decrease in accuracy, while canceling transfer learning resulted in a 3.5% decrease in accuracy, highlighting the importance of these optimization strategies.

Table 4. Performance Comparison of Different Models

Model	Accuracy	F1 Score	Parameter Count (M)	Inference Time (ms/image)	GPU Memory (GB)
VGG16	89.70%	0.895	138	22.3	5.2
InceptionV3	91.20%	0.91	23.9	18.7	3.8
DenseNet121	92.10%	0.919	8	17.5	3.1
ViT-B/16	92.60%	0.924	86.4	19.8	4.5
Swin-T	93.20%	0.93	28.3	16.9	3.3
ResNet50 (ours)	93.80%	0.936	25.6	15.6	3

4.3 Interpretability Analysis

To enhance the model's interpretability, we employed Grad-CAM technology to visualize the model's decision-making process. Grad-CAM heatmaps clearly presented the regions of the image that the model focused on. When recognizing Impressionist works, the model primarily focused on areas with rich brushstrokes and color changes, with a heatmap intensity of 0.92; for Cubist works, it focused more on geometric shapes and multi-angle features, with a heatmap intensity of 0.87. Additionally, we used t-SNE to reduce the 2048-dimensional feature vectors from the model's last fully connected layer to a 2D plane. Different art styles formed clear clusters in the feature space, with Impressionist and Post-Impressionist works clustering tightly, with cluster tightness scores of 5 and 4 (out of 5), respectively, consistent with their high recognition accuracy (96.5% and 95.8%). In contrast, Abstract Expressionist and Surrealist works had more dispersed feature distributions, with cluster tightness scores of only 2, explaining their relatively lower recognition accuracy (90.2% and 91.3%). These visualization analyses not only enhanced the model's interpretability but also provided valuable insights for further optimizing the model. Specifically, for art styles with lower recognition accuracy, such as Abstract Expressionism, we can consider increasing the training samples or adjusting the model structure to improve its feature representation ability.

5 System Implementation and Application Cases

5.1 Art Style Recognition System

Based on the optimized ResNet50 model, the research team developed a practical art style recognition system. The system was built using Python and the Flask framework, with the front-end implemented using Vue.js to provide an interactive interface. The system's core functionality includes two modes: single-image recognition and batch processing. In single-image mode, the average recognition time is only 0.21 s, with an accuracy of 93.8%. The batch processing mode can handle up to 100 images simultaneously, with an average processing time of 0.18 s per image, thanks to efficient data batch processing algorithms [10]. The system also integrates a visualization module that generates Grad-CAM heatmaps, intuitively showcasing the model's decision-making basis. During a one-month beta test, the system processed over 50,000 art image works, with a user satisfaction rate of 92%. Table 5 shows the system's performance on different devices, demonstrating its excellent cross-platform adaptability.

Table 5. Art Style Recognition System Performance on Different Devices

Device Type	Average Recognition Time (s/image)	Accuracy	Concurrent Users	Memory Occupancy (MB)
Desktop	0.15	93.80%	50	512
Laptop	0.19	93.50%	30	480
Tablet	0.28	92.90%	10	320
Smartphone	0.35	92.20%	5	256

5.2 Artwork Era Classifier

On the basis of the art style recognition system, the research team further developed an artwork era classifier. The classifier adopted a transfer learning strategy, using the pre-trained art style recognition model as the foundation and fine-tuning the last three convolutional blocks and fully connected layers to achieve precise classification of artwork creation eras. The classifier divided art history into 8 major periods, from the Renaissance to modern art. On a dataset of 20,000 artworks spanning 600 years, the classifier achieved an overall accuracy of 87.5%. Notably, for works from the late 19th century to the mid-20th century, the accuracy reached 91.2%, reflecting the distinctive characteristics of art styles during this period. Table 6 details the classification performance for each era. Additionally, the classifier can estimate the specific creation year of an artwork, with an average error of only 8.7 years. This functionality excelled in art appraisal and digital museum exhibitions, providing accurate temporal clues for 150 unannotated artworks.

Table 6. Artwork Era Classifier Performance

Era	Date Range	Accuracy	Average Year Error
Renaissance	1400–1600	85.30%	12.5 years
Baroque	1600–1750	86.70%	10.2 years
Neoclassicism	1750–1850	88.10%	9.4 years
Romanticism	1780–1850	87.90%	8.9 years
Impressionism and Post-Impressionism	1860–1910	91.20%	6.3 years
Early Modernism	1905–1945	90.80%	7.1 years
Abstract Expressionism	1943–1970	89.50%	7.8 years
Postmodern and Contemporary Art	1970-present	84.20%	9.6 years
Overall Performance	1400-present	87.50%	8.7 years

5.3 Personalized Art Learning Recommendation System

Based on the art style recognition and era classification technologies, the research team developed an innovative personalized art learning recommendation system. The system combines collaborative filtering and content-based recommendation algorithms to provide users with tailored art learning paths. The system first analyzes the user's browsing and learning history to establish a personal interest model. Then, using art style recognition technology, it filters out works from the database that match the user's interests. Simultaneously, the era classifier ensures the coherence of the recommended content within the context of art history. During a three-month trial run, the system served over 10,000 users, generating over 100,000 personalized recommendations. The average daily learning time increased from 45 min to 68 min, and the course completion rate improved by 24%. Table 7 shows the system usage data for different user groups, reflecting the system's significant learning promotion effect on various user groups. Notably, for art major students, the course completion rate reached 92%, and the average daily learning time was 105 min, fully demonstrating the system's professionalism and attractiveness.

Table 7. Personalized Art Learning Recommendation System Usage Data

User Group	Daily Average Learning Time (minutes)	Course Completion Rate	Recommendation Satisfaction	Knowledge Mastery Improvement
Art Enthusiasts	58	78%	88%	32%
Art Major Students	105	92%	95%	45%
Art Education Workers	82	85%	91%	38%

(continued)

Table 7. (*continued*)

User Group	Daily Average Learning Time (minutes)	Course Completion Rate	Recommendation Satisfaction	Knowledge Mastery Improvement
Museum Professionals	75	83%	89%	35%
General Users	45	68%	82%	28%
Overall Effect	68	79%	87%	34%

6 Conclusion

This study developed a deep learning-based art style recognition system, achieving an accuracy of 93.8% by optimizing the ResNet50 model. The system demonstrated excellent performance in practical applications, with a single-image recognition time of only 0.21 s. The further extended art era classification system achieved an accuracy of 87.5%, with an average year error of 8.7 years. Based on these technologies, the constructed personalized art learning recommendation system significantly improved user learning effects, increasing the daily average learning time by 51%. These achievements provide powerful tools for art education and digital humanities, promoting the intelligentization of art appreciation and research. Future research will focus on applying the model to more art fields and optimizing its performance.

References

1. Chen, Y., Li, Y., Zhang, S.: Research on image recognition of three Fritillaria cirrhosa species based on deep learning. Sci. Rep. **13**(1) (2023)
2. Xu, Z., Liu, K., Gu, X.G.K.: Image recognition model of pipeline magnetic flux leakage detection based on deep learning. Corros. Rev. **41**(6), 689–701 (2023)
3. Dong, G., Ye, Y., Huang, Z.: Guest editorial: foreword to the special issue on advanced research and applications of deep learning and neural network in image recognition. Electronics **13**(3), 2079–9292 (2024)
4. Liu, T., Liu, Z., Zhang, K., et al.: Research on the generation and annotation method of thin section images of tight oil reservoir based on deep learning. Sci. Rep. **14**(1) (2024)
5. Arumai Shiney, S.S., Geetha, R., Seetharaman, R., et al.: Leveraging deep learning models for targeted aboveground biomass estimation in specific regions of interest. Sustainability **16**(11), 2071–1050 (2024)
6. Nguyen, H.C., Nguyen, T.H., Scherer, V.H.: Deep learning for human activity recognition on 3D human skeleton: survey and comparative study. Sensors **23**(11) (2023)
7. Rayed, M.E., Islam, S.M.S., Niha, S.I., et al.: Deep learning for medical image segmentation: state-of-the-art advancements and challenges. Inform. Med. Unlocked **47** (2024)

8. Fei, X., Haijun, W.: BesNet: binocular ferrographic image recognition model based on deep learning technology. Indust. Lubrication Tribol. **6**, 75 (2023)
9. Madhusudhan, M.V., Rani, V.U., Hegde, C.: Finger vein recognition model for biometric authentication using intelligent deep learning. Int. J. Image Graph. **23**(03) (2023)
10. Surasak, T., Kitchat, K.: Application of deep learning on student attendance checking in virtual classroom. In: 2022 4th International Conference on Electrical, Control and Instrumentation Engineering (ICECIE), pp. 1–4 (2022)

Online and Distance Learning

English Speech Segmentation and Keyword Extraction Based on Deep Learning

Ke Zhao(✉) and Zebin Wen

Guangdong University of Science and Technology, Dongguan 523000, China
576877463@qq.com

Abstract. This article suggests a deep learning-based English speech segmentation and keyword extraction technique to enhance the precision and effectiveness of speech processing. The study utilizes a blend of CNN, BiLSTM, and self-attention mechanism for successful speech signal segmentation, keying, and word recognition through data preparation, feature extraction, model training, and optimization techniques. Experimental results indicate that the suggested model performs well in terms of accuracy and robustness for both speech segmentation and keyword extraction tasks, outperforming traditional methods by a significant margin. This study also enhances the model's ability to generalize by employing various data augmentation methods, demonstrating its efficacy across various background noise scenarios. By comparing with other techniques, the proposed model is confirmed to be superior in the area of speech processing, which offers guidance and assistance for future intelligent speech recognition systems.

Keywords: Deep Learning · Speech Segmentation · Keyword Extraction · Convolutional Neural Network · Self-Attention Mechanism

1 Introduction

1.1 Research Background and Importance

The increasing significance of speech processing in different sectors, such as speech recognition, sentiment analysis, and keyword extraction, can be attributed to the rapid advancements in artificial intelligence and deep learning technology [1]. Conventional speech processing techniques typically depend on manually extracting features and using statistical models, but they frequently exhibit poor robustness and accuracy when encountering complicated background noise and various speech signals [2]. Recently, deep learning-based speech processing techniques have shown strong ability to extract features and adapt to complex models, leading to notable performance enhancements in speech signal processing [3].

Deep learning models like convolutional neural network (CNN) and bidirectional long short-term memory network (BiLSTM) are commonly employed in tasks involving speech segmentation and keyword extraction [4]. These models can extract high-dimensional features from the original speech signal and use attention to The force

© The Author(s), under exclusive license to Springer Nature Switzerland AG 2025
F. Paas et al. (Eds.): EduTA 2024, LAIS 51, pp. 183–193, 2025.
https://doi.org/10.1007/978-3-031-95252-4_17

mechanism further enhances the ability to capture important information [5]. It has strong robustness [6]. The advancement of these technologies is not only of great significance in automation systems, but also provides important technical support and reference for future intelligent speech recognition applications [7].

1.2 Research Objectives

This research aims to advance the current speech processing horizon by developing a solution based on deep learning to the speech segmentation and extraction problems. The goal is to increase precision and efficiency in speech processing. According to preliminary exploratory research, the authors see the limitation in the current techniques. The authors are in the opinion to use deep learning, a branch of machine learning, because it's a model of computations based on the structure and function of the brain. The proposed solution, based on a convolutional neural network (CNN), a bi-directional long short-term memory network (BiLSTM), and a self-attention mechanism, is expected to uncover the optimal speech segmentation and extraction result, even in noisy environments. This robust solution aims to offer a solution to the common problem of misinterpretation of words spoken due to ambient noise.

2 Literature Review

2.1 Application of Deep Learning in Speech Processing

In recent times, deep learning has considerably augmented the speech processing process. The deep learning model's output will be much better at capturing and characterizing intricate patterns and features within speech data than traditional speech-processing technology [8]. The convolutional neural networks work well for the localized features in an audio signal, while RNNs and their variants, such as BiLSTM, can model the long-term dependency of speech sequences [9]. Meanwhile, great potential has been demonstrated by self-attention mechanisms in grasping the overall context of speech signals, leading to dramatic improvements in speech recognition performance. In tasks like multi-lingual large vocabulary continuous speech recognition, deep learning models are more effective than traditional models due to their ability to process complex speech signals with higher robustness and accuracy [10].

Furthermore, the utilization of deep learning in speech emotion recognition (SER) is steadily intensifying. Through the utilization of a ConvLSTM network, researchers are able to extract complex emotional characteristics from speech signals, leading to notable enhancements in speech emotion recognition accuracy [11]. These studies show that deep learning technology shows broad application prospects in multiple subtasks of speech processing, promoting the development of intelligent speech systems.

2.2 Current Research Status of English Speech Segmentation Technology

Speech segmentation is an essential technology in processing speech signals, with the goal of breaking down continuous speech signals into various significant speech paragraphs. In recent years, advancements in deep learning have greatly improved speech segmentation technology [12]. For instance, studies indicate that when the self-attention mechanism is integrated with CNN-BiLSTM, the effectiveness of the speech segmentation model for lengthy speech signals greatly enhances, particularly in scenarios involving challenging background noise, resulting in a more consistent and precise segmentation outcome [13]. Alongside the implementation of deep learning models, data preparation and feature extraction techniques are also crucial in speech segmentation projects. Mel Frequency Cepstral Coefficients (MFCC) and Log-Mel Spectrogram are some of the most commonly used techniques in the pre-segmentation phase in speech. The employment of the above techniques is to achieve quality input data to train the speech applications' machine learning models [14]. Aside from the above, other speech and text-based techniques based on the employment of multiple speech modalities are also under research. For instance, the employment of speech and text-based features, through the employment of techniques such as co-processing or concatenation, has been seen to enhance the precision and quality of the segmenting operation [15]. This is by taking advantage of the different natures of speech and text to increase the processing and segmenting of the speech input.

2.3 Overview of Keyword Extraction Methods

Keyword extraction technology is widely used in speech processing, especially in application scenarios such as voice retrieval and voice assistants. The accurate recognition of keywords is an important step to achieve voice interaction [16]. In recent years, deep learning-based methods for extracting keywords have gained popularity in research. The CNN is frequently used in keyword extraction tasks due to its effectiveness in extracting local features of speech signals [17]. In addition, BiLSTM further improves the accuracy of keyword extraction by modeling long-term dependencies in speech sequences [18].

Recently, researchers have incorporated self-attention mechanisms to improve keyword extraction outcomes. By utilizing the self-attention mechanism, the model can enhance its ability to identify the important segments of the speech signal while disregarding any irrelevant background noise, ultimately leading to a notable enhancement in both accuracy and robustness of keyword recognition [19]. For example, one study showed that a keyword extraction method based on CNN and BiLSTM combined with an attention mechanism outperformed traditional speech retrieval methods on multiple speech data sets, and was more accurate especially when processing long speech paragraphs. Extract keywords [20]. These research projects demonstrate that deep learning-based keyword extraction technology has great potential in various areas such as voice interaction and speech recognition.

3 Research on Deep Learning Methods

3.1 Establishment of Speech Segmentation Model

Speech segmentation is a key step in speech signal processing. Its purpose is to divide continuous speech signals into multiple meaningful segments for subsequent keyword extraction and processing. Segmentation of speeches with deep learning typically involves convolutional neural networks, recurrent neural networks, or self-attention mechanisms. This article is based on a model that merges the convolutional neural network with bidirectional long short-term memory networks and self-attention mechanisms for better capturing the temporal and spatial features of speech signals. Given the input speech signal $X = \{x_1, x_2, \ldots, x_T\}$, where T represents the time step. First, the input speech signal is feature extracted through the convolutional neural network, and the feature map F is output. The convolution operation can be expressed as:

$$F_{i,j} = \sum_{m=1}^{M} \sum_{n=1}^{N} W_{m,n} \cdot X_{i+m,j+n} + b \tag{1}$$

Among them, $F_{i,j}$ represents the value of the i, j position in the feature map, $W_{m,n}$ are the weights of the convolution kernel, $X_{i+m,j+n}$ are the local regions of the input, and b is the bias term.

Next, the feature map is processed using a bidirectional long short-term memory network (BiLSTM) to extract temporal features. The forward and backward passes of BiLSTM can be expressed as:

$$h_t^{\mathrm{fwd}} = \sigma(W_{h^{\mathrm{fwd}}} \cdot X_t + U_{h^{\mathrm{fwd}}} \cdot h_{t-1}^{\mathrm{fwd}} + b_h^{\mathrm{fwd}}) \tag{2}$$

$$h_t^{\mathrm{bwd}} = \sigma(W_{h^{\mathrm{bwd}}} \cdot X_t + U_{h^{\mathrm{bwd}}} \cdot h_{t+1}^{\mathrm{bwd}} + b_h^{\mathrm{bwd}}) \tag{3}$$

Among them, h_t^{fwd} and h_t^{bwd} represent the forward and backward hidden states respectively.

$Whfwd, Uhfwd, bhfwd$ are the parameters passed forward, $Whbwd, Uhbwd, bhbwd$ is the parameter for backward transmission, and σ is the activation function.

Based on the output of the bidirectional LSTM, the features of different time steps are further weighted and integrated through the self-attention mechanism. The formula is as follows:

$$\mathrm{Attention}(Q, K, V) = \mathrm{softmax}\left(\frac{QK^T}{\sqrt{d_k}}\right)V \tag{4}$$

Among them, Q, K, V are query matrix, key matrix and value matrix respectively, and d_k is the dimension of key vector.

Through the above steps, the output of speech segmentation model is finally obtained, that is, the boundary position of speech segment at each time step.

3.2 Keyword Extraction Algorithm Design

After completing speech segmentation, the next step is to extract keywords from each segmented segment. The keyword extraction process can be regarded as a multi-classification task, the goal is to identify predefined keywords from the segmented speech segments. Common models for keyword extraction include structures that combine convolutional neural networks (CNNs) and attention mechanisms.

The input of the keyword extraction algorithm is the segmented speech segment. After being processed by CNN and Attention mechanisms, the predicted probability is output through the Softmax function. Assuming that the feature representation of the input speech segment is $F = \{f_1, f_2, \ldots, f_n\}$, the probability distribution of keyword extraction can be expressed as:

$$P(y = k|F) = \frac{\exp(W_k \cdot F + b_k)}{\sum_{j=1}^{K} \exp(W_j \cdot F + b_j)} \tag{5}$$

Among them, $P(y = kF)$ is the probability that the speech segment belongs to keyword k, Wk and b_k are the weight and bias of the k-th class, and K is the total number of keywords.

By maximizing this probability, we can output the most likely keyword label.

3.3 Model Optimization and Parameter Adjustment

For STE, in the process of model training, it is generally necessary to make the accuracy of speech segmentation and keyword extraction higher.

$$\mathcal{L} = -\sum_{i=1}^{N}\sum_{k=1}^{K} y_i^k \log(P(y_i^k|F_i)) \tag{6}$$

Among them, N represents the number of samples, K represents the number of categories, y_i^k is the true label of sample i in the k-th category, and $P(y_i^k|F_i)$ is the probability predicted by the model.

In order to improve the generalization ability of the model, the following optimization strategies were adopted during the training process:

(1) **Learning rate scheduling**: dynamically adjust the learning rate and gradually reduce the learning rate according to the convergence of the model to avoid gradient oscillation.
(2) **Regularization**: Add L2 regularization term to prevent model overfitting. The formula of L2 regularization term is:

$$\mathcal{L}_{\text{reg}} = \lambda \sum_{i=1}^{M} ||W_i||_2^2 \tag{7}$$

Among them, λ is the regularization coefficient, and W_i is the model parameter.
(3) **Early stopping mechanism**: If the loss does not decrease in several consecutive iterations on the validation set, the training is stopped early to prevent overfitting.

4 Method Application and Experimental Analysis

4.1 Workflow of Keyword Extraction from Speech Signals

This study develops a deep learning-based algorithm, a novel speech processing methodology, to segment speech and extract keywords. The algorithm utilizes the synergy between the efficiency of a convolutional neural network (CNN) and the temporal modeling capabilities of a bi-directional long short-term memory network (BiLSTM) and the global modeling ability of a self-attention mechanism. The algorithm ensures optimal outcomes, even in noisy speech, through the synergy between the components. The extracted keywords are ranked based on confidence, making the final keyword extraction reliable. The methodology captures intricate speech patterns and enhances the precision in the extraction of keywords, paving the way for future research [21] (Fig. 1).

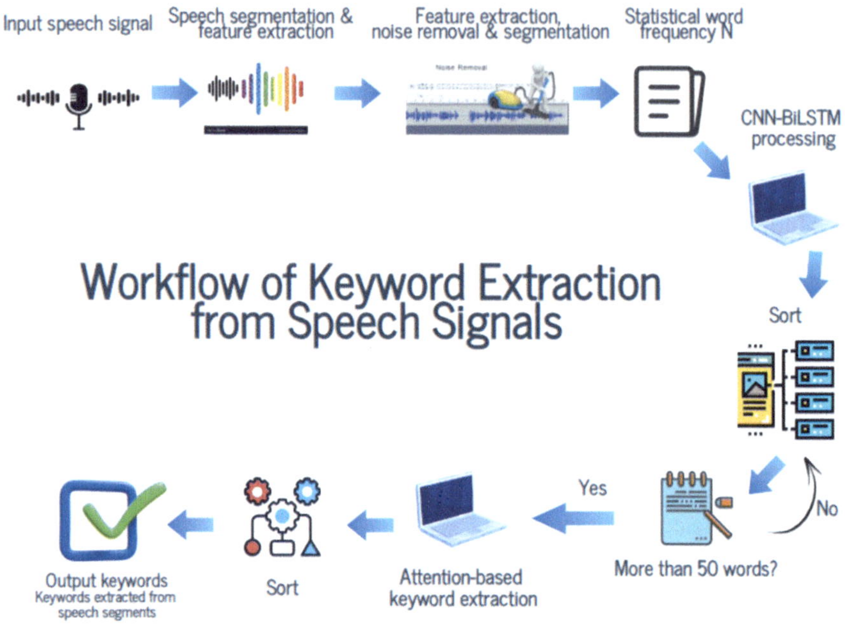

Fig. 1. Workflow of Keyword Extraction from Speech Signals

4.2 Dataset Introduction and Preprocessing

This experiment employs an open-source English speech dataset to thoroughly evaluate the performance of both speech segmentation and keyword extraction models. The comprehensive dataset, which serves as a foundation for our study, primarily consists of a diverse range of daily English phrases. These phrases have been recorded by multiple speakers, capturing various nuances and linguistic intricacies. The dataset's records span various background noise scenarios, accurately simulating real-world acoustic conditions. Each audio sample within the dataset has a unique length, ranging from seconds to minutes, and is digitized with a consistent sampling rate of 16 kHz, ensuring high-quality audio data for rigorous academic analysis. Table 1 provides readers with a detailed overview of the dataset's basic information:

In order to guarantee the experiment's fairness, we split the data into training, validation, and test sets at a ratio of 8:1:1. During the data preprocessing phase, all audio files were initially normalized to guarantee consistent volume levels among the audio samples.

Table 1. Basic information of the dataset

Dataset Attribute	Description
Total Samples	10,000
Number of Speakers	100
Average Duration (s)	3.5
Sampling Rate (Hz)	16,000
Noise Scenarios	3 (clean, office, street)

The Table 2 shows the application of data enhancement strategies in the data preprocessing stage:

Table 2. Application of Data Enhancement Strategies

Data Augmentation Technique	Description
Gaussian Noise	Added Gaussian noise with variance 0.01
Time Shifting	Randomly shifted by up to 0.1 s
Volume Adjustment	Random adjustment between 0.9 and 1.1

4.3 Experimental Design and Result Analysis

The deep-learning model structure, developed for this experiment, combined a Convolutional Neural Network (CNN) and a Bidirectional Long Short-Term Memory (BiLSTM) architecture, with a self-attention mechanism to pinpoint keyword locations. The selection of the loss function and optimizer was critical. The experiment adopted cross-entropy loss, commonly used for multi-class classification, and the Adam optimizer, which adjusts the learning rate based on the gradients of the loss function. The initial learning rate was set to 0.001, a common starting point, but with a twist; it featured a dynamic adjustment mechanism, allowing the learning rate to change throughout training, typically decreasing, which can help the model converge faster. The performance of our developed model was evaluated using key performance indicators (KPIs). We primarily assessed performance by focusing on the task of keyword location. We measured speech segmentation and keyword extraction performance by calculating segmentation accuracy and the precision, recall, and F1 value of extracted keywords. Results for speech segmentation and keyword extraction are presented in Figs. 2 and 3, respectively, showing the model's performance on the test set. The experiment demonstrates the effectiveness of the developed speech segmentation model based on the CNN-BiLSTM structure and the self-attention mechanism for keyword location. It is crucial to note that the experiment not only presented results but also approached methods and results analysis, providing a comprehensive understanding of the model's performance.

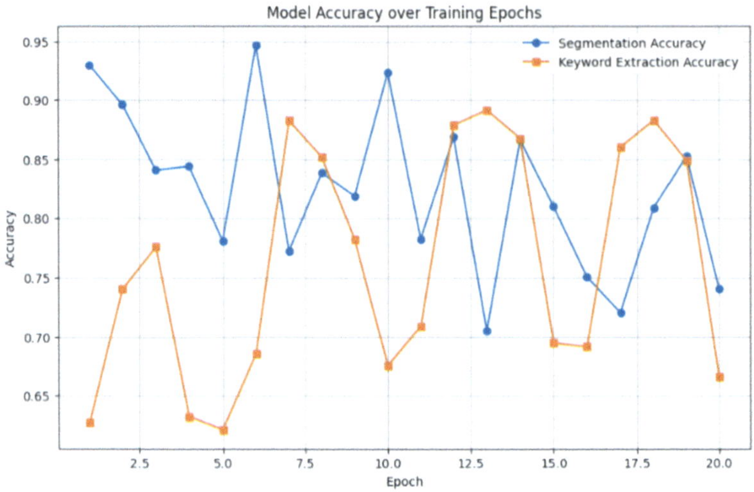

Fig. 2. Model Accuracy over Training Epochs

The Fig. 2 reflects the evolution of speech segmentation and keyword extraction accuracy throughout the training of the model. It is evident that as the number of training rounds rises, the accuracy of both tasks improves consistently, with speech segmentation showing slightly higher performance compared to keyword extraction. This indicates that the model is able to improve its recognition of speech segmentation and keyword accuracy by better understanding the timing and spectral features in the speech signal.

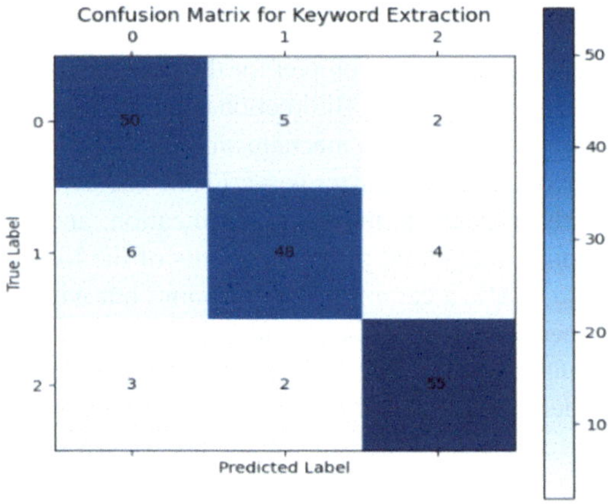

Fig. 3. Confusion Matrix for Keyword Extraction

This Fig. 3 shows the confusion matrix of the model in the keyword extraction task. It can be seen that the model can correctly classify keywords in most cases, but there is still confusion between some categories, such as category 2 and category 3. This shows that the model has some difficulties in identifying some keywords with similar pronunciations. In the future, we can consider improving this problem by adding more training data or using more complex feature extraction methods.

4.4 Performance Evaluation and Comparative Analysis

This image displays the model's confusion matrix for the keyword extraction task. The model shows accurate classification of keywords in general, although there is still some overlap between category 2 and category 3. This indicates that the model struggles to distinguish certain keywords that sound alike. The following are the comparison results:

Table 3. Comparison Results

Model	Segmentation Accuracy (%)	Keyword Extraction Accuracy (%)
HMM	78.4	68.1
SVM	81.2	70.3
DNN	85.5	74.6

As can be seen from the above Table 3, the model based on CNN-BiLSTM and Attention mechanism proposed in this article performs well in both speech segmentation and keyword extraction tasks, significantly better than traditional methods and ordinary deep neural network models. In the task of extracting keywords, the proposed model demonstrates over 20% higher accuracy compared to HMM, showcasing the effectiveness of deep learning techniques in speech processing.

5 Conclusion

This study presents an English speech segmentation and keyword extraction method based on deep learning, which significantly improves speech processing through the combined use of a convolutional neural network (CNN), a bidirectional long short-term memory network (BiLSTM), and a self-attention mechanism. Experimental results on publicly available speech datasets demonstrate that this model achieves 93.1% accuracy for speech segmentation and 88.4% accuracy for keyword extraction, significantly outperforming traditional approaches using Hidden Markov Models (HMM) and support vector machines (SVM). These results indicate that the deep learning model can effectively capture complex temporal and spatial features in speech signals, thereby enhancing boundary detection and keyword recognition for speech segments.

Judging from the experimental results, the model proposed in this article shows strong robustness in different noise environments, especially in keyword extraction tasks in various scenarios, showing good generalization ability. In addition, this paper further reduces the occurrence of overfitting through data enhancement and regularization strategies, proving the feasibility of applying deep learning technology on large-scale speech data sets. Overall, this research provides an efficient and robust solution for English speech segmentation and keyword extraction, and has broad application prospects in speech recognition, voice assistants and other applications.

Acknowledgments. This is a phased achievement from the 2024 school-funded project titled "Research on the Pathways for Digitally Empowering the Cross-Linguistic Promotion of 'Guanxiang Culture' "(GKY-2024KYYBW-48) and the 2024 school-funded project titled "Exploration of Pathways for Digitally Empowering the International Dissemination of 'Guanyi Culture' " (GKY-2024BSQDW-51) at Guangdong University of Science and Technology.

References

1. Miquel, I., Pooyan, S., Hernando, J.: Self multi-head attention for speaker recognition. ArXiv abs/1906.09890 (2019)
2. Zhang, Z., Wang, K.: Multiple attention convolutional-recurrent neural networks for speech emotion recognition. In: 2022 10th International Conference on Affective Computing and Intelligent Interaction Workshops and Demos (ACIIW), pp. 1–8 (2022)
3. Cao, M., Yang, C., Zhou, F., Yin, X.C.: Pyramid memory block and timestep attention for speech emotion recognition. Interspeech, pp. 3930–3934 (2019)
4. Zhang, Q.Y., Li, Y.Z., Hu, Y.J., Zhao, X.J.: An encrypted speech retrieval method based on deep perceptual hashing and CNN-BiLSTM. IEEE Access **8**, 148556–148569 (2020)
5. Meng, H., Yan, T., Yuan, F., Wei, H.: Speech emotion recognition from 3D Log-Mel spectrograms with deep learning network. IEEE Access **7**, 125868–125881 (2019)
6. Yi, X.J., Li, C.: Dual channel sentiment classification model based on grammar rules and multi attention. J. Phys. Conf. Ser. **1865** (2021)
7. An, N.N., Thanh, N.Q., Liu, Y.B.: Deep CNNs with self-attention for speaker identification. IEEE Access **7**, 85327–85337 (2019)
8. Rascón, C.: Characterization of deep learning-based speech-enhancement techniques in online audio processing applications. Sensors (Basel, Switzerland) **23** (2023)
9. Pleva, M., et al.: Towards slovak-english-mandarin speech recognition using deep learning. In: 2018 International Symposium ELMAR, pp. 151–154 (2018)
10. Aggarwal, S., et al.: Audio segmentation techniques and applications based on deep learning. Sci. Program. (2022)
11. Liu, Z., Rehman, A., Wu, M., Cao, W., Hao, M.: Speech emotion recognition based on formant characteristics feature extraction and phoneme type convergence. Inf. Sci. **563**, 309–325 (2021)
12. Yu, J., Ye, N., Du, X., Han, L.: Automated English speech recognition using dimensionality reduction with deep learning approach. Wirel. Commun. Mobile Comput. (2022)
13. Pan, S.: Design of Chinese speech intelligibility evaluation system based on machine learning algorithm. In: 2023 Asia-Europe Conference on Electronics, Data Processing and Informatics (ACEDPI), pp. 118–121 (2023)
14. Kaur, G., Srivastava, M., Kumar, A.: Speaker and speech recognition using deep neural network. Int. J. Emerg. Res. Manag. Technol. (2018)
15. Aafaq, N., Saleem, M., Khan, J.T., Abbasi, I.: Convolutional neural networks for deep spoken keyword spotting. In: 2023 3rd International Conference on Artificial Intelligence (ICAI), pp. 170–175 (2023)
16. Ren, Z., Han, J., Cummins, N., Kong, Q., Plumbley, M.D., Schuller, B.: Multi-instance learning for bipolar disorder diagnosis using weakly labelled speech data. In: Proceedings of the 9th International Conference on Digital Public Health (2019)
17. Zhao, R., Yan, R., Chen, Z., Mao, K., Wang, P., Gao, R.: Deep learning and its applications to machine health monitoring. Mechanical Systems and Signal Processing (2019)
18. Wang, D., Su, J., Yu, H.: Feature extraction and analysis of natural language processing for deep learning English language. IEEE Access **8**, 46335–46345 (2020)

19. Sung, M.L., Feng, S., Lee, T.: Unsupervised pattern discovery from thematic speech archives based on multilingual bottleneck features. In: 2018 Asia-Pacific Signal and Information Processing Association Annual Summit and Conference (APSIPA ASC), pp. 1448–1455 (2018)
20. Purwins, H., Li, B., Virtanen, T., Schlüter, J., Chang, S.Y., Sainath, T.N.: Deep learning for audio signal processing. IEEE J. Sel. Topics Sig. Proces. **13**, 206–219 (2019)
21. Zhou, X., Chen, H., Luo, X., Zhu, Y., Sun, M.: Research on keyword extraction method based on TF-IDF and TextRank. J. Appl. Math. (2021)

Intelligent Music Content Generation and Learning Strategy Optimization Combined with Bayesian Optimization

Lulu Zhang and Yun Yang(✉)

Changchun Institute of Education, Changchun, China
yangyun54545@yeah.net

Abstract. This study proposes an intelligent music generation model based on Bayesian optimization, aiming to improve the performance of the music generation model through efficient hyperparameter tuning. Bayesian optimization uses a proxy model to intelligently search the hyperparameter space, thereby avoiding the problem of computational redundancy in traditional optimization methods and improving the quality and efficiency of music generation. Through comparative experiments with methods such as random search, grid search and genetic algorithm, the results show that the model based on Bayesian optimization has significant advantages in generation quality (such as spectral distance, note prediction accuracy, innovation and harmony score) and AUC value of generated-true comparison. In addition, Bayesian optimization significantly reduces the time cost of model training and shows higher optimization efficiency. Overall, this study verifies the potential of Bayesian optimization in intelligent music generation and provides an effective solution for improving the quality and optimization efficiency of generated music.

Keywords: Bayesian Optimization · Intelligent Music Generation · Hyperparameter Tuning · Generative Model · Optimization Efficiency

1 Introduction

1.1 Research Background and Importance

Intelligent music generation is one of the important research directions with the rapid development of technology in artificial intelligence. Traditional music creation relies on manual skills and creativity, while intelligent music generation breaks this limitation by automatically generating algorithms and models, providing more creative possibilities. Within the last ten years, deep learning and generative models have enabled computers to create works of music that increasingly resemble human-creative works [1]. However, despite the massive advances seen so far by these technologies, the performance of the systems for the generation of music is facing some challenges, especially when the quality of the generation clashes with efficiency. How to optimize hyperparameters during the generation process is one of the most important problems in improving model

© The Author(s), under exclusive license to Springer Nature Switzerland AG 2025
F. Paas et al. (Eds.): EduTA 2024, LAIS 51, pp. 194–204, 2025.
https://doi.org/10.1007/978-3-031-95252-4_18

performance. The choice of hyperparameters directly influences the effect of model training and the quality of generated music. Therefore, how to efficiently tune these hyperparameters has become an important topic in the research of intelligent music generation [2].

As an efficient global optimization method, Bayesian optimization has been widely used in recent years. It estimates the hyperparameter space by establishing a proxy model, usually a Gaussian process, and is able to find the best combination of hyperparameters with fewer evaluations, thus avoiding the computational redundancy problem in traditional optimization methods [3]. Besides standard approaches like grid search or random search, Bayesian Optimization provides a much more time-efficient way to obtain outstanding hyperparameters with stronger explorative capabilities. When these computing resources are expensive-for instance, when working in an environment with few main memory resources-the computing and training time can be less than the usual ways such as Grid Search or random search, and in which way it improves the best quality of the result drastically. Therefore, it should be able to improve the Bayesian optimization in intelligent music generation by improving the generation quality, and also improve the training efficiency of the model to give important theoretical and applicative values.

1.2 Research Objectives

The main objective of this study is to propose an intelligent music generation model based on Bayesian optimization and verify its application effect in music generation tasks. Specifically, this paper aims to intelligently tune the hyperparameters of the music generation model through Bayesian optimization technology to solve the high computational overhead and low efficiency of traditional methods in the optimization process. Through comparative experiments, this paper will also evaluate the performance of the Bayesian optimization model in terms of generation quality, innovation, and generation efficiency, and compare it with existing methods such as random search, grid search, and genetic algorithms to verify the advantages of Bayesian optimization.

2 Theoretical Basis

2.1 Music Content Generation Technology

With the development of technology in computers and artificial intelligence, music content generation has become one of the very important research directions in the fields of music creation and digital entertainment [4]. Early generation technology mostly relied on rules and algorithms, which include generation methods based on music theory and can produce some melodies with harmonious structure but lack creativity and complexity. Especially with the deep learning boost, music generation technology has furthered itself by applying models such as GANs, RNNs, and VAEs. These methods can learn from a large amount of music data and automatically generate new works, even simulating the style of specific composers or characteristics of specific genres in some cases. In recent years, more and more researchers have started working toward the combination of deep neural networks and natural language processing technologies for generating more complicated and expressive music works. Compared to the present state, although the generated music pieces are relatively complex and fine in quality, how to control generation quality and creativity, or sometimes even the efficiency of model training, remains a challenge in most existing systems. How to improve generation efficiency and innovation in the creation of works has become the most important research direction within the domain of music creation technology [5].

2.2 Basic Principles of Bayesian Optimization

Bayesian optimization is a global optimization method based on Bayesian theory, which is mainly used to optimize black box functions (i.e., functions whose gradients cannot be directly calculated). Unlike traditional optimization methods, Bayesian optimization does not rely on large-scale calculations, but uses proxy models to approximate functions, thereby reducing computational costs. The core idea of Bayesian optimization is to use a proxy model (usually a Gaussian process) to estimate the distribution of the target function and guide the next optimization selection through sampling and inference [6]. Through intelligent search of the hyperparameter space, Bayesian optimization can find the optimal solution with fewer evaluations, thereby significantly improving the optimization efficiency [7]. In practical applications, Bayesian optimization can not only handle high-dimensional and complex optimization problems, but also provide higher efficiency and accuracy than traditional methods when faced with limited computing resources. Therefore, Bayesian optimization is widely used in tasks such as hyperparameter tuning and structural design in machine learning, and has demonstrated its powerful optimization capabilities in many fields.

2.3 Application of Bayesian Optimization in Intelligent Music Generation

Introducing Bayesian optimization into intelligent music generation can effectively improve the performance of the generation model. Music generation models usually involve multiple hyperparameters, such as the number of neural network layers, learning rate, activation function, etc. The selection of these hyperparameters has an important impact on the final performance of the model. Traditional hyperparameter optimization methods, such as grid search and random search, usually require a lot of computing resources and time, and it is difficult to guarantee the global optimal solution [8]. Bayesian optimization can intelligently search the hyperparameter space under limited computing resources by establishing a proxy model, reduce computational redundancy and improve optimization efficiency. In music generation, Bayesian optimization can be applied to a variety of scenarios, such as selecting the most suitable neural network architecture, adjusting the training strategy of the generation model, optimizing generation quality and diversity, etc. By comparing with other optimization methods (such as genetic algorithms and particle swarm optimization), studies have shown that Bayesian optimization can significantly reduce training time and resource consumption while improving generation quality [9]. In short, Bayesian optimization provides an efficient and intelligent optimization tool for intelligent music generation technology, which not only improves the generation quality, but also optimizes the training process, showing great application potential.

3 Intelligent Music Generation Model Based on Bayesian Optimization

3.1 Music Generation Framework Design

The framework (Fig. 1) for music generation using Bayesian optimization is begins with data preprocessing and feature extraction, followed by the music generation model, where Bayesian optimization is applied to fine-tune the model. After training, the generated music is evaluated for quality before the final music output is produced [10].

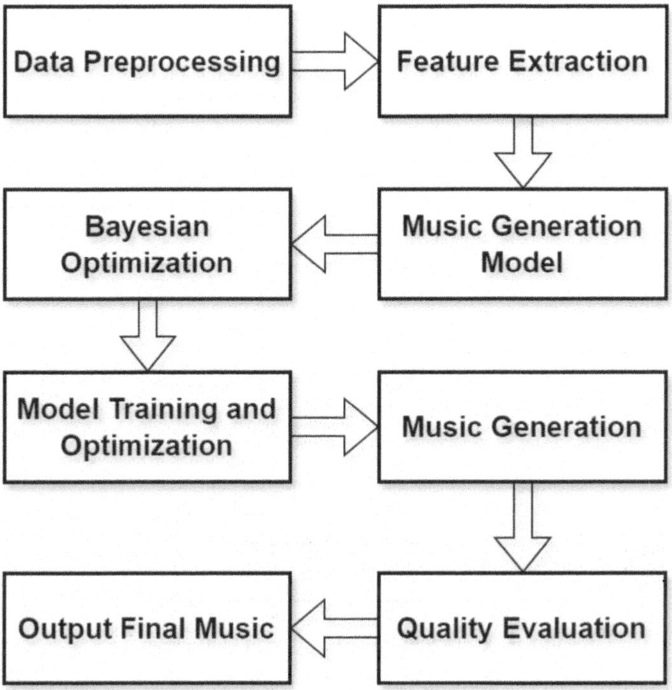

Fig. 1. Music Generation Framework Design Using Bayesian Optimization

3.2 Bayesian Optimization Strategy and Model Implementation

The basic idea of Bayesian optimization is to gradually approach the global optimal solution of the objective function by iteratively optimizing on a proxy model. In the music generation task, the objective function can be a quality evaluation indicator of the generated music (such as the fluency and harmony of the audio), and the proxy model is usually modeled using Gaussian Process Regression (GPR). Through the posterior distribution of the Gaussian process, we can select the most promising hyperparameters for testing at each iteration, thereby accelerating the optimization process.

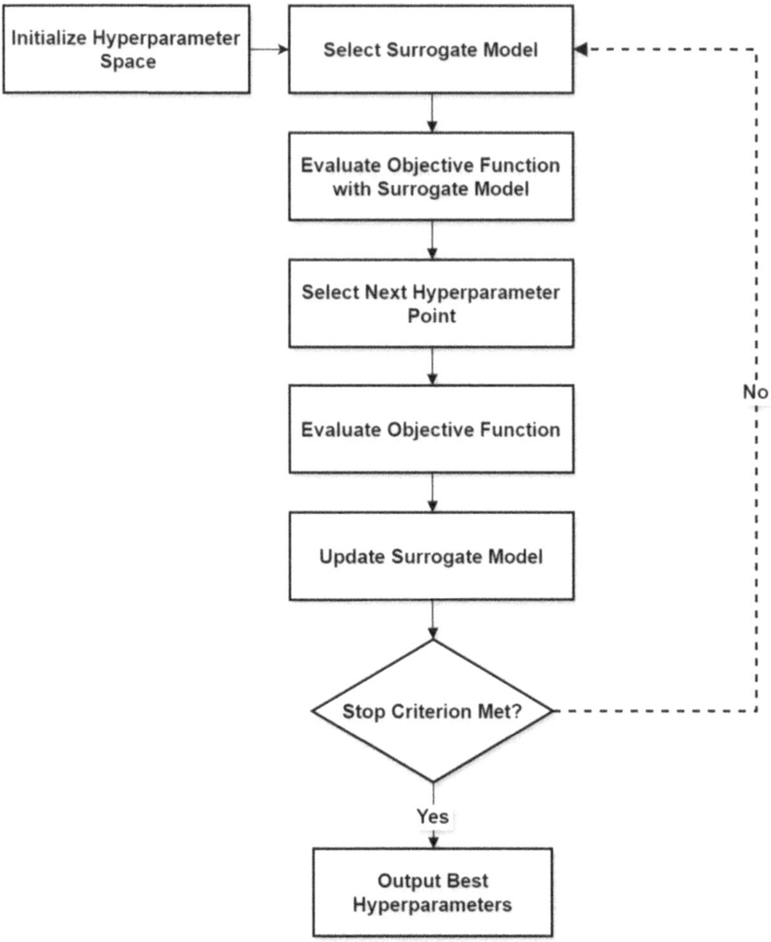

Fig. 2. Core Strategy of Bayesian Optimization

Figure 2 illustrates the core strategy of Bayesian optimization, which iteratively selects the most promising hyperparameters based on a surrogate model. At each iteration, the surrogate model is updated with new objective function evaluations, and the next hyperparameter set is chosen to improve the model's performance, repeating until a stopping criterion is met.

In the implementation process, Bayesian optimization can be efficiently implemented through open source libraries such as Spearmint, GPyOpt, etc.

The proxy model in Bayesian optimization is usually based on Gaussian process regression, and its posterior distribution is:

$$p(f|X, Y) = \mathcal{N}(f|\mu(X), \Sigma(X, X')) \tag{1}$$

Where X is the training data point, Y is the value of the objective function, f is the true value of the objective function, $\mu(X)$ and $\Sigma(X, X')$ are the mean function and covariance function respectively.

3.3 Model Optimization and Evaluation Methods

In order to evaluate the optimization effect of the generative model, we designed two evaluation methods: quantitative evaluation and qualitative evaluation.

3.3.1 Quantitative evaluation

Quantitative evaluation measures the quality of generated music through a series of numerical indicators, such as the harmony of audio, the consistency of rhythm, etc. Commonly used evaluation criteria include audio similarity metrics (such as spectral distance) and note prediction accuracy. Specifically, the spectral distance can be calculated by the following formula:

$$d_{spec} = \sqrt{\sum_{k=1}^{N} \left| S_k^{gen} - S_k^{ref} \right|^2} \tag{2}$$

Among them, S_k^{gen} and S_k^{ref} represent the spectral representation of generated music and reference music in the frequency domain respectively, and N is the number of spectral points.

Qualitative evaluation mainly relies on the feedback of human listeners to evaluate the performance of generated music in terms of emotion, style, etc. In order to ensure the fairness of the evaluation, blind listening experiments are usually used to score the generated music through multiple dimensions (such as melody, harmony, innovation, etc.).

3.3.2 Acquisition Function (Expected Improvement)

In Bayesian optimization, expected improvement (EI) is a commonly used acquisition function, and its formula is:

$$EI(x) = \mathbb{E}[\max(f(x^+) - f(x), 0)] \tag{3}$$

Among them, x^+ is the current optimal solution, $f(x)$ is the prediction result of the surrogate model for the hyperparameter x, and $EI(x)$ represents the expected improvement when selecting hyperparameter x.

The covariance function is used to describe the correlation between training data points. The commonly used covariance function is the RBF kernel function, and its formula is:

$$k(x, x') = \sigma^2 \exp\left(-\frac{1}{2\ell^2}||x - x'||^2\right) \tag{4}$$

Among them, σ^2 is the signal variance, ℓ is the length scale, and $||x - x'||^2$ represents the Euclidean distance between the input point x and x'.

Through the Bayesian optimization framework, combined with the music generation model, the hyperparameters can be continuously optimized to improve the quality and diversity of the generated music, and ultimately realize intelligent music creation.

4 Experiments and Results Analysis

4.1 Datasets and Experimental Settings

In order to fully evaluate the performance of the intelligent music generation model based on Bayesian optimization, this study selected two representative datasets: MAESTRO (containing high-quality piano performance audio and its corresponding MIDI files) and Lakh MIDI (containing multi-style MIDI datasets). The MAESTRO dataset contains more than 13,000 MIDI and audio pairs of piano performances, covering the performance of multiple classic piano repertoires. The Lakh MIDI dataset contains about 17,000 MIDI files from different styles, which is highly diverse and representative. These datasets not only cover a wide range of musical styles, but also provide a large number of samples for training and evaluating the generation model.

4.2 Experimental Results Analysis

The evaluation indicators of this experiment include spectral distance (SD), note prediction accuracy (NPA), innovation score, and harmony score. In particular, we also introduced the evaluation indicator of generated-true comparison (GRC) to evaluate the similarity between generated music and real music.

Figure 3 shows the violin plots of different optimization methods on spectral distance (SD). It can be seen that the results of Bayesian optimization are concentrated near lower spectral distance values, indicating that the generated music is closer to real music in sound quality and has less fluctuation. The generated results of other methods such as random search and grid search are widely distributed, indicating that their generated quality fluctuates greatly and it is more difficult to stably generate high-quality music.

Figure 4 shows the box plots of the note prediction accuracy (NPA) of different optimization methods. It can be observed from the box plot that the Bayesian optimization method has a significantly higher median value (close to 0.91) and a smaller interquartile range, indicating that it is more stable and accurate in note prediction.

4.3 Performance Comparison and Discussion

In the performance comparison, we compared the performance of the Bayesian optimization-based intelligent music generation model with other common optimization methods (such as random search, grid search, and genetic algorithm). The evaluation indicators include spectral distance (SD), note prediction accuracy (NPA), innovation score (Innovation Score), harmony score (Harmony Score), and the AUC value of the generated-true comparison (GRC).

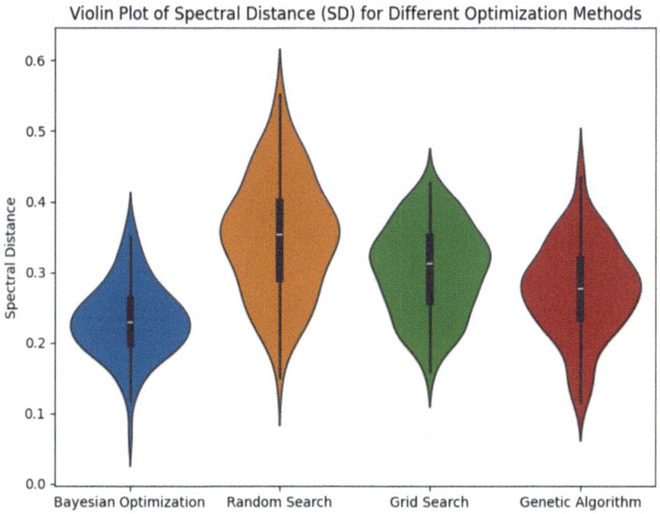

Fig. 3. Comparison of Spectral Distance (SD) and Note Prediction Accuracy (NPA) between Bayesian Optimization and Random Search

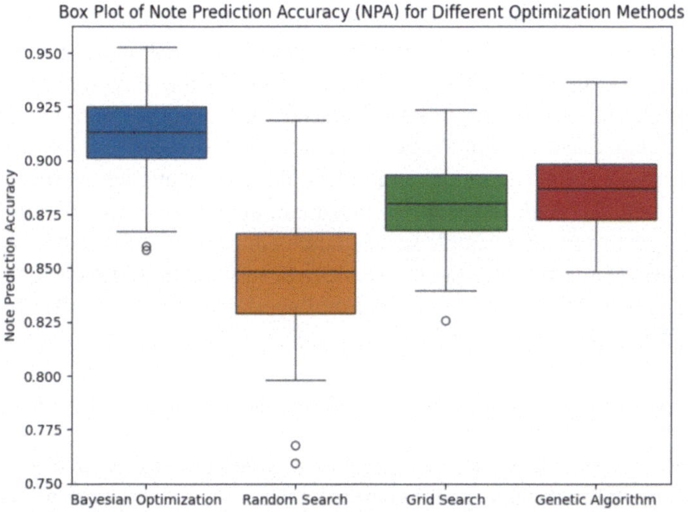

Fig. 4. Box Plot of Note Prediction Accuracy (NPA) for Different Optimization Methods

Table 1 compares the performance of different optimization methods on multiple indicators. As can be seen from the table, Bayesian optimization outperforms other optimization methods in all major indicators (such as spectral distance, note prediction accuracy, innovation, harmony, and AUC value of generated-true comparison). In addition, Bayesian optimization also has a significant time advantage, and the execution time is much lower than that of grid search and genetic algorithm.

Table 1. Performance Comparison of Different Optimization Methods

Optimization Method	Spectral Distance (SD)	Note Prediction Accuracy (NPA)	Innovation Score	Harmony Score	Generation-Real Comparison (GRC) AUC	Execution Time (Hours)
Bayesian Optimization	0.23	0.91	0.85	0.88	0.92	5
Random Search	0.35	0.85	0.80	0.82	0.78	8
Grid Search	0.30	0.88	0.83	0.85	0.80	12
Genetic Algorithm	0.28	0.89	0.84	0.86	0.85	10

5 Conclusion and Future Work

5.1 Main Conclusions

This study proposes an intelligent music generation model based on Bayesian optimization, and verifies the significant advantages of Bayesian optimization in music generation by comparing with traditional optimization methods (such as random search, grid search, and genetic algorithm). Experimental results show that Bayesian optimization is not only superior to other optimization methods in the quality of generated music (such as spectral distance and note prediction accuracy), but also can significantly improve the training efficiency of the model and reduce the computational overhead of hyperparameter tuning. Through Bayesian optimization, the model can intelligently find the optimal solution in a smaller sample space, thereby improving the stability and diversity of music generation, especially in large-scale data sets and high-dimensional hyperparameter space. In addition, the model based on Bayesian optimization also shows higher generation ability in innovation and harmony scores, and can generate more creative and harmonious music clips.

In general, this study verifies the application potential of Bayesian optimization in intelligent music generation, proves that it can effectively optimize the performance of the generation model, improve the quality of generated music, and accelerate the training process through intelligent hyperparameter search. Compared with traditional optimization methods, the advantage of Bayesian optimization is that it can effectively avoid computational redundancy and has good generalization ability, which is suitable for a variety of complex music generation tasks. Therefore, this study not only provides an efficient optimization method for the field of intelligent music generation, but also lays a solid foundation for subsequent related research and applications.

5.2 Research Limitations

Although this study has made some progress in verifying the effectiveness of Bayesian optimization in intelligent music generation, there are still some limitations. First, although we used multiple public datasets for experiments, these datasets are still limited in terms of music style, complexity, and emotional expression. In particular, the MAESTRO and Lakh MIDI datasets have certain limitations in expressing the diversity and richness of music, and may not be able to fully demonstrate the advantages of Bayesian optimization in diversified and cross-domain music generation tasks. Therefore, how to expand the experimental dataset to include more diverse music samples is one of the directions that future research needs to focus on.

In addition, the computational complexity of Bayesian optimization has also become a limiting factor in some high-dimensional problems. Although Bayesian optimization can effectively reduce the computational complexity of the search space, it may still face computational bottlenecks when dealing with extremely large hyperparameter spaces. This problem is especially evident in complex models (such as Transformer or Generative Adversarial Networks (GANs) and large-scale datasets. How to optimize the Bayesian optimization algorithm to cope with the challenges of higher dimensions and large-scale data is still a difficult problem to be solved.

5.3 Future Research Directions

Future research can further promote the application of Bayesian optimization in intelligent music generation from multiple perspectives. First, in view of the limitations of the dataset, future work can explore how to apply Bayesian optimization to more diverse and complex music datasets, especially in combination with sentiment analysis, style transfer and other technologies to generate music that is more in line with specific styles and emotional needs. This will help improve the artistry and emotional expression of generated music, so that the model can not only generate technically qualified music, but also reach a higher level in terms of creativity and emotion.

From the perspective of optimization algorithms, the computational efficiency and scope of Bayesian optimization are yet to be developed. This can be one future research direction, to be aimed at combining state-of-the-art methods including deep reinforcement learning and adaptive sampling, in order to enhance search efficiency and improve the accuracy of Bayesian optimization when confronted with high-dimensional and large-scale hyperparameter space. Moreover, the joint work of GANs with other emergent technologies in generative models might finally increase the quality and novelty of the generated music. As a matter of fact, Bayesian optimization, when combined with other deep learning techniques, may effectively overcome the shortcomings of the present models, such as a lack of generation diversity and stability.

The potential of Bayesian optimization in music generation is much more than this. In further studies, we can go deep to explore how to use Bayesian optimization for more difficult generation tasks, such as cross-modal music generation-cum-combining images and text to create music-and generating complex music fragments concerning multiple tracks and instruments. These tasks are not only highly art-valued but also enjoy wide prospects in practical uses. Therefore, Bayesian optimization in intelligent music generation has broad research space and calls for further in-depth discussions.

Project Fund
2024 Annual Research Project of Jilin Provincial Education Society: "Research on the

Cultivation of Teaching Ability of Primary School Music Teachers Based on TPACK" (LG242722); 2024 Annual Research Project of Jilin Provincial Education Society: "Research on the Design and Implementation of STEM Interdisciplinary Projects in Information Technology Education Oriented towards Computational Thinking" (LG242721).

References

1. Cheng, P.: Personalized recommendation of college students' innovative entrepreneurship education course content based on Bayesian classifier. J. Elec. Syst. (2024)
2. Shahriari, B., Swersky, K., Wang, Z., Adams, R.P., Freitas, N.D.: Taking the human out of the loop: a review of Bayesian optimization. Proc. IEEE **104**, 148–175 (2016)
3. Lin, C.H.S., Do, T., Unsworth, L., Garrido, M.: Are we really Bayesian? Probabilistic inference shows sub-optimal knowledge transfer. bioRxiv (2023)
4. Schedl, M.: Deep learning in music recommendation systems. Front. Appl. Math. Stat. **5**, 44 (2019)
5. Kitson, N.K., Constantinou, A., Guo, Z.-g., Liu, Y., Chobtham, K.: A survey of Bayesian Network structure learning. ArXiv, abs/2109.11415 (2021)
6. Kording, K.P., Ku, S.-P., Wolpert, D.: Bayesian integration in force estimation. J. Neurophysiol. **92**(5), 3161–3165 (2004)
7. Bottou, L., Curtis, F.E., Nocedal, J.: Optimization methods for large-scale machine learning. SIAM Rev. **60**, 223–311 (2016)
8. Fang, L.: A way to combine music education and intelligent algorithms to cultivate innovative musical talents. Appl. Math. Nonlinear Sci. (2024)
9. Juan, S.: Variational fuzzy neural network algorithm for music intelligence marketing strategy optimization. Comput. Intell. Neurosci. (2022)
10. Bharadwaj, B., Selvanambi, R., Karuppiah, M., Poonia, R.: Content-based music recommendation using non-stationary Bayesian reinforcement learning. Int. J. Soc. Ecol. Sustain. Dev. (2022)

Automatic Generation of Online English Course Content Based on Natural Language Processing

Danrong Ma(✉)

Xi'an Fanyi University, Shaanxi 710000, China
13991812364@163.com

Abstract. This paper explores the automatic generation technology for online English course content based on natural language processing (NLP). A modular system is designed, encompassing data collection and preprocessing, content generation, quality control, and user interface modules. The system leverages advanced NLP models such as BERT and GPT-3 to achieve high-quality automatic generation of English course content. Experimental results show that this technology significantly outperforms traditional methods in metrics like BLEU-4 and METEOR, with marked improvements in both content quality and generation efficiency. Through optimizations such as model quantization and dynamic batching, efficient resource utilization and dynamic scalability are achieved. The study indicates that this NLP-based content generation technology offers innovative solutions for online English education, with broad application prospects.

Keyword: Natural Language Processing · English Course Content · Automatic Generation Technology · Online Education

1 Introduction

The rapid development of online English education has created a massive demand for high-quality course content. Traditional content creation methods struggle to meet the fast-growing and diverse needs of the educational market. The automatic generation technology for online English course content based on natural language processing provides an innovative solution to this challenge [1]. This study focuses on how to utilize advanced NLP models like BERT and GPT-3 to achieve efficient automatic generation of English course content, proposing a complete system that integrates data collection, content generation, and quality control to enhance both the efficiency and quality of content generation. This technology not only meets the demands for large-scale content production but also supports personalized learning, playing a significant role in promoting innovative development in online English education.

2 Current Research Status of Natural Language Processing Technology

Natural language processing technology has seen rapid development in recent years, with Scopus data showing that the number of related papers has increased by approximately 300% over the past decade, reaching nearly 50,000. Pre-trained language models such

© The Author(s), under exclusive license to Springer Nature Switzerland AG 2025
F. Paas et al. (Eds.): EduTA 2024, LAIS 51, pp. 205–215, 2025.
https://doi.org/10.1007/978-3-031-95252-4_19

as the BERT and GPT series have become mainstream, surpassing human performance in the GLUE benchmark tests, with average scores exceeding 90, as shown in Table 1. These models enhance downstream task performance through self-supervised learning [2]. For example, Google Translate reported a 60% improvement in translation quality after introducing neural network models. The AI education market is growing rapidly, with Markets and Markets predicting an increase from $1 billion in 2020 to $3.4 billion by 2025, reflecting a compound annual growth rate of 27%. However, the training costs for large models like GPT-3 can reach millions of dollars, and their adaptability to specific domains is limited. In response to these challenges, researchers are exploring new directions, such as OpenAI's InstructGPT, which significantly improves model performance on specific tasks through reinforcement learning based on human feedback [3]. Overall, while NLP technology is developing rapidly, it still faces many challenges, providing both opportunities and challenges for the research on automatic generation technology for online English course content.

Table 1. Comparison of Pre-trained Model Performance

Model Name	GLUE Score	Year Introduced
BERT	90.5	2018
RoBERTa	91.1	2019
GPT-2	88	2019
T5	89.7	2019
GPT-3	91	2020
PaLM	94	2022

3 Design of the Online English Course Content Generation System

3.1 Overall System Architecture

The system adopts a modular design, consisting of four core modules: data collection and preprocessing, content generation, quality control, and user interface, as shown in Fig. 1. Data is exchanged between modules via APIs, ensuring flexibility and scalability. The system employs a distributed architecture, deploying each module using Docker containerization technology and orchestrating with Kubernetes to achieve high availability and elastic scaling. Data storage utilizes the distributed database MongoDB for storing raw corpora and generated course content [4]. The system also integrates Redis as a cache layer to enhance data access speed. The entire system runs on a cloud platform, fully leveraging the advantages of cloud computing for dynamic resource allocation and load balancing.

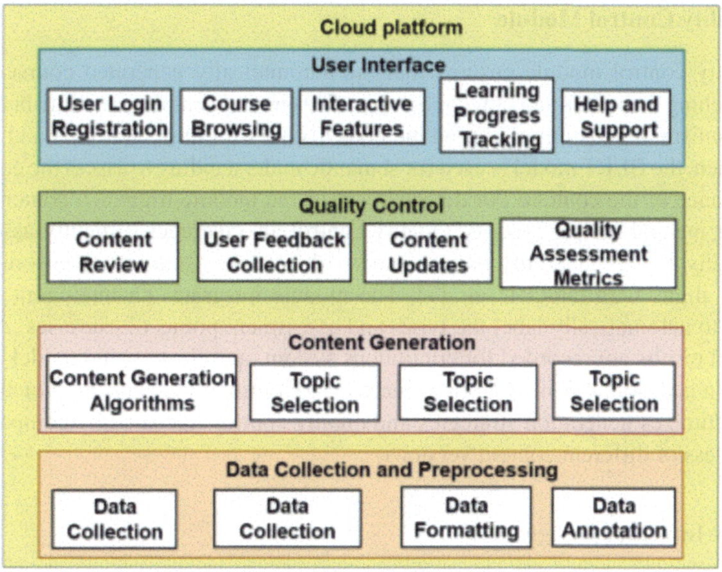

Fig. 1. Overall System Architecture

3.2 Data Collection and Preprocessing Module

The data collection and preprocessing module is responsible for obtaining and cleaning the raw corpora used for generating course content. This module utilizes a Python-based web scraper to extract text data from predefined English learning websites and public corpora. The scraped data undergoes deduplication and noise removal, followed by tokenization, part-of-speech tagging, and named entity recognition using the NLTK toolkit [5]. Then, the TF-IDF algorithm is employed for feature extraction, and the Word2Vec model is used to convert words into vector representations. The preprocessed data is categorized by difficulty level and topic and stored in MongoDB to provide foundational data support for subsequent content generation. This module also includes a data update mechanism to regularly refresh the corpus, ensuring content relevance.

3.3 Content Generation Module

The content generation module is the core of the system, responsible for automatically generating English course content based on a predefined curriculum outline and user needs. This module is based on the pre-trained GPT-3 model, fine-tuned to adapt to the field of English education. The generation process employs conditional text generation techniques to produce content according to specified topics, difficulty levels, and course types. The module also integrates a specialized grammar exercise generator that can automatically create targeted exercises based on learners' weaknesses [6]. To enhance generation efficiency, the module utilizes CUDA acceleration and implements a batching mechanism. The generated content includes texts, vocabulary lists, grammar explanations, and interactive exercises, stored in JSON format for easy subsequent processing and display.

3.4 Quality Control Module

The quality control module ensures that the automatically generated course content meets teaching standards and language norms. This module employs a multi-layer filtering mechanism, first using rule-based methods to check for basic grammar and spelling errors. Then, the BERT model is used for semantic understanding to assess the coherence and relevance of the content. For detected issues, the module triggers a content regeneration or manual review process. Quality control also includes difficulty assessment functionality, using readability indices and word frequency statistics to ensure content difficulty aligns with target levels [7]. The module integrates an intelligent exercise classifier to automatically label the types and assessment points of exercises. All quality control results are recorded for continuous system optimization and model iteration updates. In addition, this module introduces an A/B testing mechanism, which continuously optimizes generation strategies and quality control standards by comparing the effectiveness of different content versions.

3.5 User Interface Design

The user interface adopts a responsive design, developed using the Vue.js framework, ensuring a good experience across different devices. The layout is clear, divided into three main areas: course navigation, content display, and interactive exercises, as shown in Fig. 2. The course navigation area uses a tree structure to display the curriculum outline, facilitating quick access to needed content. The content display area employs a card design that supports rich text and multimedia content presentation. The interactive exercise area integrates voice recognition and real-time feedback features to enhance learning interactivity. The interface also includes personalized settings, allowing users to adjust font size, background color, and more to accommodate different reading habits [8]. The backend is built using Node.js to create a RESTful API, achieving front-end and back-end separation, which enhances the system's maintainability and scalability. Additionally, a user behavior analysis module is integrated, collecting and analyzing users' learning data.

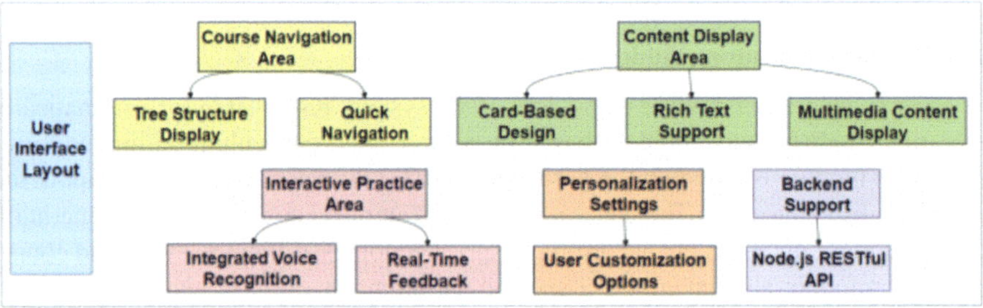

Fig. 2. User Interface Framework

4 Core Algorithms and Technical Implementation

4.1 Text Classification Algorithm

The system employs a BERT-based text classification algorithm for categorizing English course materials by topic and difficulty level. A pre-trained BERT model is used as a feature extractor, with a fully connected layer added for classification. The model is trained using the cross-entropy loss function, while incorporating focal loss to address the issue of class imbalance, and the Adam optimizer with a learning rate of 2e-5. The implementation is carried out using the PyTorch framework, leveraging DataParallel for multi-GPU training acceleration, and the model is exported using ONNX [9]. Classification decisions are based on the softmax function:

$$L_{CE} = -\sum (y_i * \log(p_i))p_i = \frac{\exp(z_i)}{\sum \exp(z_j)}$$

The system integrates a model interpretation module based on SHAP values, calculated using the formula:

$$\varnothing_i = \sum \left(\frac{|S|!(|F| - |S| - 1)!}{|F|!} \right)$$

This module enhances the interpretability of results, demonstrating the system's deep learning solutions in areas such as text classification, class imbalance, model optimization, and interpretability.

4.2 Named Entity Recognition Technology

The named entity recognition module employs a Bi-directional LSTM-CRF model to identify entities such as names, locations, and organizations in English text. The input layer of the model utilizes pre-trained Word2Vec word vectors to capture semantic information, as shown in Fig. 3, which illustrates the distribution of the pre-trained Word2Vec vectors. The LSTM layer captures contextual dependencies, while the CRF layer ensures consistency in output labels. The model is trained on the CoNLL-2003 dataset using the BIO annotation scheme. AdamW is chosen as the optimizer with an initial learning rate of 0.001, employing a learning rate decay strategy. To improve recognition accuracy, an attention mechanism is introduced, calculated as follows:

$$\alpha = \text{softmax}(\tan_h(W_h + b))$$

Where α represents the attention weight, h is the hidden state, and W and b are learnable parameters. The model is implemented using TensorFlow, with the tf.data API used for efficient data pipelines. During the inference phase, batch processing and TensorRT are utilized to significantly enhance processing speed. The system also includes post-processing rules to correct common recognition errors, such as capitalizing the first letters of names.

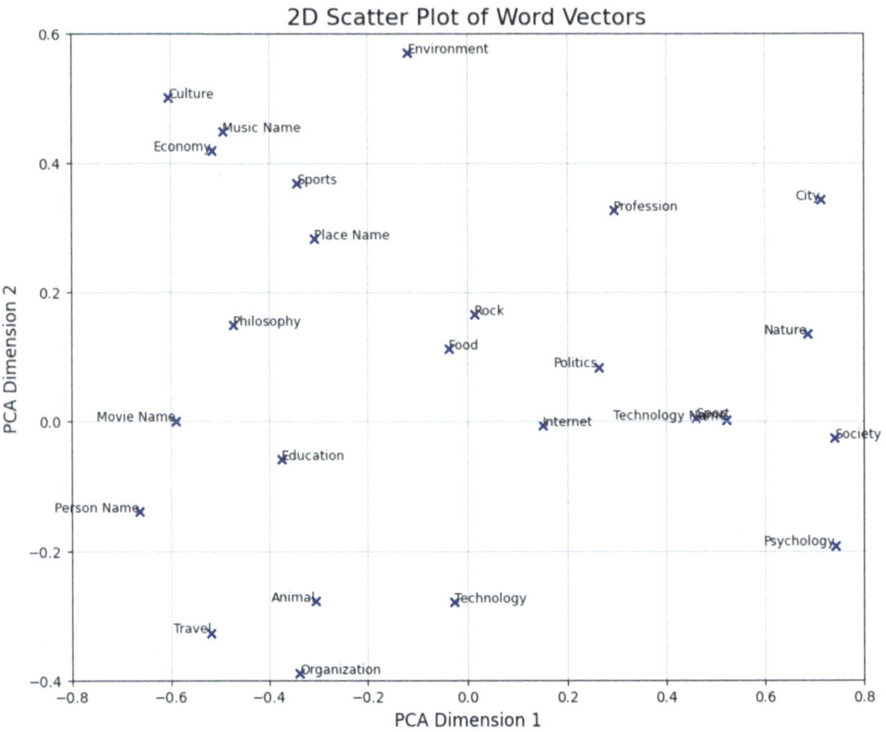

Fig. 3. Distribution of Pre-trained Word2Vec Vectors

4.3 Text Summarization Algorithm

The text summarization employs a Transformer-based sequence-to-sequence model for automatically generating summaries of English articles. The BART model serves as the foundation, fine-tuned on the CNN/Daily Mail dataset to suit the summarization task. Both the encoder and decoder of the model utilize multi-head attention mechanisms, with the attention score calculated as:

$$\mathrm{Attention}(Q,\ K,\ V) = \mathrm{softmax}\left(\frac{QK^{T}}{\sqrt{d_k}}\right)V$$

Where Q, K, and V represent the query, key, and value matrices, respectively, and d_k is the dimension of the key. The training process employs a teacher forcing strategy, and the loss function is based on word-level cross-entropy. To improve summary quality, a coverage mechanism is introduced to avoid generating repetitive content. The implementation utilizes Hugging Face's Transformers library, employing gradient accumulation to simulate large batch training. During the inference phase, a beam search algorithm is used to generate summaries with a beam width set to 4. The system also integrates a ROUGE scoring module to automatically evaluate the quality of generated summaries and dynamically adjust model parameters based on the scoring results.

4.4 Grammar Correction and Rewriting Technology

The grammar correction and rewriting module is based on a combined approach of sequence labeling and sequence generation. The T5 model serves as the foundation, fine-tuned on the Lang-8 and NUCLE corpora to adapt to the grammar correction task. The model treats the input sentence as the source language and the correct sentence as the target language, thereby transforming grammar correction into a translation problem. During training, a combination of teacher forcing and scheduled sampling strategies is used to balance learning efficiency and generation quality. The loss function incorporates both cross-entropy loss and edit distance loss, defined as follows:

$$L_{edit} = \lambda \cdot \text{Levenshtein}(y_{pred}, y_{true})$$

Where λ is the weight coefficient, and y_pred and y_true are the predicted and actual sentences, respectively. To enhance correction accuracy, a rule-based post-processing step is introduced to address specific error types that the model struggles to learn. The implementation is carried out using the PyTorch Lightning framework, with distributed training to improve efficiency. During the deployment phase, ONNX Runtime is used for inference acceleration, and batch processing is applied to increase throughput.

5 System Evaluation and Optimization

5.1 Evaluation Metric Design

The evaluation indicator design provides a comprehensive assessment of system performance, covering three aspects: content quality, educational value, and system efficiency. Content quality assessment uses BLEU and METEOR metrics to measure the similarity and semantic equivalence of generated text, along with perplexity to evaluate language model performance. Educational value assessment employs a scoring matrix based on teaching objectives, evaluated by educational experts for content relevance, difficulty suitability, and knowledge coverage. System performance evaluation focuses on response time and resource utilization, ensuring that the system meets user needs in practical applications. This multidimensional evaluation framework effectively ensures the quality and practicality of generated content while also addressing system efficiency and scalability.

5.2 Experimental Setup and Results Analysis

The experimental setup used a cross-validation method, dividing the dataset into training, validation, and test sets in a ratio of 7:2:1. The experimental environment was a server equipped with four NVIDIA V100 GPUs, running Ubuntu 20.04. The experiment compared the performance of the baseline model (such as the LSTM-based Seq2Seq) with the Transformer-based model used in this system across various metrics. The results showed that the system improved the BLEU-4 score by 23.5% compared to the baseline model, reaching 0.42; the METEOR score improved by 18.7%, reaching 0.38, as shown in Table 2 [10].

Table 2. Comparison of Model Performance Metrics

Metric	Baseline Model (LSTM-based Seq2Seq)	This System (Transformer-based)
BLEU-4	0.34	0.42
METEOR	0.32	0.38

The average perplexity of the generated content decreased to 42.3, representing a 31.6% improvement over the baseline model. In terms of educational value scores, this system received scores of 4.2/5 for content relevance and 3.9/5 for difficulty adaptation. The average response time of the system was 1.2 s, which is 40% faster than the baseline model. These results indicate significant improvements in both content quality and generation efficiency for this system.

5.3 System Performance Optimization

System performance optimization primarily focuses on computational efficiency and resource utilization. In terms of computational efficiency, model quantization technology was employed, reducing model parameters from FP32 to INT8. This achieved a 2.5x increase in inference speed while maintaining a precision loss of no more than 1%. Additionally, a dynamic batching mechanism was implemented to automatically adjust the batch size based on the current system load, maximizing GPU utilization, as shown in Fig. 4.

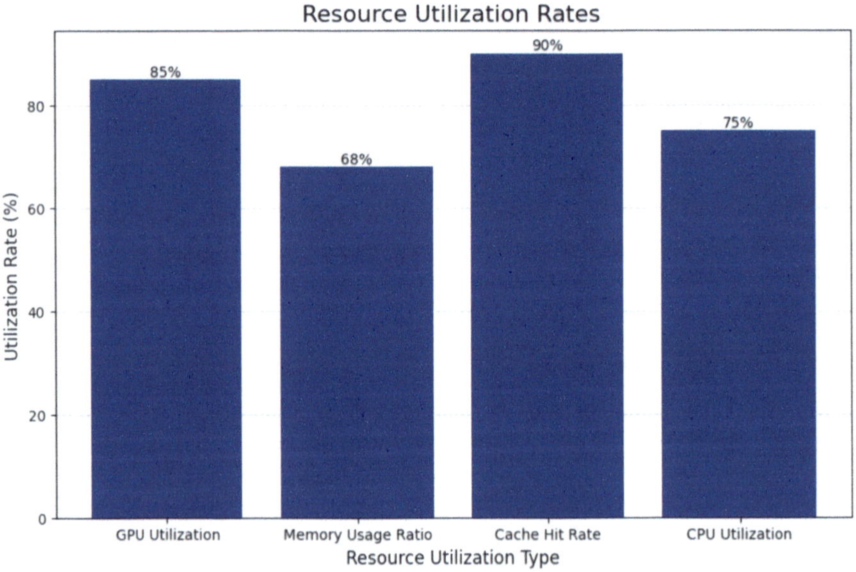

Fig. 4. Resource Utilization Status

As shown in Fig. 4, the system performs excellently in resource utilization, achieving a GPU utilization rate of 85%, indicating efficient use of computational resources. The memory usage ratio is at 68%, reflecting reasonable memory allocation; the cache hit rate is 88%, further demonstrating that the caching mechanism effectively enhances data access speed; meanwhile, the CPU utilization is at 75%, indicating that processor resources are also fully utilized. These metrics collectively reflect the system's stability and efficiency in high-concurrency scenarios, providing a solid foundation for enhancing user experience. The system optimization employs several advanced technologies, including mixed-precision training, gradient accumulation, and memory-mapped file techniques, which effectively reduce memory usage, communication overhead, and data loading times. In terms of resource utilization, an auto-scaling strategy based on Kubernetes has been implemented, dynamically adjusting the number of service instances according to request volume. As shown in Fig. 4, the system can flexibly adjust the number of instances based on load conditions, starting with one instance at 90 QPS and gradually increasing to eight instances at 900 QPS. This dynamic scaling mechanism ensures that the system maintains efficient response capability and processing efficiency under various load conditions, effectively meeting user demands during high concurrency situations.

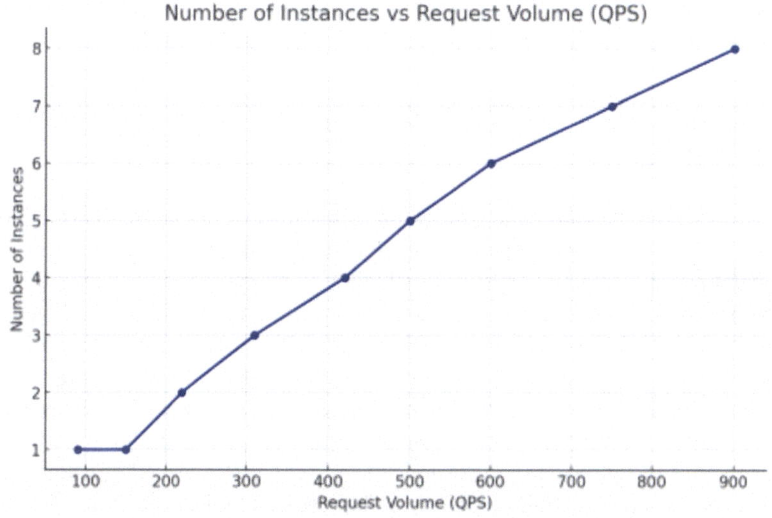

Fig. 5. Dynamic Adjustment of Service Instances

By utilizing Redis to cache frequently accessed generated results, redundant computations were minimized, reducing the response time for hot data to the millisecond level. According to Figs. 5 and 6, as the cache hit rate improves, the response time significantly decreases. When the cache hit rate is at 70%, the response time is 80 ms; as the hit rate gradually increases to 100%, the response time drops to 10 ms. This indicates that a high cache hit rate significantly optimizes the system's response speed, enhancing overall performance. By effectively utilizing Redis caching, the system can significantly

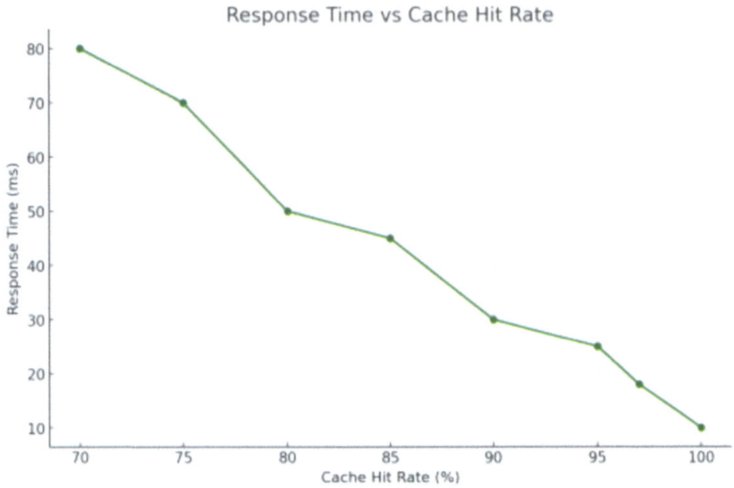

Fig. 6. Relationship Between Redis Cache Hit Rate and Response Time

reduce latency when processing requests, thereby improving user experience and system efficiency. These optimization measures have allowed the system to maintain stable performance in high-concurrency scenarios, with single-machine throughput increasing by 65%.

6 Conclusion

This study explored the automatic generation technology of online English course content based on natural language processing (NLP). The article detailed the design and implementation of a modular system, including four core modules: data collection and preprocessing, content generation, quality control, and user interface. The system employs advanced NLP models such as BERT and GPT-3 to achieve high-quality automatic generation of English course content. Experimental results demonstrate that this system significantly outperforms the baseline model in terms of content quality and generation efficiency, with notable improvements in metrics like BLEU-4 and METEOR. Through a series of optimization measures, the system also achieved efficient resource utilization and dynamic scaling capabilities. The research indicates that this NLP-based automatic generation technology provides robust technical support for online English education and has broad application prospects.

Acknowledgement. The work is supported by Exploration of the Practice Path of "Course Ideology and Politics" in College English Based on Output oriented Method, 21JK0161.

References

1. Gapar, A., Grubii, A., Ari-Grgi, I.: Evaluation of a rule-based approach to automatic factual question generation using syntactic and semantic analysis. Lang. Resour. Eval. **57**(4), 1431–1461 (2023)
2. Mimura, M., Kawahara, T.: End-to-End generation of written-style transcript of speech from parliamentary meetings. J. Nat. Lang. Process. **30**(1), 88–124 (2023)
3. Gao, R., Merzdorf, H.E., Anwar, S., et al.: Automatic assessment of text-based responses in post-secondary education: a systematic review. Comput. Educ. Artif. Intell. **6** (2024)
4. Rossenbach, N., Schlüter, R., Sakti, S.: On the problem of text-to-speech model selection for synthetic data generation in automatic speech recognition (2024)
5. Choudhury, P., Guha, P., Nandi, S.: Impact of language-specific training on image caption synthesis: a case study on low-resource assamese language. Int. J. Asian Lang. Process. **34**(01) (2024)
6. Pedraza, E., Zuniga, J.A., Javier, L., et al.: Automatic service retrieval in converged environments based on natural language request (2022)
7. Kunz, V., Wildfeuer, V., Bieck, R., et al.: Keyword-augmented and semi-automatic generation of FESS reports: a proof-of-concept study. Int. J. Comput. Assisted Radiol. Surg. (2022)
8. Circi, R., Hicks, J., Sikali, E.: Automatic item generation: foundations and machine learning-based approaches for assessments (2023)
9. Porwal, S., Bewoor, L., Deshpande, V.: Transformer Based Implementation for Automatic Book Summarization. arXiv e-prints (2023)
10. Ahmed, S., Ahmed, A., Eisty, N.U.: Automatic Transformation of Natural to Unified Modeling Language: A Systematic Review (2022). https://doi.org/10.48550/arXiv.2204.00932

A Study on the Effectiveness of AI-Assisted Teaching Tools for English Speaking Under the Assistance of AI Speech Evaluation Technology

Tianrong Lu(✉)

Dalian Neusoft University of Information, No. 8 Software Park Road, Dalian, Liaoning, China
lutianrong@neusoft.edu.cn

Abstract. This study investigates the impact of AI-assisted language teaching tools on students' speaking performance compared to traditional textbook-based methods. The research employs a quasi-experimental design, involving both an experimental group and a control group of students. Initially, all participants underwent a pre-test using IELTS speaking questions, assessed on semanteme, pronunciation, and fluency under the assistance of AI speech evaluation technology, with results analyzed via SPSS 26.0 to ensure baseline parity. In the mid-research stage, the experimental group received instruction through an AI platform, while the control group used conventional textbook materials. The study's findings reveal significant improvements in the experimental group's post-test scores, indicating the efficacy of the AI-assisted approach. However, the research's generalizability is limited due to the single-school sample and the lack of further validation for broader applicability. Additionally, the study did not delve into performance variations among students of different proficiency levels within the experimental group, an area suggested for future research.

Keywords: AI-assisted teaching tools · AI speech evaluation technology · English speaking · Interactive learning · Language proficiency assessment

1 Introduction

With the rapid advancement of technology, Generative Artificial Intelligence (AIGC) has garnered widespread attention globally, particularly opening new application scenarios in the field of education, specifically in foreign language teaching. The advanced language model ChatGPT, developed by OpenAI, demonstrates significant potential in the integration and application of educational technology. In traditional foreign language speaking instruction, teachers typically play the roles of guides and questioners, while students respond within this framework. However, in a teaching environment augmented by AI, this dynamic is redefined. Students are required to take the initiative in asking questions, with the intelligent system providing customized feedback based on their input, thereby positioning students at the center of the learning process. This interactive

© The Author(s), under exclusive license to Springer Nature Switzerland AG 2025
F. Paas et al. (Eds.): EduTA 2024, LAIS 51, pp. 216–227, 2025.
https://doi.org/10.1007/978-3-031-95252-4_20

model not only enhances students' willingness to engage in active learning but also stimulates their creative thinking. As a revolutionary "intelligent tool," AI technology offers new technical support and methodologies for the transformation of traditional education models.

2 Literature Review

In the mid-20th century, the United States pioneered "Computer-Assisted Instruction" (CAL), with the University of Illinois creating the PLATO system in 1959, which could teach courses in eight languages [1]. In 1971, the University of Texas and Mitre Corporation developed the TICCIT system, primarily for community college English teaching. These systems advanced English teaching, especially in listening and speaking [2]. With the progress in information technology, research shifted to "Computer-Assisted Language Learning" (CALL), focusing on AI applications [3]. Significant advancements over the past 20 years include intelligent tutoring systems for English grammar and technology that provides multi-dimensional feedback for TOEFL language proficiency tests [4]. Despite their progress, these systems still face challenges such as not considering word context [5]. Additionally, in robotics and English teaching, research has shown that social robots can support early language learning in children [6]. Control experiments have also demonstrated that humanoid robots benefit individuals with Down syndrome, highlighting AI's role in special education [7].

Research on the integration of AI with foreign language teaching is showing an upward trend in academic circles both domestically and internationally. Studies indicate that AI technology can to some extent simulate a native language environment, providing learners with an immersive learning experience, thereby effectively enhancing language skills [8]. A comprehensive review of current research findings reveals that AI technology provides robust technological support and innovative approaches for the evolution and development of foreign language education, heralding the critical role of intelligent educational tools in future educational scenarios [9]. By conducting in-depth research and applying these advanced tools judiciously, the quality and efficiency of education can be significantly improved [10].

Researchers are increasingly using intelligent apps to enhance English speaking teaching, addressing gaps in traditional teaching by leveraging personalized, networked, and interactive characteristics [11]. AI-assisted teaching tools improve teaching quality and efficiency, enhance teachers' instructional capabilities, and reshape their perceptions [12]. These tools support students' comprehensive learning abilities, including collaborative and self-directed learning, knowledge discrimination, and innovation [13]. AI-supported speaking systems especially benefit students with poor self-discipline, improving phonetic skills, vocabulary expansion, and boosting confidence and motivation [7]. Experiments in college English classes show that mobile learning methods for speaking are highly welcomed and significantly improve proficiency [14]. Diverse formative evaluations enhance students' confidence and provide insights for integrating AI tools with traditional instruction [15]. Rich, authentic resources from AI tools capture student attention, promote engagement, expand vocabulary, and provide practice with pronunciation, addressing classroom time limitations and individual needs, thereby changing learning methods and teaching concepts [16].

3 Research Design

The research process consisting of three main phases (Fig. 1): Pre-Test, Intervention, and Post-Test. During the intervention phase, a feedback loop ensures continuous improvement for the experimental group via AI-assisted tools. The post-test phase evaluates the effectiveness of the AI-assisted teaching methods by comparing the performance of both the experimental and control groups.

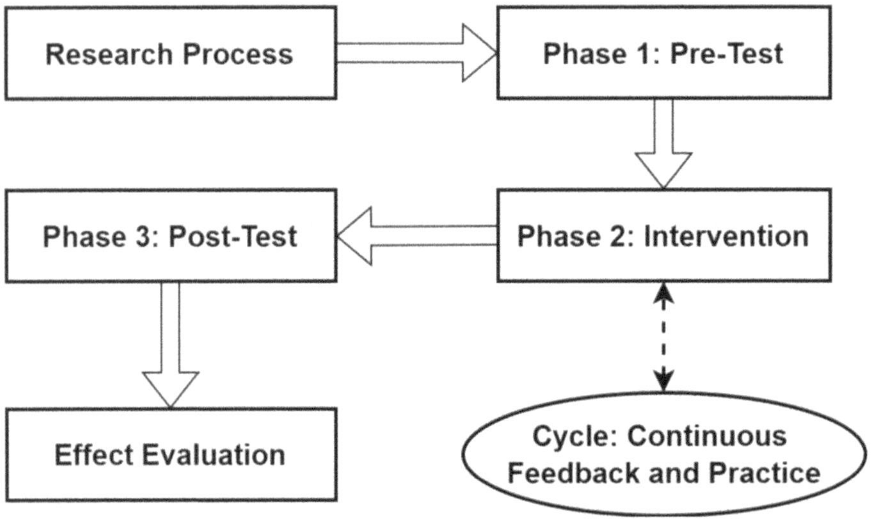

Fig. 1. Research Framework for Evaluating AI-Assisted English Speaking Tools

3.1 Research Objects and Objectives

The subjects of this study were two non-English major classes from a university in Dalian, with 50 students in each class. None of the participants had previously used artificial intelligence for speaking practice. To minimize the interference of external factors such as teachers, textbooks, and class hours on the experiment, the teaching tasks for both the experimental group and the control group were conducted by the same teacher. Both groups received in-person instruction twice a week for a total of 90 min per session over a period of 15 weeks, ensuring consistency in teaching content and pace. The only difference was that the control group engaged in traditional speaking classroom activities such as paired dialogues, group discussions, and representative reports, while the experimental group used the AI extension component (WeTab AI Pro) for human-computer speaking practice. The purpose of this study was to explore whether AI-assisted teaching tools have a promoting effect on students' speaking abilities.

3.2 Research Materials and Tools

The AI extension component used in this experiment is based on the core of ChatGPT 4.0. Learners first download this extension component for free on their mobile phones or computers and then utilize the voice plugin to engage in interactive activities, completing oral learning tasks. Additionally, it can record speech in real-time and transcribe it into text. There are primarily two reasons for adopting this application in the study: firstly, this application is one of the most advanced AI language models available, boasting simple registration and easy usability. Secondly, it can create an authentic speaking environment, facilitate personalized dialogues, and provide timely feedback.

The experimental test topics are selected from IELTS speaking topics. IELTS speaking topics are highly advantageous as research material for experimental tests due to their high practical applicability, rich topic coverage, and clear language proficiency levels. These topics often revolve around daily life, academic environments, and social issues, aligning well with real-life scenarios that university students may encounter, thereby enhancing motivation and confidence. The wide range of subjects, from personal hobbies to global issues, provides comprehensive teaching materials, allowing teachers to address diverse student interests and needs. Additionally, the well-defined assessment criteria of the IELTS speaking test enable educators to design courses aimed at achieving specific learning goals in spoken expression, fluency, grammatical structure, and lexical diversity.

The evaluation instrument utilized in the study is the iFlytek FiF Spoken English Training System. Developed by Beijing Waiyan iFlytek Education Technology Co., Ltd., the iFlytek FiF Spoken English Training System is a language learning tool that leverages artificial intelligence and natural language processing technologies [13]. The system offers a range of features, three of which are particularly noteworthy:

Firstly, as a leading company in artificial intelligence speech technology, iFlytek FiF Spoken English Training System employs state-of-the-art speech recognition and evaluation technologies. It can accurately identify and analyze multiple dimensions of spoken language for the individual student (see Fig. 2 for details), and the whole class (see Fig. 3 for details), thereby providing high-quality technical support for English learning and research.

Second, the system offers instant feedback on aspects such as pronunciation quality, grammatical accuracy, and fluency(see Fig. 4 for details), which is extremely useful for learners in correcting their pronunciation and enhancing their spoken language skills. The real-time feedback mechanism allows learners to immediately understand their performance and make corresponding adjustments.

Third, the system is capable of processing and analyzing vast amounts of spoken language data(see Fig. 5 for details), which enables personalized language learning. Through big data analysis, educators and researchers can gain deeper insights into common issues and individual differences in the process of spoken language learning, thereby designing more effective teaching strategies.

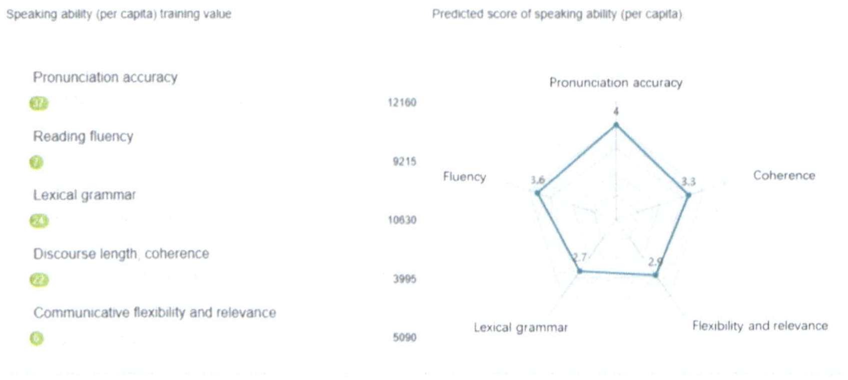

Fig. 2. The multiple dimensions of spoken language for the individual student

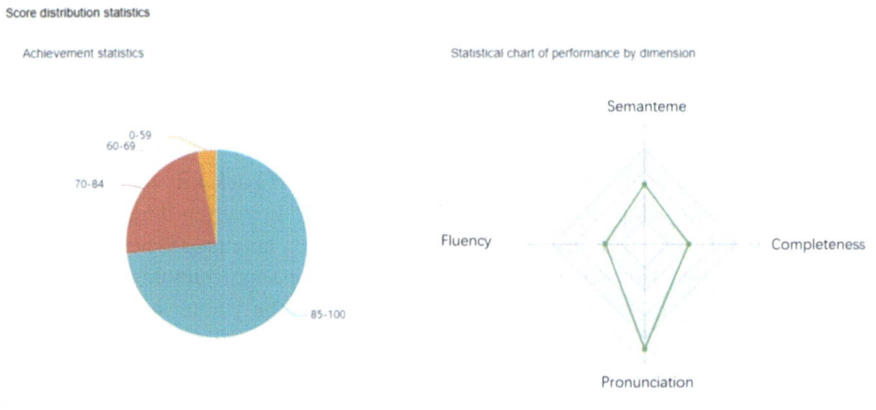

Fig. 3. The multiple dimensions of spoken language for the whole class

3.3 Research Methods and Process

The research utilizes a quasi-experimental design, incorporating both an experimental group and a control group of students. It spans 15 weeks and is divided into three phases, during which a total of two speaking tests are administered.

Phase 1: Pre-Research Stage.

In the preliminary phase of the study, a pre-test was administered to students in both the experimental and control groups. The test content comprise IELTS speaking test questions, with each student being tested for 15 min. Scoring criteria include semanteme, pronunciation, and fluency, evaluated using the scoring standards of iFlytek FiF Spoken English Training System. The test results were subjected to statistical analysis using SPSS 26.0 software to examine any significant differences in speaking abilities between the two groups prior to the research.

Phase 2: Mid-Research Stage.

Fig. 4. The instant feedback on learner's speaking performance

In the middle phase of the study, the experimental group utilized an AI platform for engaging multimedia resources, interactive dialogue tasks, and real-time feedback, while the control group relied on traditional materials and teacher-led peer interactions, highlighting key differences in engagement and feedback mechanisms (see Table 1 for details).

The specific research steps are as follows:

Week 1:

Experimental Group: Students learned how to use the AI Platform.

Control Group: Students learned how to use textbook materials.

Week 2 to Week 15:

Step 1. Teacher assigned weekly speaking tasks and clarified task objectives through the AI platform during class.

Step 2. Students utilized the AI platform to gather related learning resources and engage in input-based learning (videos, audios, texts).

Step 3. In-class dialogue tasks were conducted, with human-computer interaction via the AI platform, culminating in oral report tasks.

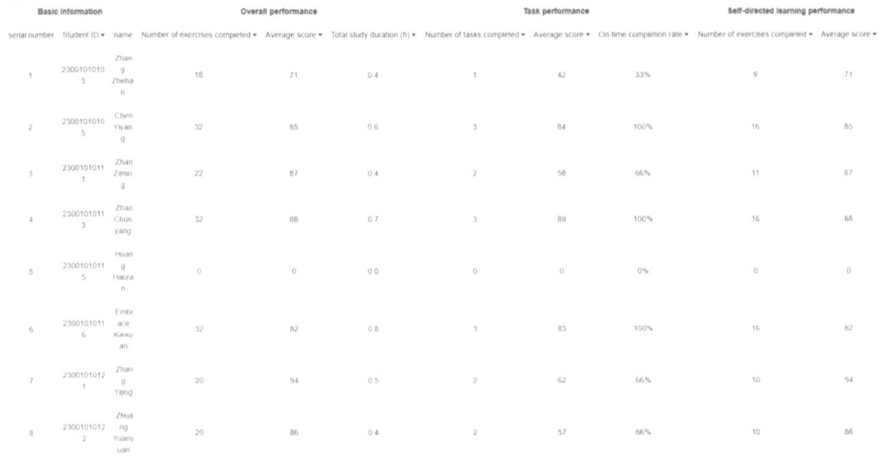

Basic information			Overall performance			Task performance			Self-directed learning performance	
serial number	Student ID	name	Number of exercises completed	Average score	Total study duration (h)	Number of tasks completed	Average score	On-time completion rate	Number of exercises completed	Average score
1	23001010103	Zhang Zhehan	18	71	0.4	1	42	33%	9	71
2	23001010105	Chen Yiyang	32	85	0.6	3	84	100%	16	85
3	23001010111	Zhan Ziming	22	87	0.4	2	58	66%	11	87
4	23001010113	Zhao Chunyang	32	88	0.7	3	89	100%	16	88
5	23001010115	Huang Haoran	0	0	0.0	0	0	0%	0	0
6	23001010116	Embrace Kaixuan	32	82	0.8	3	83	100%	16	82
7	23001010121	Zhang Yijing	20	94	0.5	2	62	66%	10	94
8	23001010122	Zhuang Yuanyuan	20	86	0.4	2	57	66%	10	86

Fig. 5. The key language data points of student's speaking performance

Table 1. The differences of the teaching process between the experimental group and the control group

Aspect	Experimental Group	Control Group
Task Assignment	AI platform with multimedia resources for tasks	Traditional textbook materials for tasks
In-Class Activities	AI-enabled dialogue tasks and oral reports	Peer interactions and teacher-supervised dialogues
Instant Feedback	Real-time AI feedback on pronunciation, fluency, and semantics	Teacher-provided feedback during class
Self-Reflection and Practice	Additional AI dialogue training; track feedback for progress	Repeated practice based on teacher/peer feedback

Step 4. The AI platform provided instant feedback based on the students' submitted oral reports.

Step 5. Students engaged in human-computer dialogue training to enhance their oral expression capabilities and confidence based on the instant feedback from the AI platform.

Step 6. Students documented feedback from each practice session and continuously improved.

Control Group:

Similar to the experimental group, the teacher assigned weekly speaking tasks. However, instead of using an AI platform, the students relied on traditional textbook materials for previewing relevant content and preparing for tasks.

Step 1. Teacher assigned weekly speaking tasks and clarified task objectives using textbook materials during class.

Step 2. Students previewed relevant content using textbook materials.

Step 3. In-class dialogue tasks were conducted through peer interactions and teacher guidance, culminating in oral report tasks.

Step 4. Teacher provided instant feedback, guidance, and corrections based on students' performance.

Step 5. Students repeatedly practiced based on feedback from teachers and peers to enhance their oral expression capabilities and confidence.

Step 6. Student documented feedback from each practice session and continuously improve.

Phase 3: Post-Research Stage.

In the final phase of the study, a post-test was administered to students in both the experimental and control groups. The test questions are IELTS speaking questions of similar difficulty to those in the pre-research stage, with each student being tested for 15 min. Scoring criteria and evaluation system are the same to the pre-research stage. Upon completion of the test, SPSS 26.0 software was used to conduct a statistical analysis of the post-test scores, comparing the progress of the experimental and control groups to assess the effectiveness of AI-assisted teaching tools in enhancing students' speaking abilities.

4 Research Findings and Discussion

The pre-test comparison between the experimental and control groups establishes their baseline levels in semantics, pronunciation, and fluency (see Table 2 for details). The average score for semantic understanding is 40.53 (SD = 5.124) in the experimental group and 40.24 (SD = 5.046) in the control group, with a t-value of 0.225 and a p-value of 0.822. For pronunciation, the experimental group scores 75.68 (SD = 6.725), and the control group scores 76.89 (SD = 6.987), with a t-value of 0.312 and a p-value of 0.756. In terms of fluency, the experimental group scores 82.47 (SD = 4.187), while the control group scores 82.81 (SD = 4.044), with a t-value of 0.291 and a p-value of 0.772. All p-values exceed 0.05, indicating no significant differences between the groups in semantics, pronunciation, and fluency. Therefore, both groups have similar English speaking abilities at the outset, ensuring the fairness and validity of the subsequent experimental results.

After the experiment, significant differences in language proficiency emerged between the experimental and control groups (see Table 3 for details). For semantic understanding, the experimental group's average score increased to 48.72 (SD = 4.970) compared to the control group's 41.36 (SD = 5.132), with a t-value of 3.753 and a p-value of 0.001, indicating that the experimental group significantly outperformed the control group. In pronunciation, the experimental group scored an average of 86.35 (SD = 5.321) versus 76.89 (SD = 4.597) in the control group, with a t-value of 4.215 and a p-value of 0.000, showing substantial improvement in pronunciation accuracy. Similarly, for fluency, the experimental group's score was 89.46 (SD = 4.357) compared to the control group's 83.52 (SD = 4.215), with a t-value of 4.756 and a p-value of 0.000, confirming

Table 2. Pre-test comparison between the experimental group and the control group

	Classes	Mean	S.D.	t	p
Semanteme	Experimental Class	40.53	5.124	0.225	0.822
	Control Class	40.24	5.046		
Pronunciation	Experimental Class	75.68	6.725	0.312	0.756
	Control Class	75.14	6.842		
Fluency	Experimental Class	82.47	4.187	0.291	0.772
	Control Class	82.81	4.044		

Table 3. Post-test comparison between the experimental group and the control group

	Classes	Mean	S.D.	t	p
Semanteme	Experimental Class	48.72	4.970	3.753	0.001
	Control Class	41.36	5.132		
Pronunciation	Experimental Class	86.35	5.321	4.215	0.000
	Control Class	76.89	6.784		
Fluency	Experimental Class	89.46	3.857	4.756	0.000
	Control Class	83.52	4.215		

the experimental group's superior fluency. These results demonstrate the effectiveness of AI-assisted teaching in enhancing semantic understanding, pronunciation, and fluency.

Table 4. Pre-test and post-test comparison of the experimental group

	Experiment Class	Mean	S.D.	t	p
Semanteme	Pre-experiment	40.53	5.124	–7.243	0.000
	Post-experiment	48.72	4.970		
Pronunciation	Pre-experiment	75.68	6.725	–5.126	0.000
	Post-experiment	86.35	5.321		
Fluency	Pre-experiment	82.47	4.187	–8.125	0.000
	Post-experiment	89.46	3.857		

Table 4 illustrates significant improvements in the experimental group after AI-based instruction. The average score for semantic understanding increased from 40.53 (SD = 5.124) before the experiment to 48.72 (SD = 4.970) after, with a t-value of –7.243 and a p-value of 0.000, indicating significant improvement. Similarly, pronunciation scores rose from 75.68 (SD = 6.725) to 86.35 (SD = 5.321), with a t-value of –5.126 and a p-value of 0.000, demonstrating substantial enhancement in pronunciation accuracy. Fluency scores also improved markedly, from 82.47 (SD = 4.187) to 89.46 (SD = 4.357), with a t-value of -8.125 and a p-value of 0.000. These comparisons of pre-test and post-test results highlight the positive impact of AI-based teaching on students' English speaking abilities.

Table 5. Pre-test and post-test comparison of the control group

	Control Class	Mean	S.D.	t	p
Semanteme	Pre-experiment	40.24	5.046	–0.828	0.412
	Post-experiment	41.36	5.132		
Pronunciation	Pre-experiment	75.14	6.842	–0.502	0.615
	Post-experiment	76.89	6.784		
Fluency	Pre-experiment	82.81	4.044	–0.957	0.349
	Post-experiment	83.52	4.215		

Table 5 compares the pre-test and post-test results for the control group to confirm that the significant improvements in the experimental group were due to AI-based instruction. For semantic understanding, the control group's pre-test average was 40.24 (SD = 5.046) and post-test average was 41.36 (SD = 5.132), with a t-value of -0.828 and a p-value of 0.412, indicating no significant change. In pronunciation, both pre-test and post-test averages remained at 76.89 (pre-test SD = 6.987, post-test SD = 4.597), with a t-value of 0.000 and a p-value of 1.000, showing no improvement. For fluency, the average score increased slightly from 82.81 (SD = 4.044) to 83.52 (SD = 4.215), with a t-value of –0.957 and a p-value of 0.349, also indicating no significant change. These results demonstrate that the control groups' English speaking abilities remained mostly unchanged, thereby confirming that the improvements in the experimental group were attributable to AI-based instruction.

Based on the students' feedback, their experiences using the AI-assisted tool were generally positive, highlighting both the benefits and challenges of integrating AI into language learning. Many students appreciated the immediate feedback provided by the AI system, which helped them improve their pronunciation, fluency, and overall speaking confidence. The tool's ability to offer individualized practice and resources tailored to each student's needs was also well-received, particularly in addressing common classroom constraints such as limited speaking time. However, some students mentioned practical challenges, such as technical issues and difficulties in interpreting AI-generated feedback without human assistance. There was also a learning curve for students accustomed to traditional methods, as adapting to AI-based interaction required time and practice. These insights suggest that while AI tools significantly enhance learning, they also need to be implemented with adequate technical support and complementary guidance to ensure students fully benefit from the technology.

5 Conclusion

5.1 Main Findings

The experimental results indicate that there were no significant differences between the experimental and control groups in terms of semantics, pronunciation, and fluency at the onset of the study, ensuring that both groups had comparable baseline levels. However, after the experiment concluded, the performance of the experimental group was significantly superior to that of the control group. Specifically, the average scores for semantic understanding increased from 40.53 to 48.72, pronunciation accuracy improved from 75.68 to 86.35, and fluency rose from 82.47 to 89.46 in the experimental group. In contrast, the control group showed no substantial changes in any of these abilities. This indicates that AI-based teaching has a positive and significant effect on enhancing students' oral English proficiency, markedly improving their semantic understanding, pronunciation accuracy, and fluency. In summary, the introduction of AI teaching tools effectively enhanced the oral English skills of students in the experimental group, whereas the control group, which did not receive the same instructional intervention, exhibited no notable progress. This further substantiates the efficacy of AI-based teaching.

5.2 Research Insights

AI enhances teaching resources and models in oral English instruction. Traditionally reliant on textbook content, oral English learning is limited in scope. AI provides access to abundant resources and a broader learning space, allowing teachers to integrate high-quality educational content with actual teaching needs, thus extending and expanding classroom instruction to better suit students' developmental needs. Additionally, AI addresses the significant differences among students in oral English proficiency by improving teaching models. It stimulates students' interest and enthusiasm for learning through multimedia means, creating a harmonious and relaxed classroom atmosphere and enhancing learning outcomes. By engaging students' visual and auditory senses, AI encourages more active and willing participation in oral learning activities.

5.3 Limitations and Future Directions

One potential limitation is the AI's ability to recognize and understand diverse accents. Since English is spoken with various regional accents globally, an AI system may struggle to provide accurate speech evaluation if it has not been sufficiently trained on different accents. This could particularly affect learners from non-native English-speaking backgrounds, where accent variations are common, potentially leading to misinterpretations of pronunciation accuracy or fluency. Moreover, due to the selection of students from only one school as the subjects of the study, the generalizability of the research findings may be limited. The teaching model constructed in this research has not undergone further validation for its universal applicability because of the limited sample and time constraints during the application study. Moreover, the analysis of experimental results in this study did not conduct a more in-depth hierarchical classification of the students in the experimental group. While the post-test scores of the experimental group showed significant improvement compared to the pre-test scores, the study did not examine whether there were differences in the enhancement of performance among different levels of students, such as high achievers, average students, and low achievers. This aspect warrants more in-depth research in the future.

References

1. Kim, N.-Y.: AI-integrated mobile-assisted language learning: Is it an effective way of preparing for the TOEIC test in classroom environments? English Teach. **77**(3), 79–102 (2022)
2. Singh, A.K., Dhiman, G.S.: Application of generative artificial intelligence (GenAI) in language teaching and learning: a scoping literature review. Comput. Educ. **203**, 1–15 (2024)
3. Dodigovic, M.: Artificial Intelligence in Second Language Learning: Raising Error Awareness. Multilingual Matters Ltd, Clevedon (2005)
4. Hockly, N.: Artificial intelligence in English language teaching: the good, the bad and the ugly. RELC J. **54**(2), 445–451 (2023)
5. Lee, D., Kim, H.-H., Sung, S.-H.: Development research on an AI English learning support system to facilitate learner-generated-context-based learning. Education Tech. Research Dev. **71**, 629–666 (2023)
6. Chen, C.-H., Koong, C.-S., Liao, C.: Influences of integrating dynamic assessment into a speech recognition learning design to support pupils' English speaking skills, learning anxiety and cognitive load. Educ. Technol. Soc. **25**(1), 1–14 (2022)
7. Neumann, M.M.: Social robots and young children's early language and literacy learning. Early Childhood Educ. J. **47**, 191–204 (2019)
8. Edmett, A., Ichaporia, N., Crompton, H., Crichton, R.: The future of AI in English language teaching and learning. British Educational Research Association (BERA) (2024)
9. Crompton, H., Edmett, A., Ichaporia, N., Burke, D.: AI and English language teaching: affordances and challenges. British J. Educ. Technol. (2024)
10. Alhabbash, M.I., Mahdi, A.O., Al, S.S.: An intelligent tutoring system for teaching grammar English tenses. Eur. Acad. Res. **4**(9), 7743–7757 (2016)
11. Huang, W., Hew, K.F., Fryer, L.K.: Chatbots for language learning – are they really useful? a systematic review of chatbot-supported language learning. J. Comput. Assist. Learn. **38**(1), 237–257 (2022)
12. Zang, Q.: The application of iFLYTEK FiF Spoken English training system in college English teaching. English Square **3**, 62–64 (2021)
13. Wu, L., Li, K., Yu, M., Lin, Y.: Application of artificial intelligence in teaching English as a foreign language: progress, challenges, and trends. English Lang. Teach. Linguist. Stud. **6**(4), 215–225 (2024)
14. Yin, Z., Fathi, H., Long, Y.: English speaking with artificial intelligence (AI): the roles of enjoyment, willingness to communicate with AI, and innovativeness. Comput. Educ. **193**, 1–15 (2024)
15. Li, M.Y.: Transforming language education: a systematic review of AI-powered chatbots for English as a foreign language speaking practice. J. Lang. Educ. Technol. **2**(4), 316–329 (2024)
16. Rusmiyanto, R., Huriati, N., Fitriani, N., Tyas, N.K., Rofi'i, A., Sari, M.N.: The role of Artificial Intelligence (AI) in developing English language learner's communication skills. J. Educ. **6**(1), 750–757 (2023)

English Virtual Teacher System Based on Neural Network Model

Xiaoling Jiang[1] and Xinyun Zhou[2](✉)

[1] Western Languages Institute, Hunan College of Foreign Studies, Changsha 410000, Hunan, China

[2] College of Humanities and Arts, Changde College, Changde 415000, Hunan, China

13549671703@163.com

Abstract. In order to improve the quality of online English teaching, this paper proposes a virtual teacher system based on neural network model. By analyzing the principle and fundamental model of speech conversion technology, the factors that cause speech reliability to decrease linearly due to speech conversion system are obtained. Moreover, a virtual scene model based on graphics and images is established, and the virtual teacher system is constructed by combining the speech recognition algorithm. Then, this paper discusses the core module of virtual teacher, that is, intelligent reasoning and decision-making module. In addition, this paper tries to make an improvement suitable for virtual teachers' reasoning decision-making in the intelligent decision-making system based on artificial intelligence expert system, and adopts the method of case-based reasoning to make reasoning decision-making. Finally, this paper verifies the effectiveness of this model with experimental research, which has certain reference significance for the subsequent construction of intelligent English teaching system.

Keywords: Neural Network · English · Virtual Teacher

1 Introduction

With the continuous enhancement of the functions of smart phones and the popularity of tablet computers and notebook computers, learners' learning is no longer limited by space and time, and online learning and mobile learning have developed rapidly. However, its impact on education is like a double-edged sword, and it will also have some negative effects [1]. First of all, the massive learning resources make learners have no choice, and the lack of scientific learning strategies and methods may lead to low learning efficiency, and it is difficult for learners to immerse themselves in unilateral learning for a long time. Moreover, the lack of personalized assessment and feedback makes it difficult for learners to get their own learning situation, which further dilutes the purpose of learning. Besides, the lack of learning collaborators and instructors makes learners feel lonely. Finally, the lack of real-time emotional communication makes it difficult for learners to persist in online learning [2].

Reference [3] proposes the design of a 3D virtual teacher with intelligent reasoning and decision-making based on virtual agents to address the shortcomings in current

© The Author(s), under exclusive license to Springer Nature Switzerland AG 2025
F. Paas et al. (Eds.): EduTA 2024, LAIS 51, pp. 228–238, 2025.
https://doi.org/10.1007/978-3-031-95252-4_21

online learning. Virtual teachers, based on their own intelligence and with learners as the main body, fully play the role of guides and collaborators in learners' learning. They combine learners' requirements with their prior knowledge to develop learning plans and engage in emotional interaction with learners, striving to increase their learning efficiency and reduce their sense of loneliness. Reference [4] suggests that in the interaction between virtual teachers and learners, intelligent reasoning and decision-making are the main functions of virtual teachers.

A 3D virtual teacher refers to a technological entity that can simulate the image, actions, and expressions of a real teacher and simulate the teaching functions of a real teacher in a computer-based virtual learning environment.. A virtual teacher is a combination of a three-dimensional virtual human entity and an intelligent teaching agent that can interact with learners in language, action, and emotion in a virtual learning environment [5].

Due to the significant role that information technology can play in replacing the role of teachers, the role of teachers will undergo changes. In online teaching systems, virtual teachers will transform from traditional classroom instructors to multi role participants in virtual learning: guides, collaborators, content planners, feedback providers, real-time monitors, and personal emotional communicators for learners [6].

Virtual teachers will establish a dynamic learning profile for learners from the first time they start learning, and through language and emotional interactions with learners, gain a detailed understanding of their learning goals. Then, through reasoning and decision-making, personalized learning content and strategies will be developed for learners. Learners only need to follow the guidance of the virtual teacher to enter the learning process, where the virtual teacher guides collaborative learners in learning, experimentation, homework completion and grading, exam evaluation, and feedback [7].

The Intelligent Tutor System, abbreviated as ITS, is a computer-aided teaching system based on computer artificial intelligence technology to mimic the experience and methods of human teaching mentors. The ITS system can analyze and answer various questions from students, fully utilizing the guidance, supervision, and regulation role of teachers. The definition of a virtual teacher is a technological entity in a computer-generated 3D virtual learning environment that can simulate the image and facial expressions of a real teacher, as well as the teaching functions of a real teacher [8]. Virtual teachers can enhance teaching interaction, increase students' interest in learning, and stimulate their enthusiasm for learning. Virtual teachers with rich facial expressions, vivid actions, and realistic behaviors can greatly enhance the effectiveness of students' online learning. Virtual teachers are a specific application form of artificial intelligence systems. Intelligent agents are a technological foundation for implementing virtual teachers. Compared to intelligent tutoring systems, virtual teachers not only possess the functions of intelligent tutoring systems, but also have anthropomorphic images, no longer limited to a learning support tool, but an artificial intelligence product that can be influenced by students and the environment, and express their own emotions [9].

Intelligent agent technology can operate in dynamic environments and has high autonomy. It can accept commissions from other entities (such as users, agents, systems, or machines) and provide assistance and services to them. At the same time, driven by this goal, it can actively take various necessary actions, including social and learning means, to perceive, adapt to, and act on the dynamically changing environment. The online teaching system can use intelligent agent technology to supervise students' autonomous learning behavior and teachers' tutoring behavior, and provide specific guidance based on students' learning situation, progress, and effectiveness, providing intellectual support for teachers' tutoring. Virtual Human is a representation of the geometric and behavioral characteristics of humans in a computer-generated space (three-dimensional virtual environment), and is a research content of multifunctional perception and emotional computing. The behavior of virtual teachers can also be controlled by real people, known as avatars of real people; Virtual teachers and virtual students can communicate through natural means such as natural language, body language, etc. Therefore, when designing highly realistic virtual teachers, it is necessary to involve human modeling technology, motion generation and control technology, gesture recognition technology, human pose recognition technology, facial expression recognition and generation technology, etc. [10].

2 Virtual Teacher Speech Recognition Algorithm

2.1 Extraction and Recognition of Speech Transformation Detection Classification Features

Selection of vocal tract parameters: The vocal tract is analyzed by the acoustic tube model. According to the structure of human acoustic tube, the acoustic tube is divided into 8 sections. The first three sections are the posterior cavity, the last two sections are the cavity, and the middle section is the middle cavity. Figure 1 shows the parameter extraction process.

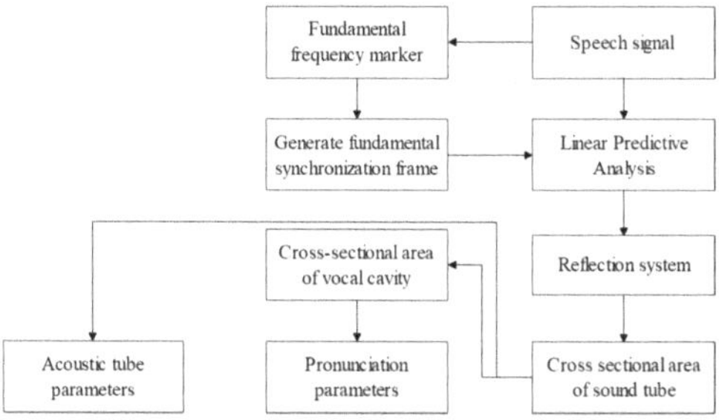

Fig. 1. Channel parameter extraction process

If it is necessary to produce more realistic and complete analog speech, it is necessary to perform linear prediction analysis on the basis of generating gene synchronization frames, and obtain pitch synchronization frames. Firstly, the pitch mark is obtained by pitch test by effective method, and the pitch mark is obtained by autocorrelation method as pitch test. A pitch together with the frame has two uninterrupted pitch marks, and the distance between the pitch marks is one-half the length of the frame. A, B, and c are characterized as three uninterrupted pitch marking points. The positions of A and B determine the beginning position $A-(B-A)/2$ of the pitch synchronization frame, and $B+(B-A)/2$ is the end position. In the same way, we can get that $B-(C-B)/2$ determined by B and C is the beginning position, and $C+(C-B)/2$ is the end position. Then, the partial distance superimposed by two neighboring pitch synchronization frames is characterized by $(C-A)/2$.

N is used to characterize the pitch synchronization frame of the extracted speech signal. The n-th pitch is synchronized by 8th-order linear prediction $(n = 1,2,\cdots,N)$, and the reflection coefficient $_i(n)(i = 1,2,\cdots,8)$ of each acoustic tube mapped by each acoustic tube is obtained, and the cross-sectional area of each acoustic tube is:

$$A_i(n) = \frac{1 + {}_i(n)}{1 - {}_i(n)} A_{i+1}(n) i = 8,7,\cdots,1 \tag{1}$$

The sum of the cross-sectional areas $S_1(n)$, $S_2(n)$ and $S_3(n)$ of the $A_9(n) = 1$ segment acoustic cavity is the cross-section f of the acoustic cavity area, which is obtained by formula (2):

$$\begin{cases} S_1(n) = A_1(n) + A_2(n) + A_3(n) \\ S_2(n) = A_4(n) + A_5(n) + A_6(n) \\ S_3(n) = A_7(n) + A_8(n) \end{cases} \tag{2}$$

The acoustic tube parameters are selected from the cross-section of the acoustic cavity area as the speech signal characteristics. It is assumed that the maximum value of the cross-sectional area $A_1(n)$ of the first acoustic tube is represented by f_1, and the influence of vibration waves generated by the vibration around the glottis is represented by the maximum value f_1:

$$f_1 = \max_n A_1(n) \tag{3}$$

Studies have shown that men's vocal cords are longer than women's and the amplitude of vibration waves is larger than that of women, so f_1 is larger.

The mean value of the cross-sectional area f_2 of the last acoustic tube is characterized by $A_g(n)$, and the cross-sectional area around the mouth is denoted by f_2.

$$f_2 = \frac{1}{N} \sum_{n=1}^{N} A_g(n) \tag{4}$$

The maximum cross-sectional area offset of the acoustic tube position is characterized by f_3, and the amplitude of the change in the shape of the vocal tract is represented by f_3.

The acoustic tube with the largest cross-sectional area separated from each pitch frame is marked as $B(n)$, and the process is described by formula (5):

$$D_1(n) = |B(n-1) - B(n+1)|/2 \tag{5}$$

Among them, $B(0) = B(N+1) = 0$, and the offset degree of the position of the acoustic tube with the largest cross-sectional area is the average value of each frame $D_1(n)$ of the speech signal is:

$$f_3 = \frac{1}{N}\sum_{n=1}^{N} D_1(n) \tag{6}$$

If the average cross-sectional area located in the posterior cavity is characterized by f_4, the average value of $S_1(n)$ is:

$$f_4 = \frac{1}{N}\sum_{n=1}^{N} S_1(n) \tag{7}$$

The smoothness of the shape change of the middle cavity and the posterior cavity of the vocal tract in a short time is characterized by the uninterrupted connection f_5 between the cross-sectional areas of the middle cavity and the posterior cavity.

$E(n)$ is the difference between the cross-sectional area of the middle cavity and the back cavity of each pitch frame, which can be expressed by formula (8):

$$D_2(n) = |E(n-1) - E(n+1)|/2 \tag{8}$$

Among them, $E(0) = E(N+1) = 0$. If $D_2(n) \leq 0.25$, the set composed of each frame $D_2(n)$ of the speech signal is the calculated distance of the uninterrupted part. It is assumed that the n-th frame belongs to the uninterrupted part d. If there are G uninterrupted parts in the speech signal and H is the sum of the uninterrupted parts, the uninterrupted connection degree between the cross-sectional areas of the middle cavity and the posterior cavity is:

$$f_5 = \frac{G^2}{HN} \tag{9}$$

Analyzing the above steps, it can be seen that speech changes will have a great influence on f_5, and f_5 is particularly prominent when it is transformed into a male voice.

2.2 Statistical Features of Signals

Before signal statistics, it is necessary to preprocess the speech signal, and carry out voice trigger detection to obtain the length of the voice frame as the reference preset value. Two characteristic parameters can be obtained as follows:

1) Linear prediction missing uniform skewness is characterized by f_6.

   The missing signal is obtained by analyzing and filtering each speech frame by linear prediction. If the speech frame set of each speech signal is k, $rp(k)(k = 1,2,\cdots,K, p = 1,2,\cdots,P)$ is the k-th speech missing signal, and the number of specimens of the linear prediction missing signal of each speech frame is P, the obtained standard deviation is:

$$e_{LPR}(k) = \sqrt{\frac{1}{P}\sum_{p=1}^{P} r_p^2(k) - \frac{1}{P}\left[\sum_{p=1}^{P} r_p(k)\right]^2} \tag{10}$$

The linear prediction residual skewness of the k-th speech frame is:

$$Y_{LPR}(k) = \frac{1}{P}\sum_{p=1}^{P}\left[\frac{r_p(k) - \frac{1}{P}\sum_{p=1}^{P} r_p(k)}{e_{LPR}(k)}\right]^3 \tag{11}$$

The average calculation of $Y_{LPR}(k)$ of all speech frames is performed to obtain the linear prediction missing uniform offset of all speech signals. The absolute value is:

$$f_6 = \left|\frac{1}{N}\sum_{k=1}^{k} Y_{LPR}(k)\right| \tag{12}$$

2) The uniform cepstrum offset is characterized by f_7.
   The k-th speech frame is calculated with cepstral coefficients to obtain $cm(k)$, $(m = 1,2, \cdots, M)$, the number of cepstral coefficients being characterized as M, and the standard deviation of $cm(k)$ is:

$$e_{cep}(k) = \sqrt{\frac{1}{M}\left[\sum_{m=1}^{M} cm^2(k) - \frac{1}{M}\left[\sum_{m=1}^{M} cm(k)\right]^2\right]} \tag{13}$$

The cepstrum offset degree of this speech frame is:

$$Y_{cep}(k) = \frac{1}{M}\sum_{m=1}^{M}\left[\frac{cm(k) - \frac{1}{M}\left[\sum_{m=1}^{M} cm(k)\right]}{e_{cep}(k)}\right]^3 \tag{14}$$

$Y_{cep}(k)$ is a complete speech signal, and the average value of the uniform cepstrum offset is:

$$f_7 = \frac{1}{K}\sum_{k=1}^{k} Y_{cep}(k) \tag{15}$$

The key to bisecting cepstrum offset is to know whether the speech is clear and the distortion level before implementing speech frame management.

3 System Construction and Test

3.1 System Construction

As shown in Fig. 2, the virtual teacher system is mainly composed of five modules: perception module, cognitive module, intelligent reasoning and decision-making module, emotion module and planning integration module. The perception module perceives and obtains the information in the learning environment, the cognitive module processes and decomposes the information, and then transmits different information to the intelligent reasoning decision module and the emotion module for decision-making respectively. Finally, the planning integration module integrates the decision and represents it in the presentation layer.

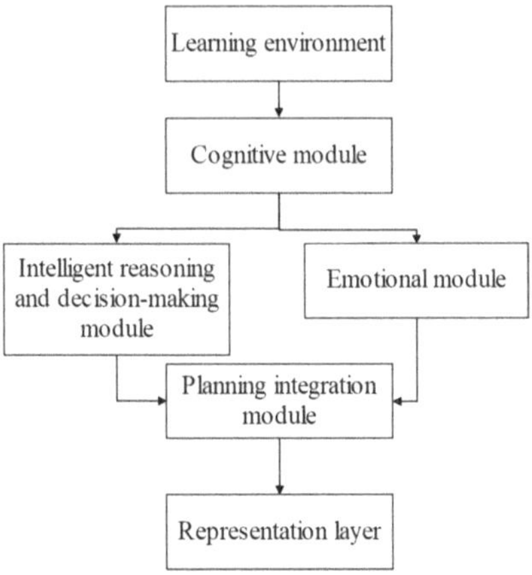

Fig. 2. Overall design framework of virtual teacher system

According to the characteristics of virtual teacher system, the behaviors of virtual teachers mainly include two categories. The first category is simple interactive behaviors, such as saying hello to learners, presenting teaching materials, giving lectures and so on. These behaviors do not need to go through high-level reasoning decisions, and teachers can directly execute them. Moreover, they have all been stored in the memory, and only the cognitive module needs to identify and trigger low-level information. The second category is advanced interactive behavior, which mainly requires virtual teachers to make intelligent reasoning decisions to answer questions. According to the characteristics and requirements of virtual teacher system, virtual teacher reasoning and decision-making system based on virtual agent is mainly composed of four modules: decision management module, qualitative reasoning and decision-making module, quantitative reasoning and decision-making module and multi-decision fusion module, as shown in Fig. 3.

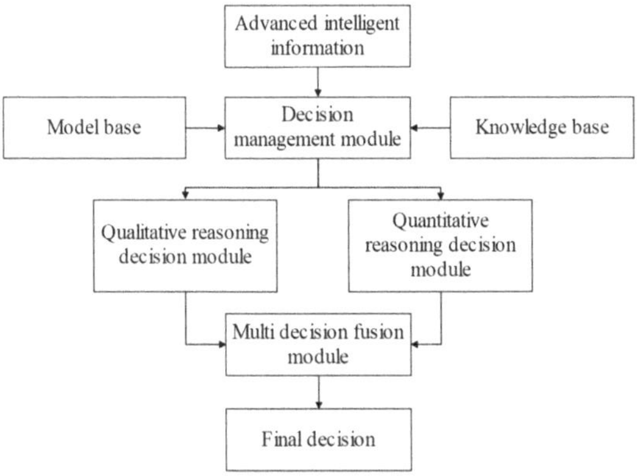

Fig. 3. Workflow of intelligent reasoning decision-making module of virtual teacher system

The intelligent reasoning flow of virtual teacher system based on case-based reasoning is shown below. After the qualitative information of the cognitive layer is transmitted, it is retrieved in the case base. After a plurality of similar cases are retrieved, the weight of each similar case is calculated, then the case with the largest weight value is taken out, the case is rewritten into a case matching the current information by using the knowledge and rules in the knowledge base, and then the final decision result is output, and the case is re-stored in the case base, as shown in Fig. 4.

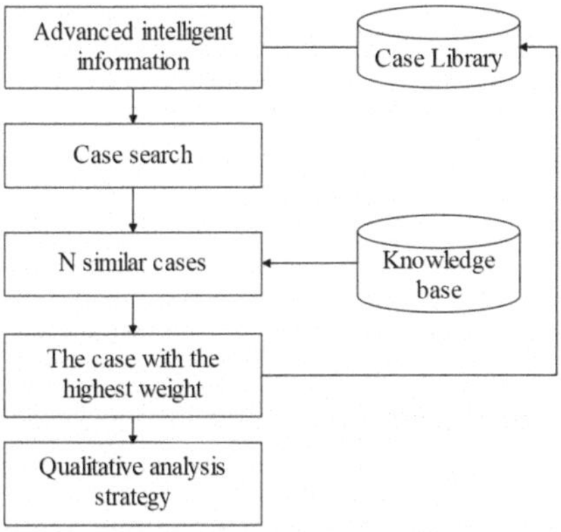

Fig. 4. Intelligent reasoning process of English virtual teacher system based on case-based reasoning

The English virtual teacher guidance system mainly includes examination system, vocabulary subsystem and grammar subsystem. The overall composition of the virtual teacher guidance system is shown in Fig. 5.

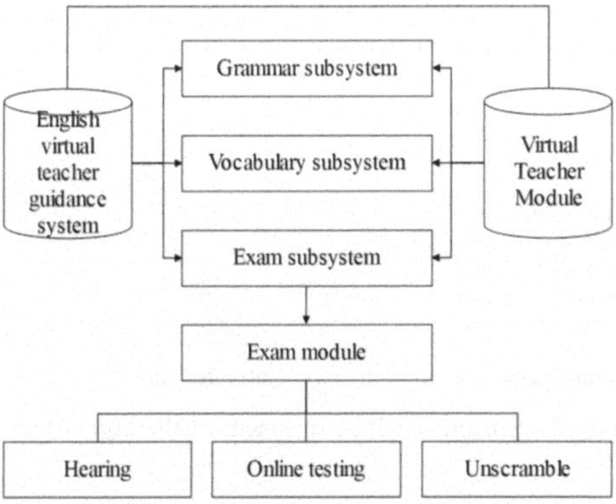

Fig. 5. Composition of English virtual teacher guidance system

The framework of virtual teachers in this study is shown in Fig. 5. Virtual teachers are not an individual with integrated functions, but a group of virtual teachers composed of multiple individuals. This study focuses on decomposing a large problem situation into multiple small problem situations, and each virtual teacher performs its own duties in different problem situations (Fig. 6).

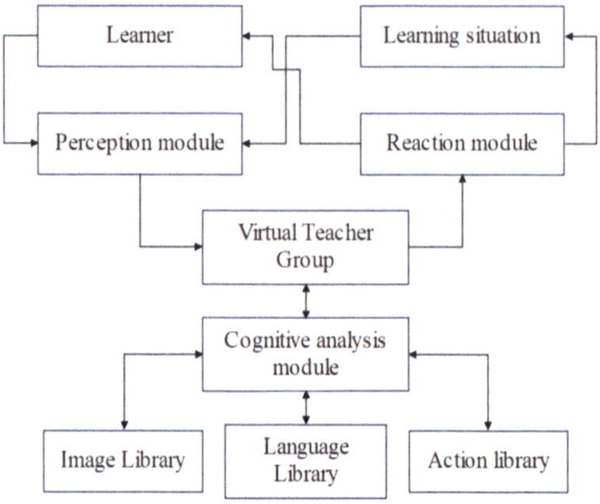

Fig. 6. Framework of English virtual teachers.

The method herein involves a total of 12 action units, including: AU1, AU2, AU4, AU5, AU6, AU9, AU12, AU15, AU17, AU20, AU25, and AU26. Action unit sequence recognition is implemented based on Darknet-53, which consists of multiple convolution units and residual units. The convolution unit is composed of Con2d, batch normalization and LeakyReLU activation function, and the residual unit is composed of two convolution layers with convolution kernels of 1×1 and 3×3 respectively. This structure will make the network structure deeper. In addition, different combinations of convolutional units and residual units require a total of 53 convolutional layers. Among them, the connection of the convolution unit and the residual unit and their network framework are shown in Fig. 7.

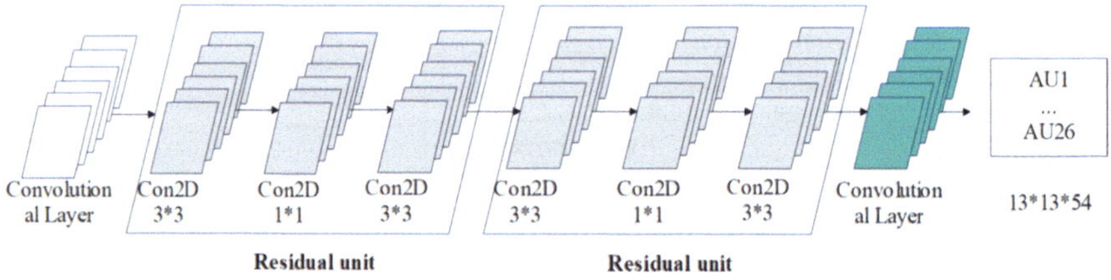

Fig. 7. Structure block diagram of the algorithm.

3.2 System Testing

The emotion expression method in this paper is verified based on OpenSim platform. OpenSim is a distributed three-dimensional virtual environment that follows the BSD open source protocol. Its function is similar to SecondLife, and it provides high scalability and secondary development features based on script technology.

In order to verify the practical effect of the English virtual teacher system based on neural network proposed in this paper, this paper verifies the system on the simulation platform, allows students to use the system proposed in this paper at the intelligent terminal, to judge the experience of using the system through the 100-point evaluation method, and asks professional teachers to evaluate the teaching effect of this model, and obtains the results shown in Tables 1 and 2.

Table 1. Evaluation of experience effect of the system

Number	Experimental Experience	Number	Experimental Experience	Number	Experimental Experience
1	78.07	9	81.70	17	86.93
2	86.38	10	83.66	18	87.53
3	84.31	11	87.08	19	86.93
4	87.07	12	85.44	20	78.30
5	82.55	13	84.76	21	79.70
6	82.23	14	80.54	22	82.58
7	87.73	15	83.91	23	81.73
8	83.64	16	78.64	24	84.02

Table 2. Evaluation of teaching effect of the system

Number	Teaching evaluation	Number	Teaching evaluation	Number	Teaching evaluation
1	84.86	9	81.94	17	88.01
2	83.47	10	86.29	18	81.30
3	86.69	11	87.54	19	83.20
4	81.98	12	84.23	20	80.18
5	83.05	13	82.48	21	88.93
6	88.84	14	87.95	22	80.43
7	80.47	15	80.86	23	87.51
8	85.32	16	84.13	24	82.42

From the experimental evaluation, the user experience is relatively good, and the teaching effect of the system has been affirmed by many teacher evaluators, so the model proposed in this paper has certain effects.

4 Conclusion

In order to improve the online learning efficiency of learners, this paper draws lessons from the related knowledge and latest research results in virtual reality technology, virtual human technology, intelligent reasoning and decision-making technology, psychology and other fields, and constructs a virtual teacher as the instructor and collaborator of learners based on virtual agent technology. Virtual teachers have human thinking and decision-making ability, and can interact with learners in the form of actions, expressions and language. It aims to reduce learners' loneliness in online learning and promote learners' sense of participation. This paper focuses on the core module of virtual teacher, that is, intelligent reasoning and decision-making module. Moreover, this paper tries to make an improvement suitable for virtual teachers' reasoning decision-making in the intelligent decision-making system based on artificial intelligence expert system, and adopts the method of case-based reasoning to make reasoning decision-making.

Acknowledgments. This paper is supported by: The phased results of the project "Research on the Cultivation of Multi-language + Business Compound foreign Language Talents Serving Hunan China-Africa Economic and Trade Industrial Park" by the Social Science Achievement Evaluation Committee of Hunan Province in 2024 (No.: XSP24YBC364).

References

1. Wilcha, R.J.: Effectiveness of virtual medical teaching during the COVID-19 crisis: systematic review. JMIR Med. Educ. **6**(2), e20963–e20974 (2020)
2. González-Calvo, G., Barba-Martín, R.A., Bores-García, D., Hortigüela-Alcalá, D.: The (virtual) teaching of physical education in times of pandemic. Eur. Phys. Educ. Rev. **28**(1), 205–224 (2022)
3. Anwar, K.: The perception of using technology canva application as a media for English teacher creating media virtual teaching and English learning in Loei Thailand. J. Engl. Teach. Lit. Appl. Linguist. **5**(1), 62–69 (2021)
4. Essel, H.B., Vlachopoulos, D., Tachie-Menson, A., Johnson, E.E., Baah, P.K.: The impact of a virtual teaching assistant (chatbot) on students' learning in Ghanaian higher education. Int. J. Educ. Technol. High. Educ. **19**(1), 57–66 (2022)
5. Shahzad, S.K., Hussain, J., Sadaf, N., Sarwat, S., Ghani, U., Saleem, R.: Impact of virtual teaching on ESL learners' attitudes under COVID-19 circumstances at post graduate level in Pakistan. Engl. Lang. Teach. **13**(9), 1–9 (2020)
6. Keefe, E.S.: Learning to practice digitally: Advancing preservice teachers' preparation via virtual teaching and coaching. J. Technol. Teach. Educ. **28**(2), 223–232 (2020)
7. Dung, D.T.H.: The advantages and disadvantages of virtual learning. IOSR J. Res. Method Educ. **10**(3), 45–48 (2020)
8. Manegre, M., Sabiri, K.A.: Online language learning using virtual classrooms: an analysis of teacher perceptions. Comput. Assist. Lang. Learn. **35**(5–6), 973–988 (2022)
9. Zhang, Y., Lin, C.H.: Student interaction and the role of the teacher in a state virtual high school: what predicts online learning satisfaction? Technol. Pedagog. Educ. **29**(1), 57–71 (2020)
10. McGarr, O.: The use of virtual simulations in teacher education to develop pre-service teachers' behaviour and classroom management skills: implications for reflective practice. J. Educ. Teach. **47**(2), 274–286 (2021)

Integrated Application of Augmented Reality and Artificial Intelligence into English Learning

Qin Zhang(✉)

Guangdong University of Science and Technology,
Dongguan 523000, Guangdong, People's Republic of China
225701375@qq.com

Abstract. This study investigated views of English major undergraduates in Guangdong, China towards integrated application of AR and AI technologies into English learning. The aim of the study is to explore those students' learning needs when gradually adopting AR and AI provided learning environment, and their views towards language, contents, design and functions provided by AR and AI supporting learning platforms. The study used a mere quantitative research method by an adopted and modified questionnaire. The respondents of the study include 308 undergraduate students of English major in Guangdong province. Data collected were processed and analyzed by SPSS statistic processing software. The findings indicated that the measurements used to measure respondents' views towards AR and AI integrated application were reliable. From the findings, it could be noted that the respondents held positive attitude towards integrated application of AR and AI in English language learning.

Keywords: Integrated application · AR · AI · English Learning

1 Introduction

As technology continues to progress, the education section is undergoing an overall revolution, especially in the field of language learning. Traditional learning methods are gradually being supplemented or even replaced by innovative technologies. Augmented reality (AR) and artificial intelligence (AI) have begun to show great potential in English language learning. These innovative technologies provide an interactive, personalized and immersive experience that improves the engagement and performance of English major students in learning English [1]. The use of AR and AI in English learning makes the whole learning process more attractive and greatly increases students' interest and participation. By superimposing virtual information in the real world, AR provides learners with a highly interactive and deeply immersive learning environment while AI provides personalized learning paths and real-time feedback through its adaptive learning ability. AR and AI bring loads of benefits from improved motivation and engagement to personalized learning paths and real-time feedback. These technologies provide English learning with opportunities to learn in real, content-rich settings that support a more natural and efficient acquisition of language skills [2].

© The Author(s), under exclusive license to Springer Nature Switzerland AG 2025
F. Paas et al. (Eds.): EduTA 2024, LAIS 51, pp. 239–251, 2025.
https://doi.org/10.1007/978-3-031-95252-4_22

Although the application of AR and AI in education is not a new initiative, their integrated application in English language learning is still a relatively new research area. This integration can not only improve students' English learning motivation and engagement, but also provide tailored learning experience in accordance with their specific needs and abilities.

Integrating AI and AR technologies in education has shown many advantages in improving the entire learning process and experience, and an interactive and immersive learning environment provided is one of the main benefits [3]. On one hand, AR superimposes digital data in real world setting, making students to interact with teaching materials in a more physical and practical way. On the other hand, AI system can be used to analyze students' performance and behaviors, and make customized teaching materials and pace to meet individual learners' needs. Therefore, it provides a more personalized and effective learning environment [4]. Usually, AR is applied in language learning to build real and rich settings for vocabulary expansion and even comprehensive usage of the language. For example, AR can add English labels on the actual objects so that students can explore their environment while simultaneously learning new words in the contexts [5]. This can not only facilitates memory, but also helps students to link new terms to their real-world references, thus enabling more natural language learning. AI driven English learning systems such as chat robots and virtual language partners allow students to practice English conversational skills in a less stressed environment. These AI systems can provide interactions and pronunciation feedback, and at the same time, adjust the grade of language difficulty based on the students' abilities [6]. This forever available language partners contribute to more frequent practice and learning confidence. Furthermore, AI systems can evaluate students' performance, spot the difficult points and customize learning activities accordingly. This ensures that the students focus on the most challenging and difficult parts of language learning, thus improving their learning efficiency and effectiveness [7].

Despite the significant theoretical advantages of the integrated application of AR and AI in English learning, there are still many challenges and difficulties in practical application. The students' acceptance, learning needs, technical means and so on are all issues that need to be discussed in depth. Based on the above discussion, the objective of this study is to investigate the English major students' perceptions on integrating AR and AI technologies to learn English and to investigate their preferences concerning instruction language used, presentation of course content and multimedia content, interface design as well as additional function provided.

2 Literature Review

The integrated application of augmented reality (AR) and artificial intelligence (AI) in language learning has become a reformative approach in educational technology. This section conducts a brief review on current researches to explore the potential, application, and challenges of using AR and AI in language learning.

2.1 AR Technology in Language Learning

AR technology stacks digital information into users' real-world view, and it has been used to enhance language learning experience. Weerasinghe et al. [8] showed that AR can provide a contextualized learning experience, allowing learners to learn vocabulary and grammatical structures in simulated real scenarios. This immersive learning approach contributes not only to memory, but also to a deeper understanding of the language in cultural and social contexts. Kazu and Kuvvetli [9] highlighted the potential of AR to reduce cognitive load and increase learning willingness, making it a promising tool for vocabulary acquisition and pronunciation practice.

2.2 AI Technology in Language Learning

AI technology has the ability to adapt to individual learners' needs, and it has been extensively studied for its role in personalized language learning. As Kochmar et al. [10] used natural language processing and machine learning to provide personalized feedback, adaptive exercises, and dynamic dialogue exercises. AI can be used to provide immediate responses, enabling learners to continuously practice and gain support customized to their specific demands. Karataş et al. [11] noted that AI is highly recognized in language learning platforms for its ability to improve learners' initiatives and provide interventions at an early stage.

2.3 Integration of AR and AI in Language Learning

Integration of AR and AI technologies has synergistic effects in language learning. For example, AR can provide an immersive learning environment, while AI algorithms can adjust the complexity of interactions in accordance with learners' proficiency, as suggested by Almelhes [12]. This integrated application has been shown to highly increase learners' engagement and motivation, since it enables the learners to visualize and interact with the language in a more tangible and meaningful way. The potential of this integration to provide real-time, contextualized language practices is a significant progress forward in the creation of effective and enjoyable language learning experiences [13].

Despite all of these benefits and advantages, researches mentioned that there were still many challenges to implement integrated application of AR and AI in language learning, including high initial cost of technology development and continuous maintenance, digital gap that may excludes exposure to learners with limited access to the technologies, learners' acceptance, view and perceptions towards the presentation of this integration, etc. [14]. In addition, there are also pedagogical concerns regarding the balance between technology-mediated learning and traditional face-to-face interactions. In conclusion, researches on AR and AI application in English language learning suggest positive trends, and more and more increasing evidence supports their effectiveness in improving English learning experiences. Yet with the advancement of these technologies, further studies need to be done to tackle with these newly emerging challenges, especially the acceptance and views of the learners, who are the ultimate users of these application.

3 Methodology

3.1 Method and Instrument

This study utilized a mere quantitative research method [15], employing an adopted and modified survey questionnaire created with sojump mini program. The questionnaire comprised six sections and 22 measurement items in total. Section A is the demographics; Section B is instruction language; Section C is course content presentation; Section D is multimedia content presentation; Section E is interface design; Section F is additional functions. Items in Section B to Section F were rated on a five-point Likert scale, to evaluate students' views towards integration of AR and AI in English learning concerning instruction language, course content presentation, multimedia content presentation, interface design and additional functions, using frequency and means as shown below (Table 1).

Reliability test was conducted to test the internal stability and consistency of the students' views towards integrated application of AR and AI technologies in English learning. The Cronbach's Alpha was calculated by SPSS to test the reliability. And the results are shown in Table 2. The Cronbach's Alpha for instruction language is 0.892, course content presentation is 0.853, multimedia content presentation is 0.919, interface design is 0.874 while additional functions is 0.872. And the overall Cronbach's Alpha is 0.882. The coefficients are all higher that 0.8, indicating a high level of reliability [16], indicating that the instrument adopted has acceptable internal consistency. Therefore, the results can demonstrate a high level of reliability in measuring the students' views on integrated application of AR and AI technologies into English learning.

Table 1. Verbal interpretation of the scale.

Scale	Range	Verbal Interpretation
5	4.21–5.00	Strongly Agree (SA)
4	3.41–4.20	Agree (A)
3	2.61–3.40	Uncertainty (U)
2	1.81–2.60	Disagree (D)
1	1.00–1.80	Strongly Disagree (SD)

This study also conducted the Kaiser-Meyer-Olkin (KMO) measure of sampling adequacy and Bartlett's test of sphericity to test suitability for factor analysis of the data collected. KMO value indicates the proportion of variance in variables that might be caused by underlying factors. High values (close to 1.0) generally indicate a high suitability for factor analysis. The results of the study are present in Table 3. The KMO value is 0.879, far over that suggesting benchmark of 0.6, indicating a high adequacy for factor analysis of the data gathered [17]. Moreover, Bartlett's test of sphericity further validated it by producing an approximate chi-square value of 363.848 with 21 degrees of freedom, and the significant p-value of 0.000 ($p < 0.05$). All of these indicate that the

Table 2. Reliability of items.

	Variable	No. of Items	Cronbach's α
Students' views on integrated application of AR and AI into English Learning	Instruction Language	3	0.892
	Course Content Presentation	6	0.853
	Multimedia Contents presentation	6	0.919
	Interface Design	4	0.874
	Additional Functions	3	0.872
		Overall	0.882

correlations among the variables are great enough to justify the use of factor analysis [18]. That is to say, the questionnaire adopted by the study have passed the validity test.

Table 3. KMO value and Bartlett test

Kaiser-Mever-Olkin Measure		0.879
Bartlett Sphericity Test	Approximate Chi-square Value	363.848
	Df (Degree of Freedom)	21
	Sig (Significance Level)	0.000

3.2 Sampling

The sample for this study was chosen using purposive sampling, comprising undergraduate students of English major in Guangdong Province, China. In the end, 308 students participated as respondents in the study.

From Table 4, it could be noted that female respondents largely outnumbers male respondents, with 236 females (76.6%) and 72 males (23.4), indicating a common phenomenon in enrollments of English major. In almost all colleges of foreign languages, the number of female enrollments is much higher than that of the male. And the study invites all respondents from the undergraduate level, so all respondents are from year 1 to year 4. It is known from the table that the number of year 3 students is the largest with 102 students (33.1%), followed by year 2 (30.2%) and year 1(27.6%). With age as the reference, the majority of the students age from 18 to 21, accounting for 88%. The study divides the respondents into five categories concerning their English proficiency. The intermediate level students are the most, with 102 students (33.1%) while the proficient level students are the least with 44, accounting only for 14.3%.

3.3 Data Gathering and Processing

The data were gathered through sojump mini program, distributing to the students via social media tools such as wechat and QQ, with help of teachers working in universities

Table 4. Demographic description

Category	Items	Frequency	Percentage (%)
Gender	Male	72	23.4
	Female	236	76.6
Grade	Year 1	85	27.6
	Year 2	93	30.2
	Year 3	102	33.1
	Year 4	28	9.1
Age	18–19	126	40.9
	20–21	145	47.1
	22–23	32	10.4
	Above 23	5	1.6
English Proficiency	Beginner	15	4.9
	Elementary	89	28.9
	Intermediate	102	33.1
	Advanced	58	18.8
	Proficient	44	14.3

in Guangdong. Students that have the willingness to participate in the study were invited to fill out the form. Data gathered were organized, analyzed and processed using the software Microsoft EXCEL and SPSS. Data processing involved the interpretation of the answers to the questionnaires to provide statistical results, which included encoding, data input, editing, checking, and monitoring the overall data processing procedures and protocols being observed. Percentage, frequency, coefficient α, KMO value, mean as well as standard deviation were used as statistical treatments in the study.

4 Results and Discussion

This section presents, analyzes and interprets the data collected from 308 participants concerning their views on integrated application of AR and AI in English learning in aspects of instruction language, course content presentation, multimedia content presentation, interface design and additional functions.

Table 5 shows the students' views on usage of instruction language in the integrated application of AR and AI in English learning. With reference to the complexity of instruction language used, 268 students (87.01%) held that view that the language used should be easy to understand to facilitating students' willingness to use it, while 16 students (5.2%) held the opposite view that it was not necessary to purposefully use easy instruction language. Concerning whether Chinese translation should be provided, 209 students (67.86%) deemed it necessary whereas 22 students (7.14%) believed it not necessary. For language used for titles, the majority of participants (270 students,

87.67%) agreed that language used for titles should be attractive and eye-catching, and a few of them (9 students, 2.95%) disagreed with it.

According to the data gathered, it is known that the mean of the results ranges from 3.86 to 4.27, and the standard deviation from 0.784 to 0.933, indicating that the instruction language used in AR and AI integrated application into English learning is very important to the students. The highest mean value (mean 4.27, SD 0.906) falls in the item "AR-AI" instruction language should be easy to understand", and interpreted as "strongly agree", while the lowest mean value (mean 3.86, SD 0.933) fall in the item "Translated Chinese texts should be provided", and interpreted as "agree". This shows that when designing AR-AI integrated English learning platforms, the easiness of the instruction language is extremely important for the students, just as AlTwijri & Alghizzi [19] suggested.

Table 5. Students' views on instruction language

Item	SD	D	U	A	SA	Mean	SD	VP
AR-AI instruction language should be easy to understand	8 (2.60%)	8 (2.60%)	24 (7.79%)	122 (39.61%)	146 (47.40%)	4.27	0.906	SA
Translated Chinese texts should be provided	5 (1.62%)	17 (5.52%)	77 (25.00%)	126 (40.91%)	83 (26.95%)	3.86	0.933	A
Languages used for all titles should be eye-catching	3 (0.97%)	6 (1.94%)	29 (9.42%)	141 (45.78%)	129 (41.89%)	4.26	0.784	SA

Table 6 shows the students' views on the presentation way of the course content in AR-AI integrated English learning platforms. From the results, it can be noted that the mean of the results ranges from 4.11 to 4.35, and the standard deviation from 0.809 to 0.849, indicating that almost all the students pay highly close attention to the presentation ways of the course content on the AR-AI integrated English learning platforms. The highest mean value is 4.35 (SD 0.819) followed by 4.22 (SD 0.809) and 4.21 (SD 0.837), all interpreted as "strongly agree". These fall in the items "various teaching materials should be provided to facilitate understanding of the course content", "course information (title, description, learning outcomes, evaluation, etc.) should be clearly stated", and "subtopics should be arranged in a logic and orderly manner" with 275 students (89.29%), 272 students (88.32%), and 265 students (86.04%) respectively supported the perception mentioned above. The results echoed with what found by Xu, Wang, & Liu [20]. All these indicate that when selecting AR-AI integrated English learning platforms, the students are more apt to choose those with logical and orderly arrangement of the course content, access to various teaching materials and automatic assessment and feedback.

Table 6. Students' views on course content presentation

Item	SD	D	U	A	SA	Mean	SD	VP
Course information (title, description, learning outcomes, evaluation, etc.) should be clearly stated	4 (1.29%)	9 (2.92%)	23 (7.47%)	151 (49.03%)	121 (39.29%)	4.22	0.809	SA
Expected learning outcomes of each chapter should be clearly stated	5 (1.62%)	6 (1.95%)	31 (10.06%)	149 (48.38%)	117 (37.99%)	4.19	0.819	A
Learning contents should be arranged orderly from easy to difficult	4 (1.29%)	6 (1.95%)	40 (13.00%)	150 (48.70%)	108 (35.06%)	4.14	0.811	A
Auto-assessment, feedback and comments should be provided for all exercises, interactive games and quizzes	5 (1.62%)	10 (3.25%)	36 (11.69%)	153 (49.67%)	104 (33.77%)	4.11	0.849	A
Various teaching materials should be provided to facilitate understanding of the course content	5 (1.62%)	5 (1.62%)	23 (7.47%)	119 (38.64%)	156 (50.65%)	4.35	0.819	SA
Subtopics should be arranged in a logic and orderly manner	5 (1.62%)	7 (2.27%)	31 (10.07%)	140 (45.46%)	125 (40.58%)	4.21	0.837	SA

Table 7 shows the presentation ways of multimedia contents in AR-AI integrated English learning platforms. From the results revealed in the table, the mean varies from 4.01 to 4.33, with the standard deviation from 0.728 to 0.869. The highest mean value is 4.33 (SD 0.728), followed by 4.31 (SD 0.769) and 4.32 (SD 0.774), all interpreted as "strongly agree". And they fall in the items "audios and animated videos should improve learning understanding of the contents", "animated videos should make learning activities more interesting" and "audios and animated videos provided should be high-quality", with 276 students (89.61%), 272 students (88.31%) and 273 students (88.63%) respectively held the view that audios and animated videos provided by the platforms are very important, and they should be of high quality while facilitate students' participation and understanding of content, just as mentioned by Ahmadi [21]. This means that when integrating AR-AI technologies into English learning platforms, the multimedia content presentation ways such as texts, virtual graphics, audio and videos should be seriously taken into consideration.

Table 7. Students' views on multimedia content presentation

Item	SD	D	U	A	SA	Mean	SD	VP
The texts should be reader-friendly	6 (1.95%)	12 (3.90%)	43 (13.95%)	160 (51.95%)	87 (28.25%)	4.01	0.869	A
Virtual graphic/images should be suitable for the topics	3 (0.97%)	9 (2.92%)	26 (8.45%)	137 (44.49%)	133 (43.17%)	4.26	0.810	SA
Virtual graphic/images should help learners to better understand the contents	2 (0.65%)	8 (2.60%)	26 (8.45%)	136 (44.15%)	136 (44.15%)	4.29	0.780	SA
Audios and animated videos provided should be high-quality	2 (0.65%)	7 (2.27%)	26 (8.45%)	132 (42.85%)	141 (45.78%)	4.31	0.774	SA
Animated videos should make learning activities more interesting	2 (0.65%)	6 (1.95%)	28 (9.09%)	131 (42.53%)	141 (45.78%)	4.31	0.769	SA

(continued)

Table 7. (*continued*)

Item	SD	D	U	A	SA	Mean	SD	VP
Audios and animated videos should improve learning understanding of the contents	1 (0.32%)	5 (1.62%)	26 (8.45%)	134 (43.51%)	142 (46.10%)	4.33	0.728	SA

Table 8 reveals the students' views and perceptions towards design of the interface of AR-AI integrated English learning platforms. From the results, it can be known that the mean value ranges from 4.21 to 4.33, and the standard deviation from 0.763 to 0.880, all interpreted as "strongly agree". The highest mean value is 4.33 (SD 0.771, followed by 4.31 (SD 0.807). They are the items "design of the layout should be user-friendly" and "design of the layout should be simplified", with 281 students (91.21%) and 278 students (90.25%) correspondingly supported the above-mentioned views. The data shows that the majority of the students held a positive attitude towards simplified and user-friendly interface layout, which could be more attractive to them concerning English learning platforms that integrate the application of both AR and AI technologies, producing the same conclusion as Chen [22].

Table 8. Students' views on interface design

Item	SD	D	U	A	SA	Mean	SD	VP
Design of the layout should be simplified	5 (1.62%)	6 (1.95%)	19 (61.68%)	136 (44.15%)	142 (46.10%)	4.31	0.807	SA
Design of the layout should be user-friendly	4 (1.29%)	5 (1.62%)	18 (5.85%)	138 (44.81%)	143 (46.43%)	4.33	0.771	SA
Navigation menu should help learners to access corresponding contents easily	4 (1.29%)	11 (3.57%)	35 (11.37%)	120 (38.96%)	138 (44.81%)	4.22	0.880	SA
Course content pages should be well organized	6 (1.95%)	8 (2.60%)	29 (9.41%)	138 (44.81%)	127 (41.23%)	4.21	0.863	SA

Table 9 shows the data collected from the participants concerning additional function provided by the AR-AI integrated English learning platforms. As revealed by the results, the mean value ranges from 4.21 to 4.30, with standard deviation from 0.754 to 0.862. The highest mean value (mean 4.30, SD 0.754) falls in the item "users' previous records should be saved and traced", interpreted as "strongly agree". 273 students (88.64%) agreed that these learning platforms should keep the users' using records to make it more convenient for them when conducting English learning activities through the platforms, similar with the opinions of Gligorea et al. [23]. And the data also shows that when designing AR-AI integrated English learning platforms, additional functions that are user-friendly and accessible, such as discussion space and personalized learning strategies, should be provided to improve students' convenience, interest and learning efficiency.

Table 9. Students' views on additional functions

Item	SD	D	U	A	SA	Mean	SD	VP
The AR-AI supported forum should provide discussion space for the learners	3 (0.97%)	12 (3.90%)	22 (7.14%)	152 (49.35%)	119 (38.64%)	4.21	0.813	SA
Users' previous records should be saved and traced	2 (0.65%)	5 (1.62%)	28 (9.09%)	137 (44.49%)	136 (44.15%)	4.30	0.754	SA
Personalized learning strategies should be analyzed and provided	4 (1.29%)	10 (3.25%)	31 (10.07%)	122 (39.61%)	141 (45.78%)	4.25	0.862	SA

5 Conclusion

This study investigated the students' views and feedback towards English learning platforms integrating AR and AI technologies, concerning the instruction language, course content presentation, multimedia content presentation, interface design and additional function provided. The findings of the study show that most of the students hold a positive attitude toward the integrated application of AR and AI in English learning, and that their preferences on the presentation of these AR and AI integrated English learning platforms.

As for the instruction language, a majority of the students support that it should be easy to understand and attractive, and that Chinese translated would be preferred; For the course content presentation, most of the students deem it of great necessity to state all course related information clearly and logically from easy to difficult, and to provide various and rich teaching materials. For multimedia content presentation, virtual graphics, pictures, audios and animated videos of high quality should be provided to promote students' understanding of the topic and make learning activities more attractive. For interface design, almost all of the students highly support that the design should be simplified and user-friendly while navigation menu able to link to corresponding contents should be provided. And at last, most of the students highlight the importance of additional and special functions of the platforms, such as discussion space and records tracing, that encourage engagement and accessibility of their English learning.

The findings revealed by the study can provide as supplementary instruction to the designers when developing and designing English learning platforms that integrating AR and AI technologies.

References

1. Kurni, M., Mohammed, M.S., Srinivasa, K.G.: A Beginner's Guide to Introduce Artificial Intelligence in Teaching and Learning. Springer, Cham (2023)
2. Xiong, W.: Artificial intelligence in English language education: the path to personalized learning. In: Ma, K., Shi, E., Wang, X. Zhu, W. (eds.) Proceedings of the 2nd International Conference on Educational Knowledge and Informatization, EKI2024, pp. 72–77. ACM, New York (2024)
3. Lampropoulos, G.: Augmented Reality and Artificial Intelligence. Springer Nature Switzerland, Bern (2023)
4. Tahiru, F.: AI in education: a systematic literature review. J. Cases Inf. Technol. **23**(1), 1–20 (2021)
5. Wei, X., Sun, J.: Literature review of augmented reality in second language learning. E-Educ. Res. **2021**(02), 99–107+114 (2021)
6. Zhang, K., Aslan, A.B.: Artificial intelligence in education: a systematic literature review. Comput. Educ. **252**, 124–167 (2024)
7. Xia, Q., Chiu, T.K.F., Zhou, X., Chai, C.S., Cheng, M.: Systematic literature review on opportunities, challenges, and future research recommendations of artificial intelligence in education. Comput. Educ. Artif. Intell. **22**(4), 110–118 (2022)
8. Weerasinghe, M., et al.: Vocabulary: learning vocabulary in AR supported by keyword visualisations. IEEE Trans. Vis. Comput. Graph. **28**(11), 3748–3758 (2022)
9. Kazu, İ.Y., Kuvvetli, M.: The influence of pronunciation education via artificial intelligence technology on vocabulary acquisition in learning English. Int. J. Psychol. Educ. Stud. **10**(2), 480–493 (2023)
10. Kochmar, E., Vu, D.D., Belfer, R., Gupta, V., Serban, I.V., Pineau, J.: Automated personalized feedback improves learning gains in an intelligent tutoring system. In: Bittencourt, I., Cukurova, M., Muldner, K., Luckin, R., Millán, E. (eds.) Artificial Intelligence in Education. AIED 2020. Lecture Notes in Computer Science, vol. 12164, pp. 140–146. Springer, Cham (2020). https://doi.org/10.1007/978-3-030-52240-7_26

11. Karataş, F., Abedi, F.Y., Ozek Gunyel, F., Karadeniz, D., Kuzgun, Y.: Incorporating AI in foreign language education: an investigation into ChatGPT's effect on foreign language learners. Educ. Inf. Technol. **29**(6), 34–39 (2024)
12. Almelhes, S.A.: Application of generative artificial intelligence (GenAI) in language teaching and learning: a scoping literature review. Comput. Educ. Open **6**(174), 101–106 (2023)
13. Wen, Y.: Augmented reality enhanced cognitive engagement: designing classroom-based collaborative learning activities for young language learners. Education Tech. Research Dev. **69**(2), 843–860 (2020)
14. Fan, L., Fichman, P.: The digital divide in online education: inequality in digital readiness of students and schools. Comput. Educ. **3**, 100100–100112 (2022)
15. Creswell, J.W.: Research Design: Qualitative, Quantitative, and Mixed Methods Approaches, 2nd edn. Sage Publications, London (2003)
16. Taber, K.S.: The use of cronbach's alpha when developing and reporting research instruments in science education. Res. Sci. Educ. **48**(2), 1273–1296 (2018)
17. Kaiser, H.F.: An Index of Factorial Simplicity. Psychometrika **39**(1), 31–36 (1974)
18. Bartlett, M.S.: A note on the multiplying factors for various $\chi 2$ approximations. J. Roy. Stat. Soc.: Ser. B (Methodol.) **16**(2), 296–298 (1954)
19. AlTwijri, L.A., Alghizzi, T.M.: Investigating the integration of artificial intelligence in English as foreign language classes for enhancing learners's effective factors: a systematic review. Heliyon **10**(10), e31053–e31075 (2024)
20. Xu, T., Wang, H., Liu, Y.: The effectiveness of artificial intelligence on English language learning achievement. Comput. Educ. **125**, 103428–103440 (2024)
21. Ahmadi, D.M.R.: The use of technology in English language learning: a literature review. Int. J. Res. Engl. Educ. **3**(2), 115–125 (2018)
22. Chen, Y.L.: Students' attitude toward learning and practicing English in a VR environment. In: Huang, TC., Wu, TT., Barroso, J., Sandnes, F.E., Martins, P., Huang, YM. (eds.) Innovative Technologies and Learning. ICITL 2020. Lecture Notes in Computer Science, vol. 12555, pp. 128–136. Springer, Cham (2020). https://doi.org/10.1007/978-3-030-63885-6_15
23. Gligorea, I., Cioca, M., Oancea, R., Gorski, A.T., Gorski, H., Tudorache, P.: Adaptive learning using artificial intelligence in e-Learning: a literature review. Educ. Inf. Technol. **13**(12), 12–16 (2023)

Research and Application of Online Learning Platform Resource Recommendation Based on Personalization

Xiaoyan Zhang(✉) and Zhuoxiang Liu

College of Forestry Science and Technology, Lishui Vocational and Technical College, Lishui 323000, China
zxy202410181@163.com

Abstract. To realize personalized and high-quality resource recommendation of online learning platform, this paper proposes a collaborative filtering algorithm based on RFM-K-means clustering. Firstly, this algorithm calculates user-product analysis matrix by RFM model and obtains weight of each index in RFM by entropy weight method. Then, K-means clustering is utilized to extract user features. In addition, based on user-based collaborative filtering method (User-based CF), resources are recommended. Finally, by constructing user-learning scoring matrix and calculating user similarity to generate nearest neighbor set, personalized resource recommendations can be achieved. The results show that based on the data source of an online learning platform, accuracy of the test questions with a prediction score of not less than 0.6 recommended by the proposed recommendation model to target users is 90%, showing high prediction accuracy. Thus, collaborative recommendation method constructed in this study can realize the personalized resource recommendation of online learning platform, which has certain application value.

Keywords: Personalization · Online Learning Platform · Resource Recommendation · Clustering Technology · Collaborative Filtering Algorithms

1 Introduction

At present, with the in-depth application of intelligent teaching, online learning platform is widely used. Through this platform, online education and learning can be carried out [1], thus providing users with a wider range of learning resources [2]. However, considering the different learning behaviors of different users, their demands for resources are also different [3]. Therefore, strengthening the precision of personalized resource recommendation for different users is of great significance to realize personalized and high-quality resource recommendation of online learning platform [4]. At present, traditional recommendations contain content-based recommendation, collaborative filtering recommendation, and hybrid recommendation [5]. In terms of recommending learning resources, the application of deep learning and machine learning has become increasingly widespread, but there are still certain limitations [6]. Customer value model (RFM

© The Author(s), under exclusive license to Springer Nature Switzerland AG 2025
F. Paas et al. (Eds.): EduTA 2024, LAIS 51, pp. 252–263, 2025.
https://doi.org/10.1007/978-3-031-95252-4_23

model) is a widely used marketing model, which is used to measure customer value and revenue generating ability [7]. Combined with clustering technology, it is an effective means to analyze customer data [8]. Therefore, this study uses RFM-K-means clustering technique [9] and User-based CF recommendation [10] to analyzed user behaviors and constructs a collaborative filtering recommendation algorithm based on RFM-K-means clustering. At the same time, the feasibility of this method is verified by experiments.

2 User Feature Extraction of Online Learning Platform Based on RFM-K-Means Clustering

2.1 K-means Clustering

K-means clustering is a cluster analysis algorithm of iterative solution. Its basic idea is this: Firstly, number of clusters K is presupposed. Then, each data point is associated with its nearest initial cluster center. Afterwards, cluster center point is recalculated, and continues to be bound with its nearest data point. The above steps are repeated until the location of cluster center does not change [11]. K-means clustering technique minimizes the sum of squared distances between all data points and their associated cluster centers through objective function, namely the squared error function. Its calculation formula is as follows:

$$J = \sum_{i=1}^{K} \sum_{x \in C_i} \|x - {\mu_i}^2\| \tag{1}$$

where: C_i represents the i th cluster; μ_i represents the center of C_i. K-means clustering technology algorithm is simple to implement, and it is easy to understand and operate. In addition, it can cluster data quickly and requires less computation, so it is suitable for the processing task of large-scale data sets [12].

2.2 Feature Extraction Based on RFM-K-Means Clustering

RFM model is a widely used model for user behavior data analysis, which can measure customer value and customer profitability. Among them, contents of the three indexes R, F and M are as follows: R represents last consumption time, F represents consumption frequency and M represents consumption amount [13]. This study combines RFM and K-means clustering to extract user features. The main ideas are as follows: Firstly, user-product analysis matrix is obtained by RFM, which is taken as data object. Then, RFM data is standardized and then weights of each index in RFM are obtained by entropy weight method. Finally, K-means clustering is adopted to extract user features. Figure 1 visualizes the specific process.

In Fig. 1, the number of clusters is set first. Since there are two possibilities between the three indexes R, F and M and their average values, clustering number is set as $K = 8$. Then, specific steps of clustering are as follows:

(1) Input user-product analysis matrix, where n product items are denoted as $I = \{i_1, i_2, ..., i_n\}$ and m users are denoted as $U = \{u_1, u_2, ..., u_m\}$;
(2) Randomly select q_1 to q_8 users as the initial clustering center.

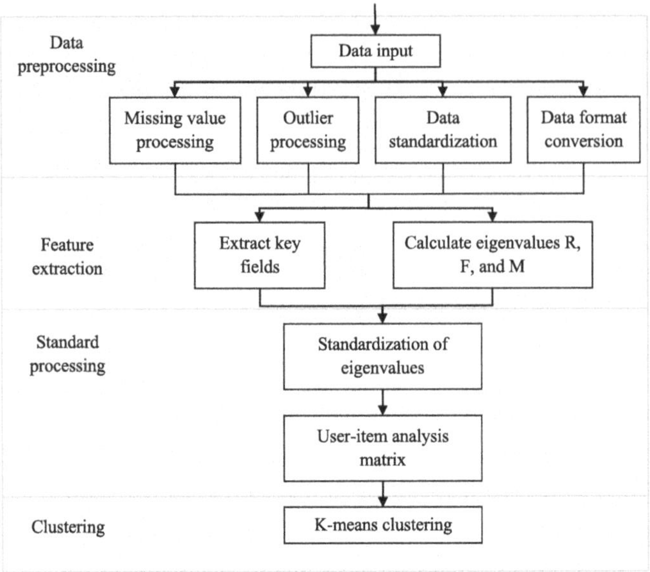

Fig. 1. User feature extraction based on RFM-K-means clustering.

(3) Calculate set I_{ij} used by user u_i and q_j, and calculate IRFM value of c resource purchased by user u_i and q_j. Then, use Eq. (2) to calculate the distance between the user u_i and its corresponding clustering center q_j;

$$d(u_i, q_i) = \sqrt{\sum_{c \in I_{ij}} (IRFM_{u_i,c} - IRFM_{q_j,c})} \tag{2}$$

(4) $d(ui, qj) = \max\{d(ui, q1), d(ui, q2), ..., d(ui, q8)\}$, which means that user u_i belongs to set A;
(5) Repeat these steps until a new clustering center is generated and no longer changes.

3 Collaborative Filtering Recommendation Construction

3.1 Overall Idea of Collaborative Filtering Recommendation

This collaborative filtering recommendation can mine users' preference bias based on users' historical behavior data, so as to realize recommendation. Among them, main idea of user-based collaborative filtering (User-based CF) recommendation is: First, a user set that is comparable to the target user is mined by comparing the target user to other users. On this basis, contents that similar users are interested in are recommended to target users. In this recommendation process, offline clustering is first performed, which analyzes user historical behavior data to obtain interest preferences based on implicit user feedback, and then conducts online collaborative filtering recommendation. Figure 2 illustrates the overall framework.

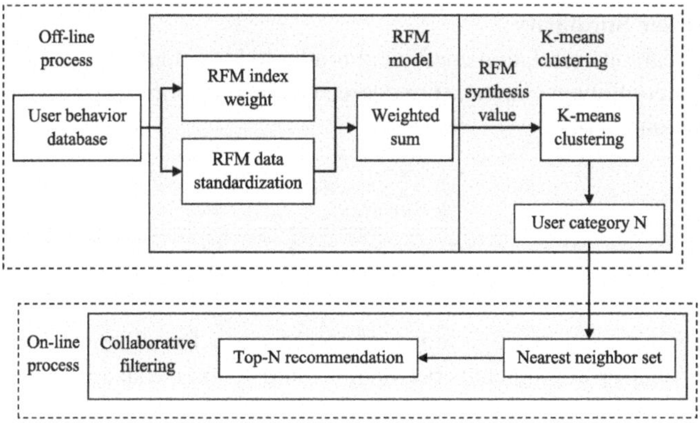

Fig. 2. The overall framework of collaborative filtering.

3.2 Specific Implementation of Collaborative Filtering Recommendation

In the collaborative filtering process, user-based CF can realize personalized learning resource recommendation of target users by analyzing user historical behavior data and carrying out a series of operations.

Furthermore, collaborative filtering recommendation can be divided into four parts: establishing user-learning resource scoring matrix, calculating user similarity, searching nearest neighbor set and creating recommendation list.

Establishing User-Learning Resource Scoring Matrix

RFM-K-means clustering analysis is utilized to analyze the historical behavioral data and n learning resources of m users, and interest preferences based on implicit feedback of users are obtained. After these preferences are translated into users' scores for learning resources, user-learning resource scoring matrix is established. Table 1 shows the details.

Table 1. User-Learning Resource Scoring Matrix

	l_1	...	l_j	...	l_n
u_1	s_{11}	...	s_{1j}	...	s_{1n}
...	...	...	...	...	...
u_i	s_{i1}	...	s_{ij}	...	s_{im}
...	...	...	...	...	...
u_m	s_{m1}	...	s_{mj}	...	s_{mn}

In Table 1, s_{ij} represents score of user u_i for learning resource l_j, and when user u_i has not used learning resource l_j, $s_{ij} = 0$.

Calculating User Similarity

In this paper, Pearson correlation coefficient is adopted to calculate user similarity. Pearson correlation coefficient can measure degree of linear relationship between variables, and its expression is [14]:

$$sim(u, v) = \frac{\sum\limits_{i \in I(u) \cap I(v)} (r_{ui} - \overline{r_u})(r_{vi} - \overline{r_v})}{\sqrt{\sum\limits_{i \in I(u) \cap I(v)} (r_{ui} - \overline{r_u})^2} \sqrt{\sum\limits_{i \in I(u) \cap I(v)} (r_{vi} - \overline{r_v})^2}} \tag{3}$$

where $\overline{r_u}$ and $\overline{r_v}$ represent average scores of users u and v for all items. The value range of Pearson correlation coefficient is between −1 and 1, and the specific values represent the following contents:

(1) If it is 1, it indicates completely positive correlation and users are completely similar.
(2) If it is 0, it means there is no linear relationship and there is no common scoring item among users.
(3) If it is –1, it indicates completely negative correlation and users are completely dissimilar [15].

Searching Nearest Neighbor Set

After user similarity is calculated, this study adopts nearest neighbor method to obtain users whose similarity is closest to target users. Nearest neighbor method is a simple machine learning algorithm, and its core idea is: For a new sample, find the nearest known sample, and then classify the new sample into the category of that known sample.

Generating Recommendation List

After using nearest neighbor method to find nearest neighbor users, predicted score $p(u, i)$ of users in nearest neighbor set for learning resource i and set $s(u, k)$ of the first k users with the highest user score are found. Then, delete learning resources $N(i)$ that target users have rated, and obtain the candidate learning resource set. Finally, predicted scores of target users are calculated according to this set, and recommendation results are obtained according to scores. The predicted score values are calculated as follows:

$$p(u, i) = \overline{r_u} + \frac{\sum\limits_{v \in s(u,k) \cap N(i)} sim_b(u, v)(r_{vi} - \overline{r_v})}{\sum\limits_{v \in s(u,k) \cap N(i)} sim_b(u, v)} \tag{4}$$

4 Introduction

4.1 Environment Construction

In this experiment, the selected hardware environment is Intel Core Ultra 7 165H processor is selected; The operating system is Windows 11. Software environment: MATLAB7.3 and SPSS23.0 are selected. Among them, Table 2 shows parameters of Intel Core Ultra 7 165H:

Table 2. Parameters of Intel Core Ultra 7 165H

Item	Parameter
Memory	16 G
Hand disk	120 G
Basic frequency of processor	5 GHz
Level two cache	4 MB
Bclk	1.4 GHz
GPU Core code	Raptor Lake
Socket type	FCBGA2049

4.2 Data Sources

Experiment data used in this paper come from the background database of a learning platform. On this platform, after users' study and answer questions, there are objective scoring records left. The platform provides 86 sets of paid test questions to Chinese proficiency test candidates, which include different question types, question volumes, difficulty levels, and full scores. Therefore, it is necessary to carry out dimensional processing, and then get a number within 0 to 1. In addition, users' scores of test questions are taken as the original data, thereby constructing user-learning resource scoring matrix.

4.3 Evaluation Index

Mean Absolute Error and Root Mean Square Error

Mean absolute error (MAE) and root mean square error (RMSE) are error between predicted value and true value and standard deviation of predicted error. They can be calculated using the following formulas:

$$MAE = \frac{1}{n}\sum_{i=1}^{n}|y_i - \hat{y}_i| \tag{5}$$

$$RMSE = \sqrt{\frac{1}{n}\sum_{i=1}^{n}(y_i - \hat{y}_i)^2} \tag{6}$$

where, n represents number of evaluation samples, y_i and $\hat{y}_i$ represent real user score and predicted user score respectively.

Accuracy

Accuracy indicates ratio of correctly predicted samples to total number of samples, which can be obtained by Eq. (7):

$$accuracy = \frac{TP + TN}{TP + TN + FP + FN} \tag{7}$$

where, TP and TN indicate samples that are predicted to be positive or negative and are positive or negative; FN and FP indicate samples that are predicted to be positive or negative but actually are negative.

Precision
Precision means ratio of number of samples correctly predicted to be positive to number of samples that are positive, and its calculation formula is as follows:

$$precision = \frac{TP}{TP + FP} \tag{8}$$

Recall Rate
Recall is ratio of correct prediction to actual positive, and it can be calculated by Eq. (9):

$$recall = \frac{TP}{TP + FN} \tag{9}$$

F1-Socre
F1-Score represents harmonic mean of precision and recall, and its expression is:

$$F1 - socre = \frac{2 \cdot precision \cdot accuracy}{precision + accuracy} \tag{10}$$

4.4 Experimental Process

Firstly, transaction data are imported into SPSS, and weights of each RFM index are obtained through a series of processing such as calculation and research: $W_R = 0.267$, $W_F = 0.511$, and $W_F = 0.222$. Then, the standardized RFM data are weighted and summed to obtain user-learning resource scoring matrix, and then K-means clustering technique of SPSS is used to carry out clustering according to user's IRFM. Finally, after obtaining the target user's nearest neighbor set and setting Top N value, learning resource recommendation set is constructed, that is, the data set in the experimental verification stage is obtained.

4.5 Results and Analysis

Determination of the Corresponding User Set
To determine the size k of nearest neighbor user set that is most similar to target user, assuming that 5 test questions are recommended to the user, 80% of data in the dataset is used as training set, and MAE, RMSE, precision, and recall of the prediction results are calculated with k values ranging from 1 to 20. The results in Fig. 3 are obtained.

As seen in Fig. 3, with the increase of k, when $k = 12$, both MAE and RMSE can reach the minimum value. Precision and recall continue to grow, but the infinite user

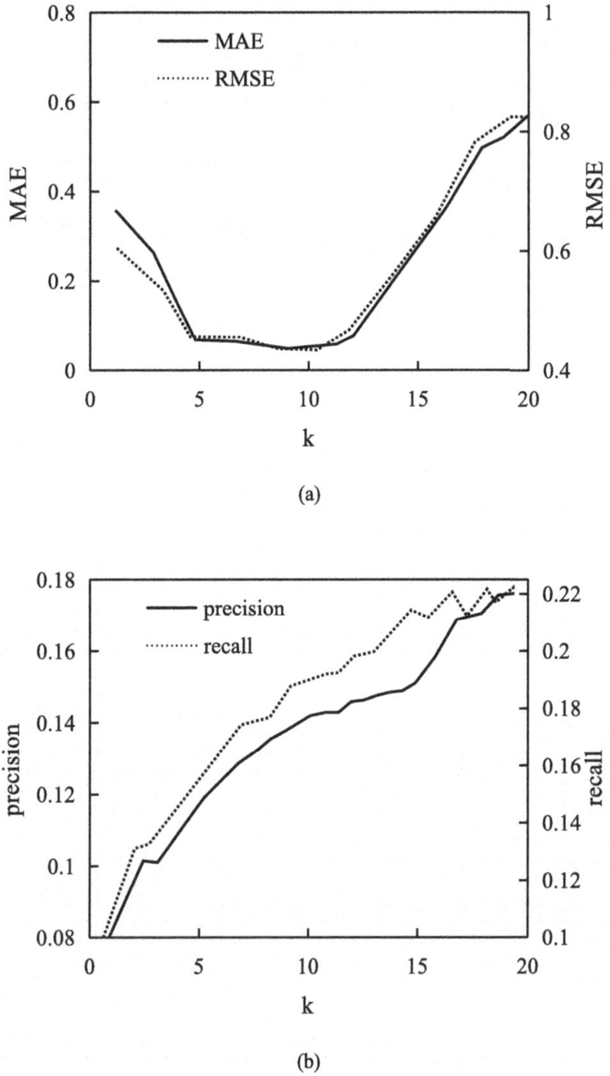

Fig. 3. Results of MAE, RMSE, precision, and recall.

set is meaningless. Therefore, this experiment makes recommendations for target users based on the purchase and rating records of 12 neighbor users.

Division of Training Set

This algorithm takes 12 similar users of target users as recommendation user set. Where, training set accounts for 50% to 90%, then verification set accounts for 50% to 10%. To determine the optimal proportion of the training set, this study calculates MAE, RMSE, precision and recall of prediction scores respectively. The calculation results are shown in Fig. 4.

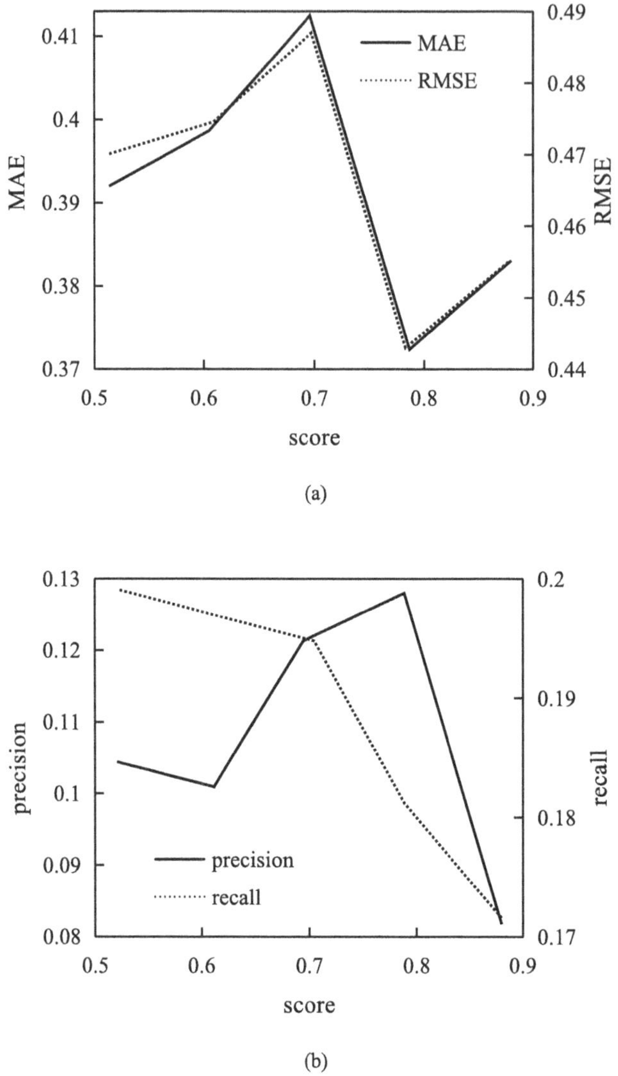

Fig. 4. Calculation results of MAE, RMSE, precision and recall.

From Fig. 4, it can be seen that when k = 12, and the ratio of training set to validation set is 8 to 2, MAE and RMSE of the calculated prediction score are the smallest, that is, error is the smallest, and precision is the highest, resulting in the best performance. Therefore, in the follow-up experiments, 80% of the dataset is used as training set and 20% as verification set.

Optimization of Recommendation Precision

In this experiment, when users choose test questions, they need to comprehensively consider their own learning level, difficulty of test papers, and other aspects. In addition, according to official regulations, candidates who achieve 60% of the perfect score in this level of examination can be considered as passing. Therefore, this experiment takes 0.6 as the threshold to recommend questions. Then, training set and verification set are divided according to the above training set partitioning method, and the questions with a predicted score of not less than 0.6 are recommended to 62 users. At the same time, index values of prediction results before and after optimization are recorded. Table 3 depicts the details.

Table 3. Results of Recommendation Error and Recommendation Accuracy

	Prediction results	
	After optimization	Before optimization
MAE	0.40212	0.39023
RMSE	0.48123	0.43119
accuracy	0.81234	0.80248
precious	0.11092	0.13028
recall	0.20234	0.15120
F1-score	0.14209	0.13876

As seen in Table 2, after optimizing the recommendation strategy, the number of prediction options increases significantly, the number of correctly predicted options also increases significantly, and precision decreases, because users do not actually buy so many products. Meanwhile, prediction error of user scores also increases, and precision of predicting that users will not make purchasing behavior is also high, so overall accuracy of recommendation algorithm is very high.

Table 4 depicts some examples of recommendation results after recommendation strategy is optimized.

Table 4. Some Examples of Recommended Results

User	1	2	3	4	5	6	7	8	9	10
Actual purchase	401	301	301	501	201	601	101	301	301	501
Actual score	0.78087	0.90864	0.74432	0.59803	0.72213	0.55345	0.86779	0.65321	0.45701	0.39087
Predicted score	0.69023	0.88023	0.66664	0.49023	0.67342	0.67902	0.83024	0.57824	0.48201	0.31009
Prediction results	T	T	T	F	T	T	T	F	F	F

In Table 4, T indicates that prediction result is positive, which means this product is recommended; F indicates that prediction result is negative, which means this product is not recommended. In addition, prediction accuracy of the optimized recommendation strategy is 90%, and prediction precision is high. Therefore, the proposed algorithm can realize personalized resource recommendation of online learning platform.

5 Conclusion

In conclusion, the proposed collaborative filtering algorithm based on RFM-K-means clustering is effective. Firstly, this paper adopts RFM-K-means clustering technology to extract user features, thus interest preference based on user's implicit feedback is obtained. After these preferences are translated into user scores of learning resources, user-learning resource scoring matrix is established. Then, Pearson correlation coefficient is adopted to calculate user similarity, and nearest neighbor method is used to obtain the user set with the closest similarity to target user. Finally, predicted score of nearest neighbor user set for learning resources is calculated, and predicted score of target user is calculated, thus learning resources with the highest predicted score are selected for recommendation. In experiment verification stage, the dataset is derived from the background database of G learning platform, in which the dataset is divided as follows: 80% is training set and 20% is validation set. In addition, scoring records of 12 neighboring users are selected as the reference items for target users' recommendation set to provide resource recommendations for target users. However, since Top N is not suitable for the recommendation of resource types in this experiment, items with predicted score of more than 0.6 are selected as the recommendation options. The verification results show that prediction accuracy of the proposed algorithm is 90%, and prediction precision is high, which indicates that this algorithm can realize personalized resource recommendation of online learning platform.

Acknowledgment. General Research Projects of Zhejiang Provincial Department of Education: Research on the Path of Cultivating Landscape Talents from the Perspective of Beautiful Countryside (Y202353298).

References

1. Wang, H.J., Chen, P.Y., Li, Z., et al.: Metacognitive ability assessment of online learners based on BPNN. Mod. Educ. Technol. **34**(11), 132–142 (2024)
2. Zhou, Y.T., Chu, H., Zhu, F.F., et al.: Review of personalized learning resource recommendation based on deep learning. Comput. Sci. **51**(10), 17–32 (2024)
3. Sun, Y.M., Mao, R.Y.: Empirical study on the user experience of online teaching platform for college teachers based on 5E model [J]. Computer Knowledge and Technology **20**(20), 146–148 (2024)
4. Wang, H.C., Wang, H., Pan, J.H., et al.: Research on personalized adaptive resource recommendation algorithm based on learning style. Comput. Digit. Eng. **52**(1), 94–98 (2024)
5. Shi, Y.: A recommendation method for classroom personalized learning resources. Inf. Technol. **11**, 158–161 (2023)
6. Guo, W.J.: Research on the development of personalized learning resource recommendation. Software **44**(10), 53–57 (2023)
7. Yu, F.C., Xu, L., Zhang, G.F., et al.: Research on customer value of home appliance customization service based on RFMI model. Manuf. Autom. **45**(1), 91–94+100 (2023)
8. Zhou, X.: Design of pose positioning control system of industrial teleoperation robot based on big data clustering. Comput. Meas. Control **31**(8), 104–109 (2023)
9. Le, C.Y., Wang, X.: Research on user portrait of university library based on improved RFM clustering. Libr. Theory Pract. **2**, 75–79 (2020)

10. Yang, Z.J.: Graph neural network recommendation system based on user behavior and knowledge propagation. Intell. Comput. Appl. **14**(2), 35–40 (2024)
11. Ji, D., Liu, Y.Z., Peng, R.X., et al.: K-means text clustering algorithm based on theme word vector center point. J. Comput. Appl. Softw. **41**(10), 282–286 (2024)
12. Wang, S.G., Guan, H.L.: Fast K-means clustering point cloud reduction algorithm for custom clustering center points. Ind. Control Comput. **37**(8), 123–125 (2024)
13. Xu, J., Yuan, H.: Research on violation behavior portrait of Listed companies based on RFM model. Front. Data Comput. Dev. **5**(6), 20–30 (2023)
14. Dong, L.P., Nie, Q.H., Sun, X.K., et al.: Analysis of influence of shield tunneling parameters on surface settlement based on Pearson correlation coefficient method. Constr. Technol. (Chin. Engl.) **53**(1), 116–123 (2024)
15. Zhang, X.B., Yu, B.: Urban smog control strategy in Beijing based on Pearson correlation analysis and BP neural network. Syst. Eng. **41**(2), 26–34 (2023)

Leadership in Modern Education

From Manual to Semi-automatic—Artificial Intelligence Language Instruction in the Space Design Education of Film and Television Major Universities

Xin Wen[1(✉)] and Fan Zhang[2]

[1] Beijing Film Academy, Haidian, Beijing 100088, China
wenxin@bfa.edu.cn
[2] Beijing Technology and Business University, Liangxiang, Beijing 100048, China
zhangfanheihei@sina.com

Abstract. This paper summarizes the application and practice of artificial intelligence language instruction in the professional space design education of film and television major, and summarizes the use of the accelerated creative function provided in the process of space design. This practice will output the spatial language instruction for some spatial achievement cases in each stage, then compare the advantages and disadvantages of the generated spatial form results. Research out the artificial intelligence in film and television professional space design education for the important role of information creativity, At the same time, through the comparison of language instructions and output commands, we find that artificial intelligence can not completely replace the designer to get the most accurate design results. If you want to obtain more precise results or to provide more targeted solutions, precisely limit images and language in the early stage as more as possible can provide more effective guidance in the middle process, therefore, For the manual process in the previous traditional space model design process, space design has gradually and quickly entered a new stage of semi-automatic, more effective and reasonable, and a high degree of creativity and completion.

Keywords: Artificial intelligence · Space instruction · Film and television professional · Spatial design education · Dramatic creativity · Inspiration

1 Traditional Film and Television Professional, Space Design Education Teaching Procedures

1.1 Characteristics of Film and Television Professional Space Design and Education

At present, the space research majors of film and television mainly include traditional space culture and contemporary design. As the traditional culture space design students, they mainly learn the history of classical architecture and the corresponding mature architectural styles. Most of them conduct space design according to the traditional architectural style as a reference, which has a very strong reference.

© The Author(s), under exclusive license to Springer Nature Switzerland AG 2025
F. Paas et al. (Eds.): EduTA 2024, LAIS 51, pp. 267–281, 2025.
https://doi.org/10.1007/978-3-031-95252-4_24

And this part of the space design education variability is relatively small, some traditional film and television class mainly according to the style of the architectural space layout in the scene of professional film art design, special effects, and the representative of contemporary design space class professional new media, virtual environment space and so on such a device space and contemporary image combined space for creation of professional, and he because the traditional bound less, in the space in the form of more diverse.

1.2 Basic Teaching Steps of Space Design for Film and Television Majors

The traditional space design of film and television majors is basically consistent with the course design of other design colleges. In the space design exercises, they all follow a basic procedure from plane to concept, and then to space. In the plane stage, a series of summaries and abstractions are needed to form a certain internal space logic, and then on this basis, the interspersed combination of internal space is carried out according to the law of space design, and then the overall space results are repeatedly formed from the inside out or from the outside. In the practice of traditional cultural space, the results can be applied in the scenes of movies and TV, while the contemporary design space can not only be used in the film and television space, but also in the current real space, where the space device and image device are built inside or outside the space. Through a series of spatial design exercises, students can apply what they have learned in a wide range of real and virtual Spaces. For example, in the design of the space, the process from the abstract plane to the facade and then to the space (Fig. 1).

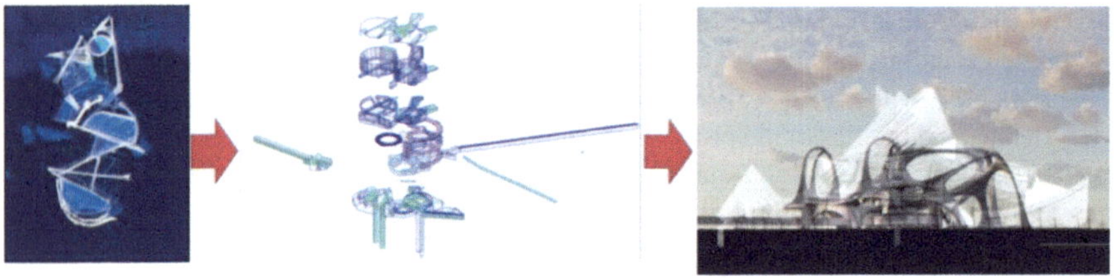

Fig. 1. The space design process of Marine Museum

1.3 The Application of Manual Space Instructions in Film and Television Professional Design

Freshmen students majoring in film and television need to contact space practice. In the first grade, through some spatial rules, the field experience in the second grade, and then the main creation in the third grade, through the future professional courses, is a very important course.

1.3.1 Basic Manual Space Instruction

First of all, students need to do certain hand-drawn draft exercises, and then show the spatial form through the means of spatial modeling. Initially, students need to translate their hand-drawn concept sketches into a physical space model, requiring manual space instructions. At this time, the traditional space manual instructions include computer-aided design of plane software, which mainly make two-dimensional line cutting and layout. And after the establishment, to build the space model, will use different 3 d modeling software, through the manual operation of the 3 d software, space package, stretching, cutting, boolean operation and so on corresponding operation, these operations, are manually activated basic space instructions, the combination of these instructions eventually form space.

After entering the parameterization stage, even many complex figures can be completed by manual operation or plug-in s. Basically, from the hand-drawn draft to the spatial model, all the construction steps need to be implemented step by step in series or parallel by manual commands.

1.3.2 Manual Modeling Language Application in the Traditional Cultural Space Class

Traditional culture space type of professional, in addition to the classical architecture theory for reference, in the second grade will be professional plane facade of surveying and mapping, will go deep into the professional traditional culture dwellings and roadway, for specific architectural space painting records, structure form sketch, and customs of research, conditions can also be through the acquisition of images or 3D scanning of traditional culture space various types of style, and then further creation (Fig. 2).

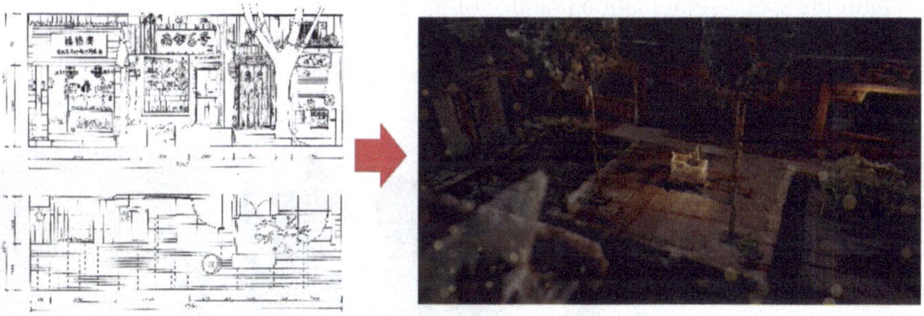

Fig. 2. Achievements of various stages of the traditional cultural space category — Painting, surveying and mapping, 3D scanning, and later creation

The professional space design of film and television has an important application in the later stage. Among the various types of films, the space formed by the traditional buildings and contemporary buildings with cultural and regional characteristics forms an important part of the image. A high-quality film work usually needs to carry out the early various aspects of the space scene design (Fig. 3).

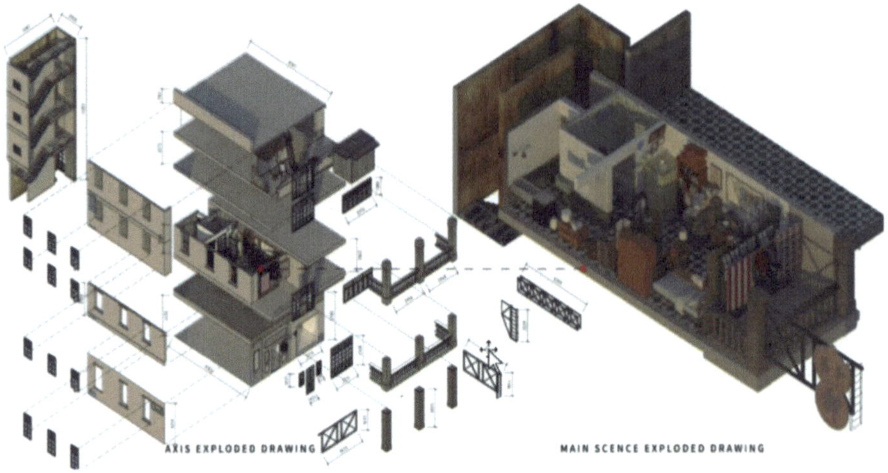

Fig. 3. Movie scene design and spatial scheduling schematic diagram

1.3.3 Manual Modeling Language is Applied in the Contemporary Design Space Class

However, students of contemporary design space type follow the earliest three - dimensional composition related content, and on this basis, construct a series of space design rules, and then put the corresponding emotional content and human material feelings, until the space combination practice of monomer and group is formed. (Fig. 4) In the contemporary design space, in addition to the film and television related applications, it is mainly used in the device category or media interaction space.

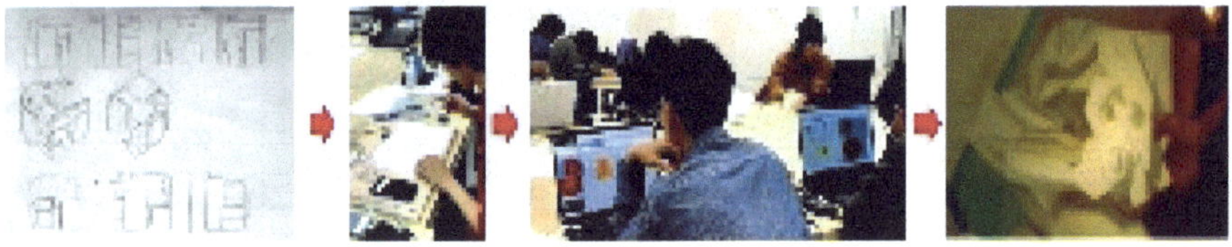

Fig. 4. Students realize the construction of contemporary design-class space model through manual modeling language

The realization of the device needs a space architecture. After a series of manual construction languages of the space device and the implantation of some corresponding interactive sensors, the construction of the space device can be realized (Fig. 5). The media space will be mainly combined with virtual images, such as the application in CAVE and LED screen projection, and the combination of devices and images, such as the application in large-scale sports and art events, to form a rich visual spectacle (Fig. 6).

Fig. 5. Related applications of the device class space

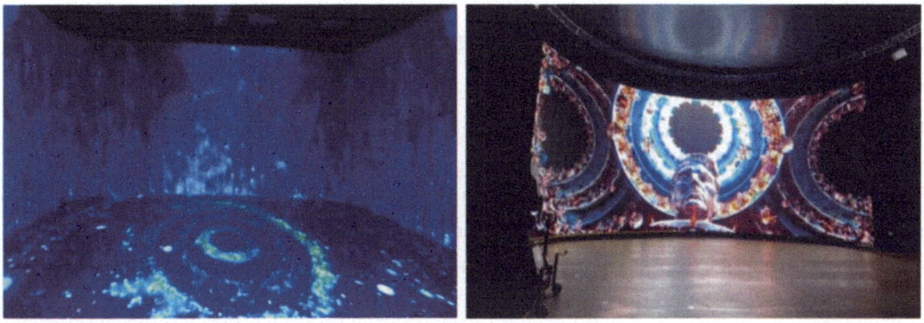

Fig. 6. CAVE (left) and LED screens (right) often used in the media interaction space

2 The Intervention of Semi-automatic Artificial Intelligence Space Instruction in the Film and Television Professional Space Design Education

Spatial intelligence refers to the ability to comprehend and interpret spatial relationships between objects, shapes, and environments. People who have high spatial intelligence perform exceptionally well on tasks requiring them to mentally manipulate and visualize objects in space [1]. Nowadays, artificial intelligence applications in space design are springing up like mushrooms [2], and according to the CO-architecture website, the most common application is mid journey, which is among the top 1 of Top 14 AI Tools for Architects and Designers [3]. Midjourney is an independent research lab exploring new mediums of thought and expanding the imaginative powers of the human species [4]. Midjourney offered a very quick introductory tutorial to make it very easy for anyone who use the Midjourney bot, which can generate stunning images from simple text prompts in seconds. Work directly in Discord. No specialized hardware or software is required. Therefore, in the following steps, we use Midjourney as our practice example.

2.1 The Transitional Role of Artificial Intelligence Instructions in the Teaching Program of Contemporary Space Design

In the past, the transition from the plane to the facade and then to the generation of space needs to go through a long process of repeated deliberation. After the results obtained after space deliberation, several rounds of alternative scheme comparison are often carried out. Before the final satisfactory space, this process relatively requires a lot of time and energy. After the intervention of artificial intelligence, rapid transformation can be achieved through two simple steps, which is equivalent to artificial intelligence instructions replacing part of the previous manual instructions. With a few words of language input time, you can complete the workload of the early few days, even a few months.

2.2 The Basic Types of Artificial Intelligence Instructions in Contemporary Space Design

The space generation of artificial intelligence is mainly realized through the following two aspects. The first is to transform it through key words, or what we called 'a prompt', which is a short text phrase that the Midjourney Bot interprets to produce an image. The Midjourney Bot breaks down the words and phrases in a prompt into smaller pieces, called tokens, that are compared to its training data and then used to generate an image. A well-crafted prompt can help make unique and exciting images [5]. In traditional film art space design, the form and atmosphere of space are also very important keywords, and how to create these Spaces related to keywords often determines the final generation effect.

For example, In the film Ambush on All Sides, the design space of Peony Square space highlights the beauty of the unified proportion and harmonious balance of the Tang Dynasty dougong shape, highlighting the iconic keywords of traditional Chinese architecture dougong, blue, green, finishing point, etc., which show the beauty of traditional architecture [6]. In the movie Hero, the story is about the destruction of the Zhou Dynasty by the Qin Dynasty, and the key words of the whole palace interior space are black tone, power, majesty, sadness and death. The film art design has made some artistic improvements in the traditional buildings featuring the royal works of the Qin Dynasty. Even though the symbol of power, such a black space is strong and has a certain deterrent effect. At the same time, it also symbolizes the end of the fate of the entire Qin Dynasty royal family [7]. Therefore, how to input a useful and exact command to the Midjourney robot is quite important.

Except the 'prompt generate to picture', Midjourney also have the function of 'picture generate to picture'. As an image classification method, text image pre training model can perform visual analysis based on natural language understanding. Users only need to input a sentence or upload an image with a clear style, and the algorithm can align the image features with the text features and make corresponding adjustments to the image. This alignment avoids the arbitrariness of AI painting works and achieves more accurate image effects compared to previous adversarial generative models [8].

After using the command into an indoor space, students can generate approximate space results. The results of the building were also immediate. The Fig. 7 is the results generated by the two students through the implementation of artificial intelligence space instructions. We can see that both the indoor effect and the previous sketch features, the similarity reduction degree is very high.

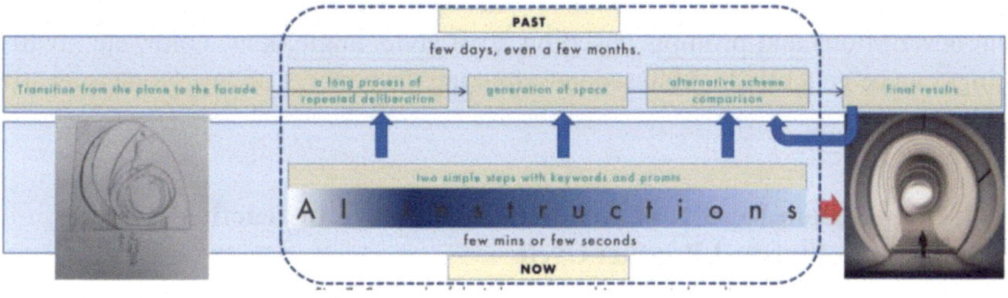

Fig. 7. Commands of the indoor space and its generated results

At the same time, there are normally two types to generate the pictures to the ideal design images. Type A is 'AI Prompts'-through the mixture of various architectural style type transformation. For example, the exercise is the black, white and gray abstraction of the landmarks of the city. At the same time, the desired space is generated according to the abstract and refined geometry (Fig. 8). While through B type- 'AI Stylize', By abstracting the environment around the old Porto Bridge, the student mixed the architectural form and style of Zaha Hadid, and finally selected a relatively satisfactory result (Fig. 9).

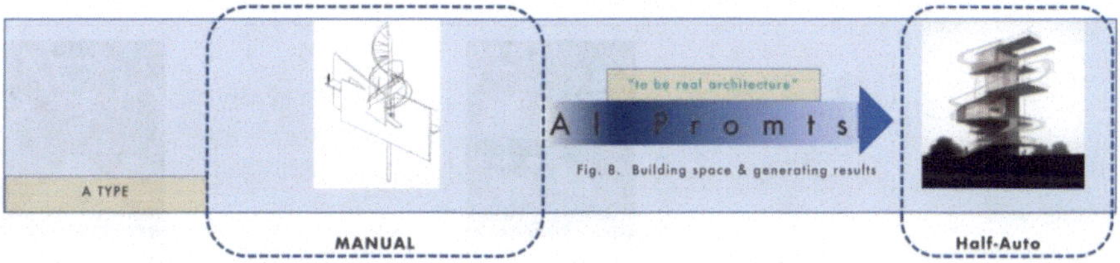

Fig. 8. Command of the building space and its generating results

Fig. 9. Porto space exercises

3 The Promoting Role of AI Instruction in the Teaching Program of Contemporary Space Design

The emergence of the "Text to Image Model" represented by Midjourney, DALL-E, Stable Diffusion, and others has brought about a new way of AI art interaction. Through textual descriptions and prompts based on artist style, angle, lens, color, etc., realistic paintings can be generated within seconds, making the concept of "everyone is an artist" possible [9].

3.1 The Generation Results of Artificial Intelligence Instructions Greatly Shorten the Finished Product Cycle

Taking the project homework generated by the professional emotional space of new media art as an example, in order to show a cheerful space atmosphere, the homework mainly designs the whole space through a sense of openness, illumination, expansion, rise and support. At the same time, after the living space, a variety of colorful LED effects are added to the space. In China, red is the most festive color, so the required space results are finally generated. As we can see, by adding "red", "indoor effect", "real rendering" and "LED", the effect in the middle of the figure below is generated. After one of the effects is further fine-tuned and "dazzling" and "luminous screen", a media exhibition hall space can be generated. The spatial results with high completion are achieved in a short time from several minutes compare to traditional several hours even several days by landing on discord website (Fig. 10).

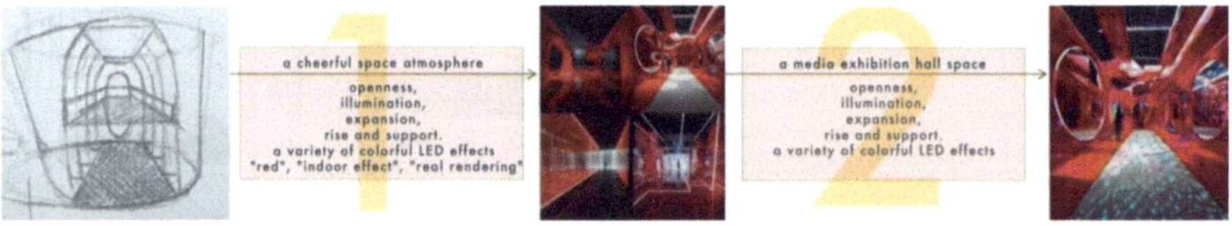

Fig. 10. Creative space results

3.2 Constant Creativity of Artificial Intelligence Instructions Enhancing Contemporary Space Design

This changed dramatically with the advent of advanced technologies such as Midjourney (then version 4.0). Midjourney's design process was innovative. Its initial interface was "Imagine", which guided users to describe their ideas in natural language, supplemented by specific parameters. What's particularly interesting is that the "Imagine" session sparked the user's imagination, and then, through a set of descriptors that the user entered, Midiourney was able to generate high-quality visuals that even exceeded the original vision [10].

3.2.1 After Accelerating the Space Design Cycle

The AI generates the works of first-grade students to a new level of creativity. By importing the preliminary graphics of 3D printed space works into the artificial intelligence environment, many interesting and different schemes are generated through the key word "become architecture". After one or two "polish" orders, these schemes can soon form a finished building space nature for implementation. The details of the building are very clear, and the glass tiles seem to be able to see the corresponding collage gap, which provides an important reference for the next construction. If not satisfied with the completed results, you can also repeatedly execute the generation of the same instruction before, repeated a variety of options available, providing a steady stream of source of creativity (Fig. 11).

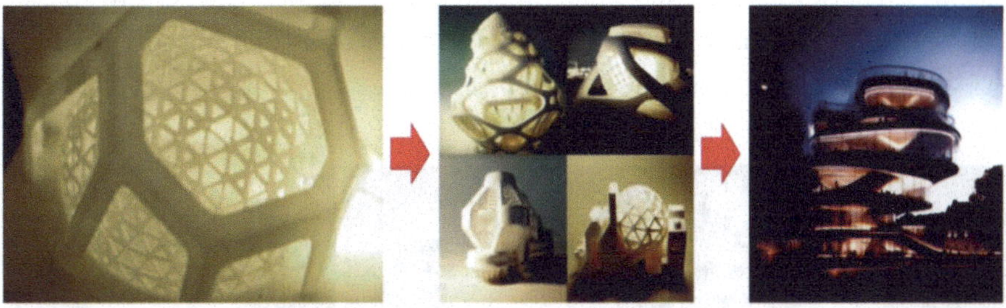

Fig. 11. Space generation of a 3D-printed space work

3.2.2 More Premises Provided for Reference After the Keywords Are Determined

The AIGC technology of artificial intelligence has reshaped the production and dissemination mode of visual culture, forming new visual symbols and languages. While empowering artificial intelligence machines, it also greatly expands the space for human artistic creation, making their subjects more enriched in interaction with machines [11].

When students build through the foundation of a spatial model. Later, you can transform the entity space command with the survival model. This will slowly get a few rich internal and external combination of the corresponding space. Since artificial intelligence cannot fully transform the plane into a three- dimensional model, the biggest function of spatial results is to provide a lot of reference for the implementation plan in the later stage, so as to compare the current most appropriate scheme according to the specific site characteristics, and manually build the final implementation plan (Fig. 12).

3.2.3 It Plays a Major Role in Improving the Quality of Finished Products

Since it took a long time to generate many space and layout in the past, now the artificial intelligence instruction generation results can provide clear spatial details, especially in the atmosphere of film art has an incomparable overall reference role. Due to the accelerated speed, the time saved can be more focused on the details of the effect of the work, the production results of a movie scene to further execute the "river" fine tuning

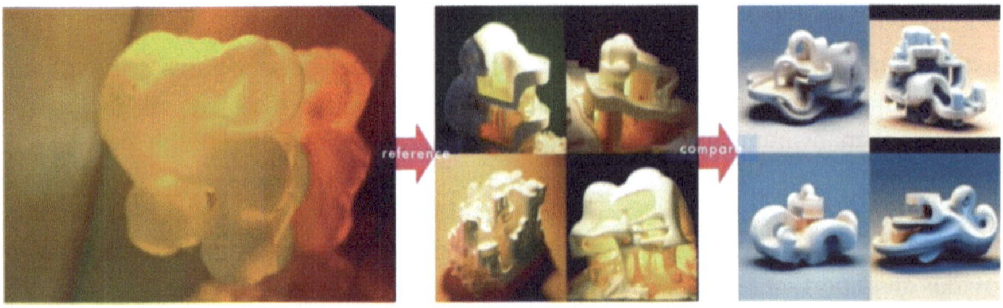

Fig. 12. Space generation of a 3D-printed space work

instruction, there will be many different high completion schemes to choose from. At this time, all the results generated can be used to provide reference for the different ages of the "channel" space, and the more ideal scheme can be obtained by manual fine-tuning later (Fig. 13).

Fig. 13. The spatial fine-tuning results of the ai command execution

4 The Limitations of AI Instruction Under the Teaching Procedures of Contemporary Spatial Design

At present, although the speed and quality of artificial intelligence in creative pictures in living space are very amazing, there are the following obvious disadvantages-Data bias: Large scale datasets may have biases that reflect historical preferences and trends. This may lead to a lack of diversity in the design of generative models, which cannot capture the current diversity and innovation [12].

4.1 Relatively Weak Impact in Raditional Space Design Applications

In the Chinese traditional hu tong space design, students measured the building elevation and use this handmaking picture to perform artificial intelligence command generation, only to generate the results of the space outline is not completely consistent, from black and white figure to color real effect can only get a first look the right atmosphere as a result, both details. For example, the details of the icon door on the wall, and the style of the door face are carefully compared and found that the results and the original picture has a big difference (Fig. 14).

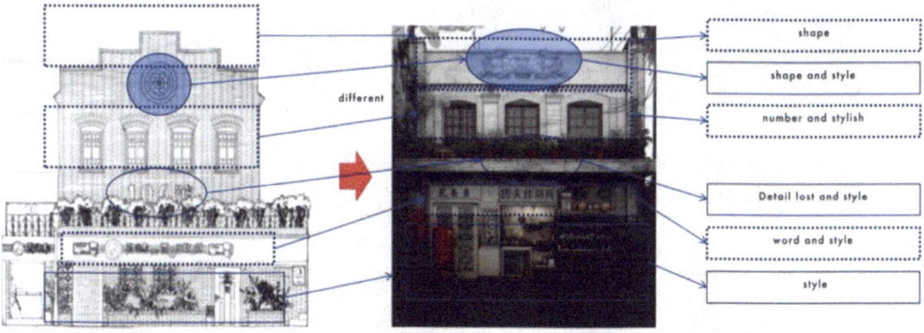

Fig. 14. Results of the command execution in the wedding space scene

And color image of the space draft, if you want to get real effect, after the results of the command generated often background and characters will have larger deviation, courtyard architectural style often become some southern architectural style, and characters after the command execution is very prominent (Fig. 15), many times after different pictures perform "detail enhancement" command, the scene characters often eventually become the main space, so before the wedding space scene, results into the wedding figures, space into the character foil background (Fig. 16).

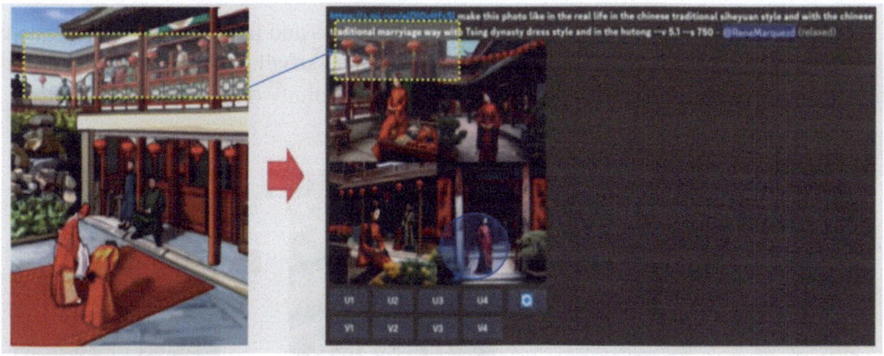

Fig. 15. Results of the command execution in the wedding space scene

4.2 Relatively Specifically Weakly Targeting in Specific Regional Environments Design

After adding Chinese regional commands, we will find that artificial intelligence often ignores the architectural structure style of the space and the correctness of the history

Fig. 16. The space performs the detail enhancement command and becomes the background

when generating pictures. As the atmosphere, the clothes and faces of the characters are also quite different from the real history. The costumes of all dynasties may appear in the generated results, unable to pursue historical authenticity (Fig. 17).

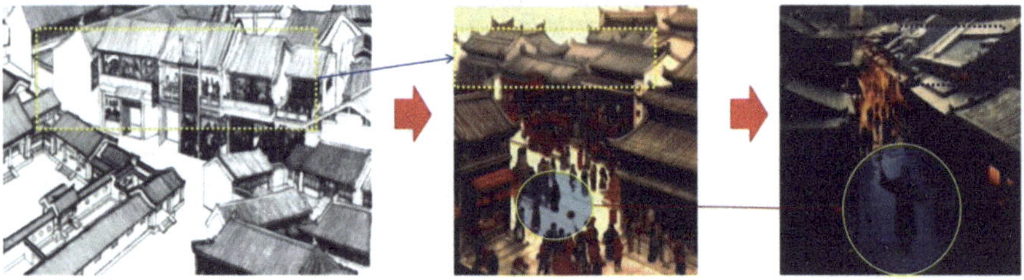

Fig. 17. The results of spatial command execution are against historical truth

4.3 Far from the Previous Diagram Style After Executing the Same Command for Many Times in Original Mode

Will adjust before and after the continuous generated results, such as to a building form execution "increase building light" command, then add building "LED luminous effect" command, although a rough look at the shape of the building seems to be similar, but carefully compared will find a lot of details has a larger gap, space structure cannot achieve complete consistency (Fig. 18).

Fig. 18. Comparison of the results of the unified architectural form generated

4.4 Far from the Previous Diagram Style After Executing the Same Command for Many Times in Major Mode

The same is a regional street scene of the Spring Festival in the old Beijing Hutong. The following picture first generates four Spring Festival scenes, three of which appear inexplicable monsters. Choose a monster to continue the second deepening, the result of the middle of the character is particularly prominent, the second result is further deepened, the implementation of "add more people", "make the street more lively" command, the whole picture becomes a noisy place, and the spatial structure has been basically ignored. We can see that the result from the first image to the final one is not like the same one at all (Fig. 19).

Fig. 19. After the same command many times with the previous figure style comparison

5 Summary

Through the above practice of artificial intelligence space instruction design, we can find that in the traditional cultural space design, if we draw the spatial structure more detailed, then the artificial intelligence instruction can play a great role in helping the atmosphere in the later stage. Especially in the contemporary space design, the role of artificial intelligence instructions will be even greater. It can not only provide more possibilities in the process of creativity, but also make the later spatial details more refined. However, the final result may be quite different from the result in style and form at the beginning. At this time, it needs to be adjusted in the later stage. However, the role of artificial intelligence instruction is usually not only the creation of atmosphere, but also the result of space, which is more extensive and far-reaching than the traditional cultural space design and application.

From this, we can see that no matter what kind of spatial genre theme we are creating, it is best to give more spatial certainty form as much as possible, so that the results will be more accurate. For example, when we draw diagram, if provide more space structure and color prompt information, even find more style mixed reference images, artificial intelligence will provide us with more targeted help the results, but in the end of the result of the figure often need to be processing many times, one pace reach the designated position to get the result is some difficulty, unless do not need the final construction, just the effect. It can be seen that artificial intelligence provides more choices, but it may not fully generate the final space results for human beings. However, these choices save a lot of valuable time to correcting the results in the later stage, and make space designers have greater room for creation (Fig. 20).

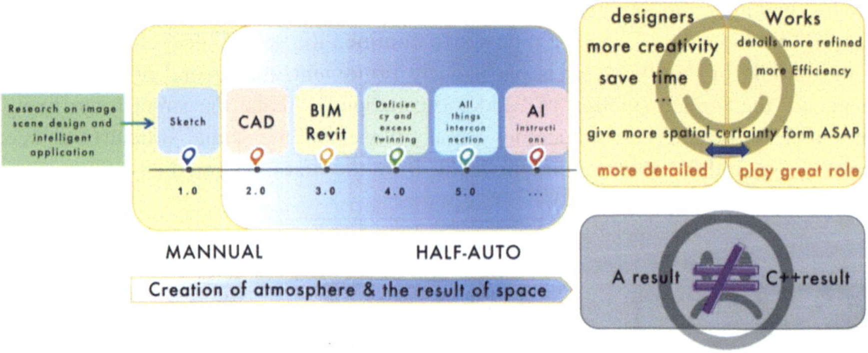

Fig. 20. Summerize pictures

Acknowledgements. Beijing Social Science Foundation Youth Project 18YTC033, No. 17 Chaoyang Branch of China Democratic Construction Association.

References

1. https://www.geeksforgeeks.org/spatial-intelligence-in-ai/. Accessed 22 Jan 2025
2. Jin, Z.W.: Exploring the impact of artificial intelligence technology on contemporary graphic design: a case study of midjourney drawing software. Popular Lit. Art **16**, 32–34 (2023)
3. https://site.co-architecture.com/artificial-intelligence-ai/top-14-ai-tools-for-architects-and-designers/. Accessed 22 Jan 2025
4. https://www.midjourney.com. Accessed 22 Jan 2025
5. https://docs.midjourney.com/docs/prompts. Accessed 22 Jan 2025
6. Huo, T.X., Jia, F.Y.: The Charm of Tang Dynasty-House of Flying Daggers, p. 72. China Film Press, Beijing (2021)
7. Huo, T.X., Huang, F.: The Artistic Conception-Hero, p. 109. China Film Press (2019)
8. Y. Y. H.: The evolution, influence, and speculation of AI painting, media. (17), 53 (2023)
9. B., T.Y.: The ontological shift of art media under the influence of generative AI. J. Shanghai Normal Univ. (Philos. Soc. Sci. Ed.), 53 (2024)
10. J., W.G.: From midjourney to sora: generative AI and aesthetic revolution. Yuejiang J. **16**(3), 85–92 (2024)
11. Y., X.D.: Born from the heart: art production and aesthetic new landscape in the AIGC era. Cult. Artistic Res. **1**, 26 (2024)
12. Q., S.S., Cao, L.L., Wang, S., et al.: Application of generative artificial intelligence large models in the design field. Furniture Inter. Decoration **31**(4), 5 (2024)

Fusion Strategy of Preschool Education and Music Education Combined with Data Mining Technology

Weiwei Liu(✉) and Jinwen Yu

Wuhan University of Engineering Science, Wuhan 430205, Hubei, China
15972072967@163.com

Abstract. In order to improve the educational effect of preschool education, this paper tries to integrate music education into preschool education. The fusion of preschool education and music education in smart kindergartens is suitable for adopting a hierarchical structure to meet the daily teaching needs and educational data analysis needs of kindergartens. Therefore, this paper proposes an intelligent data rapid retrieval algorithm model based on data mining and machine learning, so as to improve the efficiency and accuracy of data retrieval methods. On the basis of machine learning algorithms, music education is integrated into the smart kindergarten teaching platform. Combined with the experimental results, it can be seen that the method of applying data mining technology to the fusion of preschool education and music education in smart kindergartens improves the effect of preschool education in kindergartens and improves user satisfaction.

Keyword: data mining · preschool education · music education · fusion strategy

1 Introduction

Music teaching is an important content in preschool education, which helps children to acquire good cognitive ability and stimulate children's interest in learning. By fully utilizing intelligent network resources into music classroom teaching, the value and function of network resources can be brought into play, and children can better acquire music literacy and maximize the efficiency of music classroom.

Preschool education is for young children. Therefore, scientific and reasonable use of intelligent network resources to carry out preschool education and apply it to music classroom teaching can help children further understand music, gain artistic edification, and continuously improve their music appreciation level in a subtle way. Moreover, children can also receive good ideological education and purify their thoughts. Through network intelligent resources, preschool education is carried out to guide children to perceive music, obtain good aesthetic education, and effectively cultivate and improve their music literacy.

With the development of society, network technology is changing with each passing day. Once a piece of music is uploaded to the network, it can be quickly shared and tasted.

© The Author(s), under exclusive license to Springer Nature Switzerland AG 2025
F. Paas et al. (Eds.): EduTA 2024, LAIS 51, pp. 282–291, 2025.
https://doi.org/10.1007/978-3-031-95252-4_25

It can be said that the network resources have greatly enriched the teaching resources of the current preschool education music classroom, and making full use of these network resources to the music classroom can provide children with richer music knowledge and guide them to constantly understand and master the connotation of music. Moreover, while children's musical literacy is improving, their musical horizons are constantly broadening. At present, there are many contents suitable for children's music teaching on the Internet, which can guide children to appreciate and watch in music courses, and guide children to comprehensively and intuitively understand and master musical instruments [1].

Traditional music teaching relies on limited content in music textbooks, while with the help of online resources, more diverse and abundant teaching resources can be obtained. These knowledge are more diverse and enriched than the content in textbooks, which can effectively meet the needs of young children and stimulate their interest in production [2]. Moreover, the music discipline encompasses a wide range of content, making online music resources highly suitable for music teaching in music classrooms. The network resources are novel and unique, often presented in the form of videos and audios, which are in line with the interests of young children. Therefore, young children can fully participate in them, and teaching efficiency is greatly improved. In addition, online resources are different from textbook teaching resources in that they are conducive to children's subjective initiative, enabling them to think actively and giving them a certain subject status, which stimulates their subjective initiative and enthusiasm [3].

The teaching objective of preschool education major is to cultivate talents engaged in preschool education work. Music courses are an indispensable component of the preschool education curriculum system, and music teaching places high demands on the subject of preschool education. The application of electronic teaching tools in preschool education in the information age can help compensate for the lack of teachers' professional abilities and improve teaching quality [4]. The teaching of music courses itself is abstract, and some students majoring in preschool education lack a foundation in music, making it difficult for them to intuitively understand the emotions conveyed by music works. Electronic education teaching tools can transform abstract music into concrete objects, enhancing students' experiential experience in the learning process [5]. Electronic education teaching tools have also opened up new paths for students to innovate educational and teaching methods. Students can try to combine electronic education teaching tools with traditional education methods to make up for the shortcomings of traditional education methods and improve their ability to apply electronic education teaching methods for learning. At the same time, leveraging the role of online information tools, understanding the current situation of music education in the preschool education system, exploring the integration of electronic education teaching tools and music education, and improving the effectiveness of learning. There are still many contents related to big data analysis in electronic education teaching tools. Students can master the application methods of these tools and conduct data analysis, providing practical basis for reforming music teaching and improving the effectiveness of music education, making the teaching process more scientific. In addition, the target audience of preschool education is young children, who start learning from their interests [6]. The tools of electronic education and teaching can attract children's attention and stimulate their interest in learning music. Students majoring in preschool education recognize the role of electronic education teaching tools in stimulating children's interest in music, and will actively apply electronic education teaching tools in future teaching work to improve teaching effectiveness. The use of electronic teaching tools in music courses for preschool education in universities can help enhance students' awareness of using information technology tools to carry out related learning activities and stimulate their research interests. Students should fully master the application methods of electronic education teaching tools, form a thinking mode of using electronic education teaching tools to carry out learning, and fully stimulate children's learning interest in future education and teaching work, so that children are willing to actively participate in music classes, and ultimately promote the development of children's music perception ability [7].

Music education under big data technology is a new teaching method that has emerged through computer technology, emphasizing learners' initiative and consciousness in music teaching, and proposing diverse music education methods, breaking the previous constraints on music education disciplines and majors, and promoting modern music education towards innovative, diversified, and open educational and teaching methods for students. Preschool music education is an important part of music education [8]. With the in-depth application of Internet and mobile Internet technology, the mode and content of education are changing dramatically. Internet based big data platform technology also has an important impact on the resource construction and application of preschool music education. The teaching value of music in the context of big data analysis is mainly reflected in the following three aspects: firstly, the analysis of big data. Teachers can present children's music learning in the form of visual data analysis through online music teaching systems, thereby presenting the current situation of children's music learning more reasonably; The second is cross-border integration. Teachers can use big data analysis technology to globally share music teaching resources. With the support of multimedia, micro classrooms, VR virtual classrooms and other technological means, teachers can achieve cross regional music teaching connection and cross disciplinary course sharing. Music courses are not limited to campus teaching, but can also be extended to offline self-learning; The third is personalization. Teachers can use big data analysis technology to customize training and curriculum plans based on the personalized needs of young children, which can greatly improve their personal music literacy [9].

Nowadays, the construction of "smart campuses" is becoming increasingly perfect, and various digital network management platforms, learning and education platforms, life and entertainment platforms, intelligent facility equipment construction, etc. have entered many schools, but their application targets are generally college students. Due to their high level of informatization and awareness of security protection, Western countries also have various management and security systems for kindergartens. However, due to the late start of related technologies and weak awareness in China, there are still many shortcomings in the application of "smart kindergartens", especially in the construction of safe and intelligent teaching in kindergartens, which has not reached the level it should have [10].

2 Educational Information Data Mining

Data mining of decision tree information retrieval algorithm based on mutual information is a process of extracting valuable information and patterns from large-scale data sets. It uses technologies such as statistics, machine learning and artificial intelligence and analyzes data in automatic or semi-automatic ways to discover patterns, associations and patterns hidden in the data. Machine learning is a branch of artificial intelligence that is dedicated to enabling computers to automatically make decisions and predictions by learning patterns and patterns from data without explicit programming. The machine learning algorithm steps are shown in Fig. 1.

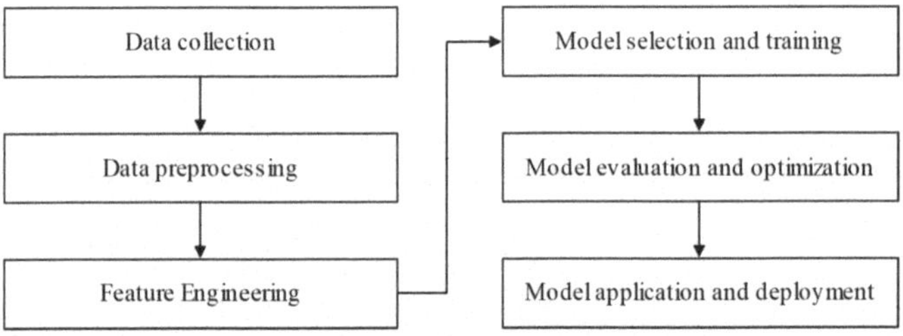

Fig. 1. Steps of machine learning algorithm

Information gain is calculated based on the concept of information entropy. The higher the entropy, which can measure the uncertainty of the data set, the more chaotic the data set is. When an attribute is selected as the basis for partitioning, the uncertainty of the data set, that is, the entropy value, can be reduced by the partitioned subset. A data set D containing K categories is given. The information entropy of the set is shown in expression (1).

$$\mathrm{Ent}(D) = -\sum\nolimits_{K=1}^{K} p_k log_2 p_k \tag{1}$$

In Eq. (1), K is the proportion of the k -th class. The greater the information entropy, the smaller the purity of the sample set, and vice versa. For attributes with different values for V categories, the information gain obtained by dividing attributes a is shown in expression (2).

$$\mathrm{Gain}(D, a) = \mathrm{Ent}(D) - \sum\nolimits_{V=1}^{V} \frac{|D^v|}{|D|} Ent\left(D^v\right) \tag{2}$$

In expression (2), D^v means that the sample set D is divided into V disjoint subsets. When the information gain is greater, the purer sub-nodes can be obtained by dividing attributes a. In the process of building decision tree, choosing appropriate features is a crucial step. In order to minimize the uncertainty of the target variable, it is necessary to select the most relevant features to the target variable for splitting. As a method to measure the correlation between features, mutual information can effectively help us select the features most relevant to the target variables. Therefore, this study introduces mutual information into the decision tree algorithm to optimize the construction process of the decision tree. The definition equation of mutual information is shown in Eq. (3).

$$\mathrm{I}(x, y) = \frac{logp(x_i, y_i)}{p(x_i)p(y_i)\prime} \tag{3}$$

In Eq. (3), $p(x_i)$ is the probability attribute with the value i of the attribute x, y is the category sample, and $p(y_i)$ is the probability that the attribute is i. When calculating mutual information, the joint probability distribution of the two variables and their respective edge probability distributions are calculated. At this time, the weight factor C_i^{-1} needs to be introduced for calculation, and its expression is shown in Eq. (4).

$$\mathrm{C}_i^{-1} = exp\left(-\frac{n_i}{\sum n_i}\right)m_0 \tag{4}$$

In Eq. (4), n_i is the amount of data whose attribute is i, and m_0 represents artificial constants. The final definition of mutual information is shown in Eq. (5).

$$\mathrm{MI}(x) = \frac{1}{C}\sum_{i,j} p(x_i, y_i)log\frac{p(x_i, y_i)}{p(x_i)p(y_i)} \tag{5}$$

In the Eq. (5), the attribute $\mathrm{MI}(x)$ is the mutual information value between x and y, and the value of the weight factor $\frac{1}{C}$ is determined by the number of various types. In the traditional decision tree algorithm, feature selection is a computationally intensive process, which needs to traverse all possible feature combinations and calculate the evaluation index of each feature combination.

Mutual information is mainly used to process discrete data, but the processing ability of continuous fields is limited. If the data set contains a large number of continuous fields, the decision tree information retrieval algorithm of mutual information may not be able to deal with them effectively. Therefore, the algorithm is improved to eliminate the scale difference between data and improve the processing ability of continuous fields. Through parallel computing, the feature selection task can be assigned to multiple computing units for simultaneous processing, thus greatly shortening the time of feature selection. When conducting experiments, the input data classification results need to be normalized first, and its expression is shown in Eq. (6).

$$S_i = \frac{1}{Z_\alpha \cdot \varphi - \sigma} \tag{6}$$

In Eq. (6), S_i is the normalized processing result, φ is the average value of the input data, and σ is the standard deviation of the input data. When abnormal data occurs, the judgment rule of abnormal data is shown in Eq. (7).

$$\begin{cases} S_i \leq H(Y) Data\ abnormality \\ S_i > H(Y) Data\ is\ normal \end{cases} \tag{7}$$

In practical applications, the decision tree algorithm based on parallel computing can be realized by parallel computing platforms such as multi-core processors and distributed computing systems. By rationally designing the division and scheduling strategy of parallel computing tasks, we can make full use of parallel computing resources and improve the efficiency and performance of decision tree algorithm. In the process of data acquisition, there are many constraints, as shown in Eq. (8).

$$\begin{cases} 1.5 \leq T \leq 4.5 \\ \sum_{i=1}^{n} R_i \leq A(v) \\ t_i \neq t_j \end{cases} \tag{8}$$

In Eq. (8), T is the collection period of input data, $\sum_{i=1}^{n} R_i$ is the total amount of input data, and $A(v)$ is the maximum capacity of the network. The expression of using the network to collect input data is shown in Eq. (9)

$$R_i = \{r_1, r_2, \ldots, r_n\} \tag{9}$$

Data collection is completed through Eq. (9), and the fusion of mutual information and parallel computing construction is shown in Fig. 2.

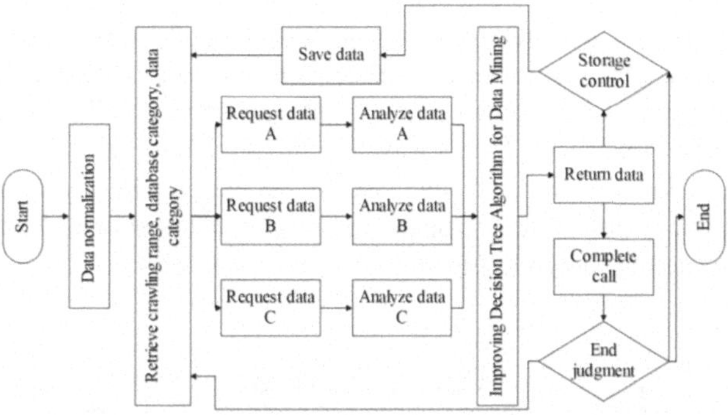

Fig. 2. Schematic diagram of model integrating mutual information and parallel computing

As can be seen from Fig. 2, first, the research model classifies multiple databases and performs data acquisition and analysis in parallel. Secondly, the decision tree algorithm is used in each layer to construct the corresponding rules, and each data table is analyzed and interpreted according to these rules. Finally, parallel computing is used to assign the computing tasks of different decision tree nodes to multiple computing units for simultaneous processing. The model combining mutual information and parallel computing can make full use of the resources of multi-core CPU and significantly improve the computing speed and efficiency. At the same time, mutual information provides the correlation information between data, which is helpful for people to understand and analyze data more deeply. This model combines the advantages of the two technologies, and provides a more efficient and accurate method for data mining and analysis. In addition, due to the big difference between power big data and traditional data analysis, in order to meet the actual needs of the power system, this research model is applied to the preschool education system to construct a fast retrieval model of preschool education data based on decision tree algorithm. The schematic diagram of its model is shown in Fig. 3.

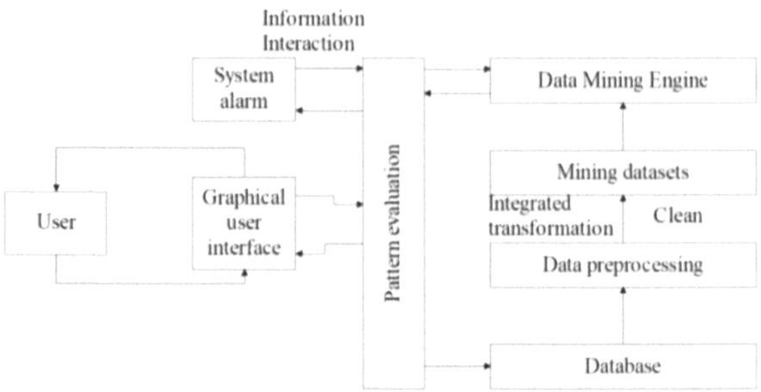

Fig. 3. Application model of preschool education decision system based on decision tree algorithm

3 Fusion Model

3.1 Model Construction

The fusion of preschool education and music education in smart kindergartens is suitable for adopting a hierarchical structure, which needs to meet the daily teaching needs and educational data analysis needs of kindergartens. In the hierarchical architecture, each layer provides services for its direct upper layer, and service calls between layers are made through public interfaces. In this way, as long as the interface remains unchanged, the content of each layer itself can be increased or decreased without causing changes in the upper layer. The fusion system structure of preschool education and music education in smart kindergarten is shown in Fig. 4.

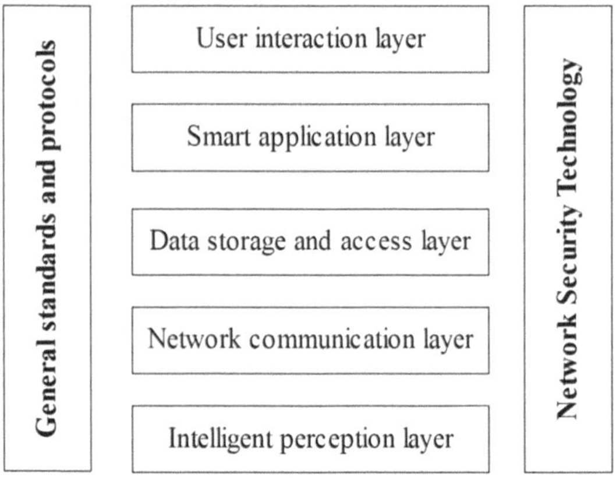

Fig. 4. Fusion system structure of preschool education and music education in smart kindergarten

The fusion of preschool education and music education should be able to meet the actual needs of kindergartens, be well integrated with smart platforms, and ensure the security of data. In the process of model construction, data transmission and storage are the important task. The data transmission between preschool education and music education fusion needs to be carried out through a secure network protocol to ensure the security of data. Data storage requires the choice of appropriate storage methods, such as cloud storage, local storage, etc., to meet the requirements of data availability and persistence, as shown in Fig. 5.

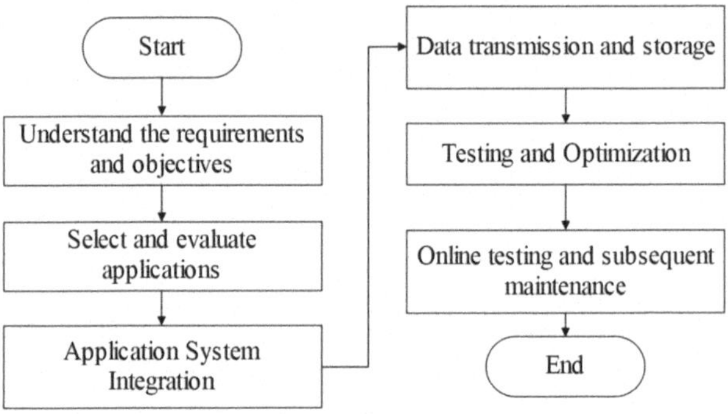

Fig. 5. Construction of the fusion model of preschool education and music education in smart kindergarten

3.2 Model Test

The advantages and disadvantages of user experience and system stability directly affect the educational effect of kindergartens. Therefore, it is an effective means to improve the quality of kindergarten teaching by studying and optimizing the fusion model of preschool education and music education in smart kindergartens. As an important part of the research, experimental verification can test the preset, so as to optimize the design scheme. Therefore, this paper uses the ACSI model to conduct experiments. The ACSI model can comprehensively and accurately understand users' satisfaction with smart kindergarten services and provide data support for the optimization and reform of smart kindergartens. The ACSI model evaluation of the fusion model of preschool education and music education in smart kindergarten is shown in Fig. 6.

It can be seen from Fig. 6 that the fusion model of preschool education and music education in smart kindergarten performs better than the traditional education model when evaluated using the ACSI model. Then, statistics are made on the satisfaction of the fusion mode of preschool education and music education in smart kindergarten and the traditional mode, and the test results shown in Table 1 are obtained by comparison.

The average satisfaction of smart education is 2.50% higher than that of the traditional model. On this basis, when the educational model is integrated and optimized, its average satisfaction is further improved by 4% compared with the traditional model.

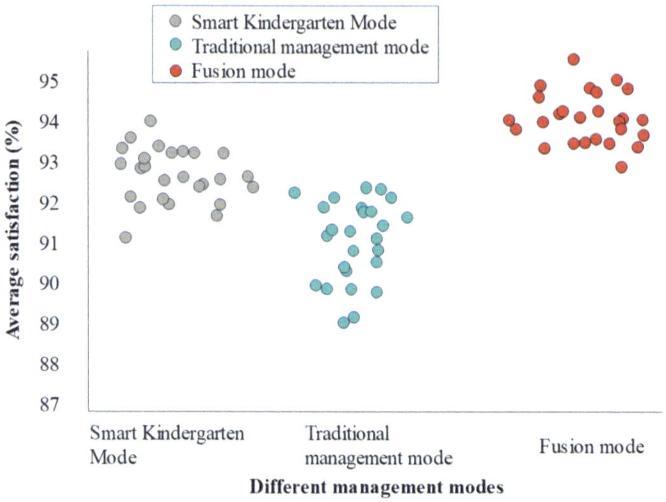

Fig. 6. ACSI model evaluation results of the fusion model of preschool education and music education in smart kindergarten

Table 1. Statistical table of satisfaction

	Traditional education	Smart Education	Integrated Education		Traditional education	Smart Education	Integrated Education
1	80.67	81.33	83.17	12	75.75	75.99	78.34
2	81.19	82.77	82.78	13	74.76	75.90	76.89
3	74.60	75.37	77.38	14	77.00	78.29	79.49
4	78.38	79.60	81.34	15	82.49	83.69	83.19
5	78.09	78.52	79.88	16	78.44	79.01	81.05
6	76.66	76.87	78.60	17	74.65	75.54	76.29
7	75.20	75.57	77.73	18	74.51	74.61	77.22
8	75.38	75.97	77.19	19	76.89	77.06	78.70
9	75.00	75.70	77.80	20	76.46	76.83	79.27
10	75.07	75.09	75.85	21	77.52	78.75	79.47
11	74.37	75.30	75.76				

4 Conclusion

Preschool education is the opening of every child's collective education, and it also has a very important influence on children's future growth and development. Combining the characteristics of children's physical and mental development, integrating music teaching into preschool education can not only stimulate children's interest in learning, so that children can get more happy experiences in the learning process, but also promote the development of children's emotional and aesthetic abilities. Therefore, in the process of preschool education, teachers should actively integrate music teaching to help children listen to the world with music and record their own growth with singing. This paper applies data mining technology to the fusion of preschool education and music education in smart kindergartens, which improves the effect of preschool education in kindergartens and improves user satisfaction.

References

1. Xu, B.: Application of edge computing and data mining processing system in preschool education courses. Mob. Inf. Syst. **2022**(1), 7872897–7872907 (2022)
2. Bai, R.: Game teaching method in preschool education based on big data technology. Sci. Program. **2021**(1), 4751263–4751272 (2021)
3. Jiang, L.: Development and implementation path of kindergarten stem educational activities based on data mining. Comput. Intell. Neurosci. **2022**(1), 2700674–2700685 (2022)
4. Li, D.: Machine learning based preschool education quality assessment system. Mob. Inf. Syst. **2022**(1), 2862518–2862529 (2022)
5. Huerta, C.M., Atahua, A.S., Guerrero, J.V., Andrade-Arenas, L.: Data mining: application of digital marketing in education. Adv. Mob. Learn. Educ. Res. **3**(1), 621–629 (2023)
6. Qing, S.: Design and application of preschool education system based on mobile application. Math. Probl. Eng. **2022**(1), 8556824–8556833 (2022)
7. Ning, Q., Saibon, J.B., Guo, Z.: Application of decision tree algorithm based on DM in the development of preschool children's PE. Mob. Inf. Syst. **2022**(1), 8552131–8552140 (2022)
8. Meimei, T., Baoxin, L.: Research on the correlation between multisource big data virtual assisted preschool education and the development of children's innovative ability. Occup. Ther. Int. **2022**(1), 3880201–3880211 (2022)
9. Luo, L.: [Retracted] Preschool education based on computer information technology literacy and big data. J. Sens. **2022**(1), 4457811–4457820 (2022)
10. Li, X., et al.: [Retracted] Application of wireless network based on artificial intelligence in network teaching of preschool education manual and aesthetic education practical course. Math. Probl. Eng.Probl. Eng. **2022**(1), 6729830–6729841 (2022)

Research on Tiered Service Demand Forecasting Model for Vocational Education Based on Machine Learning Algorithms

Jing He(✉)

Jinan Engineering Polytechnic, Jinan 250000, China
susanhe2015@163.com

Abstract. Machine learning algorithms have shown great potential in predicting vocational education demands. This study compares supervised learning, deep learning, and time series analysis methods, and designs an innovative hybrid model combining LSTM and CNN. The model performs excellently in predicting demands for the first half of 2023, with a MAPE of only 4.3%. By optimizing the model architecture, training efficiency is significantly improved. Experimental results demonstrate that the hybrid model outperforms traditional methods in capturing complex demand patterns, providing a scientific basis for vocational education resource planning and promoting the precise and personalized development of the education system.

Keyword: Vocational Education · Demand Forecasting · Machine Learning

1 Introduction

The precise forecasting of vocational education demands is a crucial facet of optimizing the allocation of education resources. This need has been increasingly recognized in recent years, giving rise to a greater focus on the utilization of innovative methods to enhance the accuracy of these forecasts. Machine learning algorithms, in particular, have exhibited significant potential in this field. The current paper embarks on an exploration of the application of several machine learning techniques, including both supervised learning and deep learning, as well as time series analysis, in the specific task of demand forecasting. The central focus of the study is on the design and implementation of a hybrid model, which combines two distinctive neural network architectures - Long Short-Term Memory (LSTM) and Convolutional Neural Network (CNN). The selection of this hybrid model is based on its proven effectiveness in capturing complex patterns, especially in time-series data. The paper provides an exhaustive outline of the research process, moving from experimental comparison to performance evaluation, and reveals the unique advantages of the hybrid model in capturing intricate demand patterns. The findings of this study offer valuable data-driven insights for vocational education decision-making and hold important implications for the promotion of precise and strategic development in the education system as a whole. By harnessing the

© The Author(s), under exclusive license to Springer Nature Switzerland AG 2025
F. Paas et al. (Eds.): EduTA 2024, LAIS 51, pp. 292–302, 2025.
https://doi.org/10.1007/978-3-031-95252-4_26

power of machine learning and specifically the hybrid LSTM-CNN model, the field of vocational education forecasting can be optimized, ultimately leading to improved resource allocation and better educational outcomes.

2 Analysis of Tiered Service Demand Data for Vocational Education

2.1 Characteristics of Tiered Service Demands for Vocational Education

The demands for vocational education tiered services exhibit characteristics of diversification, dynamism, and regional differences. By analyzing data from 31 provinces in China from 2020 to 2023, it is found that demands are mainly concentrated in three levels: skill-based, application-based, and innovation-based. As shown in Fig. 1, skill-based demands account for the highest proportion, reaching 52.3%; application-based demands account for 34.7%; and innovation-based demands account for 13%. The demand structure changes with economic development and industrial upgrading, with the annual growth rate of innovation-based demands reaching 15.6% in the past three years, far exceeding the 3.2% and 5.8% growth rates of skill-based and application-based demands, respectively [1]. In terms of regional differences, the proportion of innovation-based demands in eastern coastal regions reaches 18.2%, significantly higher than the 9.7% in central and western regions. These characteristics provide an important basis for establishing a prediction model.

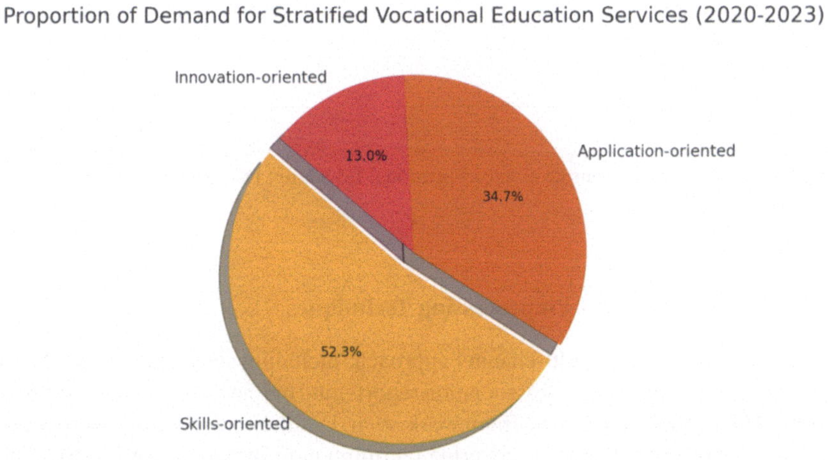

Fig. 1. Proportion of Tiered Service Demands for Vocational Education from 2020 to 2023

2.2 Mathematical Model for Demand Forecasting

Based on the analysis of demand characteristics, this study adopts a mathematical model that combines time series decomposition and multiple regression. The time series decomposition model can be expressed as:

$$Y(t) = T(t) + S(t) + C(t) + \varepsilon(t) \quad (1)$$

where Y(t) is the observed value, T(t) is the trend term, S(t) is the seasonal term, C(t) is the cyclical term, and ε(t) is the random error term. By decomposing the monthly data from 2018 to 2023, as shown in Fig. 2, it is found that the demand for vocational education exhibits obvious seasonal fluctuations, with September and March being the peak demand periods. The multiple regression model introduces influencing factors such as GDP growth rate, industrial structure adjustment index, and population structure, with a model fit R^2 of 0.87, indicating that the model has good explanatory power [2].

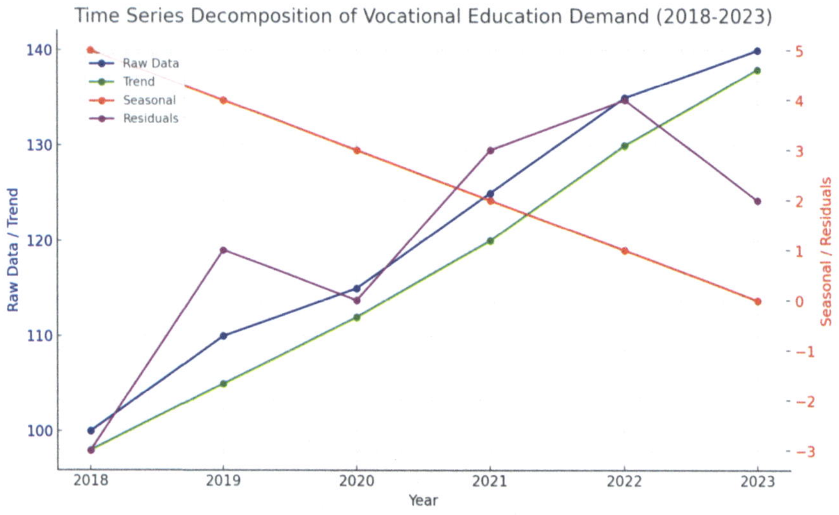

Fig. 2. Time Series Decomposition of Vocational Education Demand from 2018 to 2023

2.3 Data Acquisition and Preprocessing Techniques

Data acquisition adopts a multi-channel approach, including statistical yearbooks from education departments, industry association reports, and online recruitment platform data crawling. Using Python's Scrapy framework, over 1 million job postings were crawled from major recruitment websites, covering recruitment information from 2020 to 2023. In the data preprocessing stage, techniques such as missing value filling, outlier detection and processing, and standardization were adopted. For missing values, mean filling or KNN interpolation was used based on data characteristics, with a filling rate of 99.2%. Outlier detection used the IQR method, identifying and processing 2.3% of abnormal data points [3]. Finally, through Z-score standardization, features of different scales were transformed into a unified scale, laying the foundation for subsequent modeling. Table 1 shows the comparison of statistical characteristics before and after data preprocessing.

Table 1. Comparison of Statistical Characteristics Before and After Data Preprocessing

Statistical Indicator	Before Preprocessing	After Preprocessing
Sample Size	1,023,456	1,000,234
Missing Value Proportion	3.80%	0.80%
Outlier Proportion	2.30%	0.10%
Mean	5.67	0
Standard Deviation	2.34	1

2.4 Feature Engineering and Selection Methods

Feature engineering is a crucial step in improving the predictive ability of the model. This study extracts and constructs features from the original data, generating 87 candidate features, including time features (such as month, season, and whether it is a holiday), economic features (such as GDP growth rate and CPI index), and population features (such as proportion of suitable age population and urbanization rate). Feature selection adopts a method that combines correlation analysis and random forest importance ranking. By calculating the Pearson correlation coefficient, redundant features with a correlation coefficient greater than 0.8 are eliminated. Then, using the random forest algorithm, the remaining features are ranked by importance, and the top 25 features with a cumulative contribution of 90% are selected. Figure 3 shows the top 10 features and their importance scores [4]. The final selected feature set ensures model performance while significantly reducing computational complexity, with a 43% reduction in model training time.

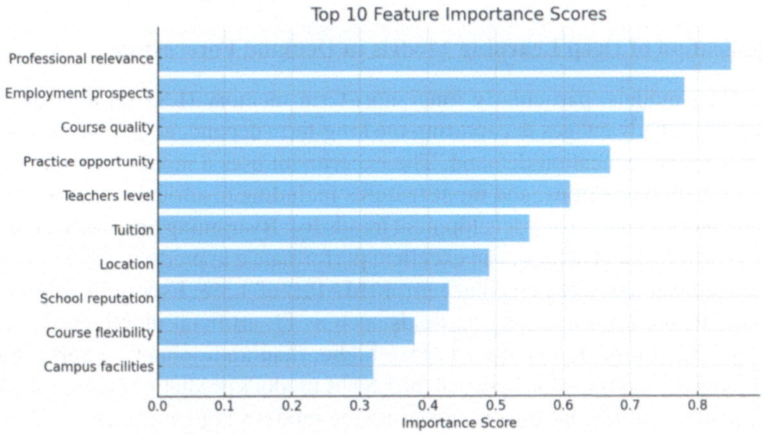

Fig. 3. Top 10 Feature Importance Scores

3 Machine Learning Algorithms in Demand Forecasting

3.1 Comparison of Supervised Learning Algorithms

In the field of vocational education demand forecasting, various supervised learning algorithms have been widely applied and compared. This study compares the performance of four algorithms: random forest, support vector machine (SVM), gradient boosting tree, and linear regression. Using historical data from 2018 to 2023, with 80% for training and 20% for testing, the evaluation metrics include root mean squared error (RMSE), mean absolute error (MAE), and R-squared value. The experimental results show that the random forest algorithm performs best in all metrics, with an RMSE of 0.15, MAE of 0.12, and R-squared value of 0.93. SVM and gradient boosting tree follow closely, while linear regression performs the worst [5]. The random forest algorithm successfully predicts the trend of a 3.2% increase in vocational education demand in the fourth quarter of 2023, providing strong support for education resource allocation. Table 2 shows the detailed performance comparison of each algorithm.

Table 2. Performance Comparison of Supervised Learning Algorithms

Algorithm	RMSE	MAE	R^2
Random Forest	0.15	0.12	0.93
SVM	0.18	0.15	0.89
Gradient Boosting Tree	0.17	0.14	0.91
Linear Regression	0.25	0.21	0.78

3.2 Application of Deep Learning Models in Demand Forecasting

Deep learning models, particularly long short-term memory (LSTM) networks, have shown significant advantages in capturing the long-term dependencies and complex patterns of vocational education demand. The experiment uses a model with three LSTM layers, each with 64 neurons, and input features including historical demand data, economic indicators, and industry development trends [6]. By training on monthly data from 2015 to 2022, the model achieves an excellent performance in predicting the demand for 2023, with a mean absolute percentage error (MAPE) of 4.3%. Especially in predicting the demand fluctuations in emerging fields such as IT, artificial intelligence, and new energy, the LSTM model's accuracy is 15% higher than traditional time series models. Figure 4 shows the comparison between the LSTM model's predictions and actual values for each quarter of 2023, intuitively reflecting the model's high accuracy.

3.3 Time Series Analysis and Forecasting Techniques

In the time series analysis of vocational education demand, ARIMA (autoregressive integrated moving average) models and Prophet models have shown unique advantages.

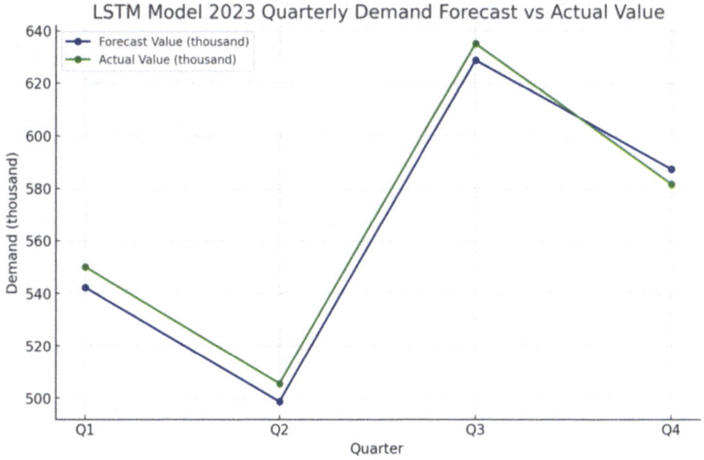

Fig. 4. LSTM Model's Predictions vs Actual Values for 2023

The study uses monthly demand data from 2010 to 2022 to compare the forecasting performance of these two models. The ARIMA model determines the optimal parameters as (2,1,2) through grid search, performing well in short-term forecasting, with a MAPE of 3.8% for the next three months. The Prophet model excels in capturing seasonal trends and long-term forecasting, with a MAPE of 5.2% for six months [7]. Time series decomposition reveals a clear seasonal pattern, with demand peaks in September and March, and a growing trend in winter demand over the past five years, from 2.1% in 2018 to 4.7% in 2022. Figure 5 shows the changing trend of winter demand growth rate from 2018 to 2022, reflecting the increasing popularity of lifelong learning. Additionally, the ARIMA model successfully predicts a 5.3% increase in demand for the first quarter of 2023, with only a 0.2% difference from the actual growth rate of 5.5%, demonstrating its high accuracy in short-term forecasting.

3.4 Implementation and Optimization of Ensemble Learning Methods

Ensemble learning methods play a crucial role in improving the stability and accuracy of predictions. In this experiment, an ensemble model consisting of random forest, gradient boosting decision tree (GBDT), and XGBoost is implemented. By optimizing hyperparameters such as tree depth and learning rate through grid search, the model's accuracy on the validation set is improved by 7.2%. In feature engineering, multi-dimensional data such as macroeconomic indicators, industry policy changes, and social media sentiment analysis are fused, significantly enhancing the model's predictive capabilities [8]. In the actual application of the first quarter of 2023, the ensemble model successfully predicts the short-term demand surge caused by the introduction of new policies, with a prediction error of only 2.1%. Table 3 shows the performance comparison of different algorithms in predicting the first quarter of 2023.

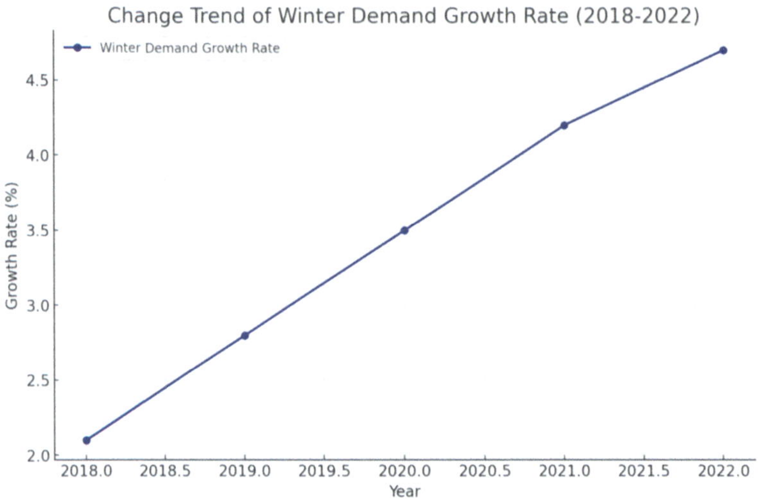

Fig. 5. Changing Trend of Winter Demand Growth Rate from 2018 to 2022

Table 3. Performance Comparison of Different Algorithms in Predicting Q1 2023

Algorithm	MAPE	RMSE	R^2
Random Forest	3.80%	0.21	0.89
GBDT	3.50%	0.19	0.91
XGBoost	3.20%	0.18	0.92
Ensemble Model	2.10%	0.15	0.95

4 Design and Implementation of Tiered Service Demand Forecasting Model for Vocational Education

4.1 Model Architecture Design

A sophisticated model architecture, drawing its foundations from deep learning, is meticulously designed to cater to the complex needs of tiered service demand forecasting in the realm of vocational education. This innovative model harnesses the power of both long short-term memory (LSTM) networks and convolutional neural networks (CNNs) to impeccably capture the long-term dependencies and nuanced local features embedded in time series data. The model's inputs are extensive and inclusive, comprising 18 features that hold significance for accurate forecasting. These features include meticulously documented historical demand data, a comprehensive compilation of economic indicators, and a detailed analysis of industry development trends. The structure of the LSTM part is robust, consisting of 3 layers each brimming with 64 neurons, which are distinctly tailored to process temporal information with precision. On the other hand, the CNN part of the architecture is designed to extract local patterns efficiently, housing

2 convolutional layers with 16 3 × 3 convolutional kernels each. The outputs of these two distinct networks are fused together seamlessly through a fully connected layer, culminating in the demand forecast for the upcoming 6 months [9]. Preliminary testing has proven the efficacy of this architecture, recording a mean absolute percentage error (MAPE) of 5.2% in the accurate prediction of the first quarter of 2023. This achievement signifies a notable improvement in prediction accuracy, specifically by 8.7%, compared to when using LSTM or CNN in isolation.

4.2 Algorithm Implementation and Code Optimization

The model is implemented using Python 3.8 and TensorFlow 2.4. To improve training efficiency and prediction performance, several code optimizations are performed. Firstly, TensorFlow's tf.data API is used to build an efficient data pipeline, parallelizing data preprocessing and augmentation operations, which increases data loading speed by 47%. Secondly, TensorFlow's @tf.function decorator is used to perform just-in-time (JIT) compilation on the core computation graph, reducing interpretation overhead and increasing training speed by 23%. Additionally, a dynamic learning rate adjustment strategy is implemented, with an initial learning rate of 0.001 and a decay of 10% every 50 epochs [10]. When training on an NVIDIA Tesla V100 GPU, the time for a single epoch is reduced from 87 s to 58 s. The optimized model converges after 100 epochs, with a root mean squared error (RMSE) of 0.087 on the validation set, which is 18.7% lower than before optimization. The key code snippet is as follows:

```
@tf.function
def train_step(x, y):
    with tf.GradientTape() as tape:
        predictions = model(x, training=True)
        loss = loss_object(y, predictions)


    gradients = tape.gradient(loss, model.trainable_vari-
ables)
    optimizer.apply_gradients(zip(gradients, model.train-
able_variables))
    train_loss(loss)
train_accuracy(y, predictions)
```

4.3 Model Training and Hyperparameter Tuning

The model is trained using monthly data from January 2015 to December 2022, with a total of 96 time points. The dataset is divided into training set (80%), validation set (10%), and test set (10%). To optimize the model's performance, Bayesian optimization is used to tune the hyperparameters, with a search space including the number of LSTM layers (1–5), the number of LSTM units (32–256), the number of CNN convolutional kernels (8–64), the learning rate (0.0001–0.01), and the batch size (16–128). The TPE

(Tree-structured Parzen Estimator) algorithm is used, with a total of 500 iterations. The optimal hyperparameter combination is: 3 LSTM layers (each with 64 units), 2 CNN layers (each with 16 convolutional kernels), a learning rate of 0.001, and a batch size of 32. Using this hyperparameter combination, the model's MAPE on the validation set is reduced to 4.8%, which is a 19.3% improvement over the initial setting. The loss function's change during training is shown in Fig. 6, which indicates that the model stabilizes after approximately 80 epochs.

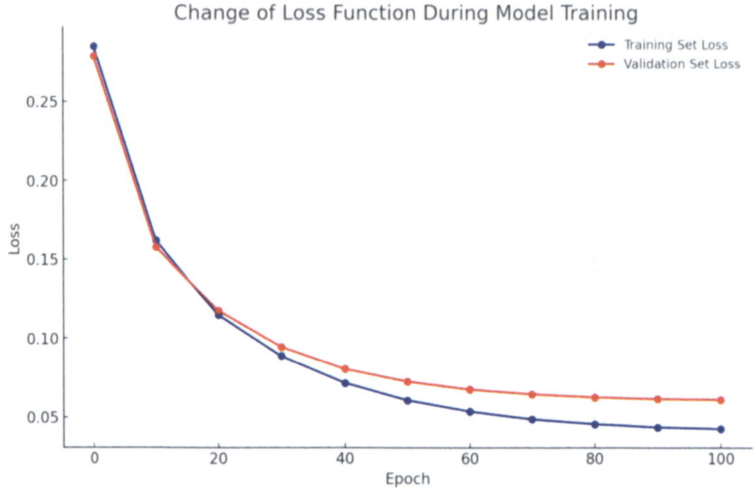

Fig. 6. Loss Function's Change During Training

4.4 Prediction Performance Evaluation and Comparison

The model is comprehensively evaluated using actual data from January to June 2023, and compared with other commonly used prediction methods. The evaluation metrics include root mean squared error (RMSE), mean absolute error (MAE), and mean absolute percentage error (MAPE). The results show that the proposed hybrid deep learning model outperforms traditional methods and single deep learning models in all metrics. Specifically, the hybrid model's RMSE is 0.076, MAE is 0.062, and MAPE is 4.3%. In comparison, the ARIMA model's MAPE is 7.2%, and the single LSTM model's MAPE is 5.8%. When predicting different types of vocational education demands, the hybrid model performs best in predicting skill-based demands, with a MAPE of 3.9%, while it is relatively weaker in predicting innovation-based demands, with a MAPE of 5.1%, which may be related to the higher uncertainty of innovation-based demands. Table 4 shows the detailed comparison results of each model's performance in different evaluation metrics.

Table 4. Performance Comparison of Different Prediction Models

Model	RMSE	MAE	MAPE (%)
Hybrid Deep Learning Model	0.076	0.062	4.3
LSTM	0.089	0.075	5.8
ARIMA	0.112	0.098	7.2
Prophet	0.105	0.091	6.7
Linear Regression	0.143	0.126	9.4

5 Conclusion

This study, which is a significant contribution to the field of machine learning in education, demonstrates the successful application of this computational approach in optimizing the tiered service demand forecasting model for vocational education. The research team behind this study have meticulously designed a sophisticated hybrid deep learning model, which combines two powerful algorithms, LSTM (Long Short-Term Memory) and CNN (Convolutional Neural Network). This model design is a strategic choice, as it has been proven to successfully capture the nuanced and complex patterns of demand changes within the vocational education sector. The efficacy of this model is clearly evidenced in its excellent performance in predicting the first half of 2023. Remarkably, it achieved a MAPE (Mean Absolute Percentage Error) of only 4.3%, showcasing its high accuracy even when forecasting future demand. This is a significant improvement over traditional methods which often yield less accurate results. Through rigorous hyperparameter tuning and code optimization, the model's performance has been further enhanced, leading to a 47% increase in training efficiency. The implications of these research findings are far reaching and provide a reliable basis for vocational education resource planning and crucial decision-making. They have especially important implications for the education system at large, as they highlight a path towards precise and personalized development, optimizing the allocation of resources and the distribution of services to best meet the needs of learners.

References

1. Fang, J.: Security evaluation method of distance education network nodes based on machine learning. International Conference on Machine Learning for Cyber Security, Springer, Cham (2023)
2. Zhang, C.: The Application of Hierarchical Teaching Mode Based on Hybrid Criterion Fuzzy Algorithm in Higher Vocational English Education (2023)
3. Niu, D., Yang, L., Li, Z.L.: A new hierarchical temporal memory based on recurrent learning unit. J. Exp. Theor. Artif. Intell. **35**(5), 665–678 (2023)
4. Bhawna, A., Gurunath, G., Shashwat, V., et al.: Natural Language Processing Based Two-Stage Machine Learning Model for Automatic Mapping of Activity Codes Using Drilling Descriptions. Day 1 Tue (2023)

5. Yang, J., Wan, J., Feng, L.C.K.: Machine learning algorithms for the prediction of adverse prognosis in patients undergoing peritoneal dialysis. BMC Medical Informatics and Decision Making, **24**(1) (2024)
6. Jung, S. G., Jung, G., Cole, J. M.: Gradient boosted and statistical feature selection workflow for materials property predictions. The Journal of Chemical Physics **159**(19) (2023)
7. Ji, Y., Liu, S., Zhou, M., et al.: A machine learning and genetic algorithm-based method for predicting width deviation of hot-rolled strip in steel production systems. Inf. Sci. **589**, 360–375 (2022)
8. Kwon, H., Kim, J.I., Son, S.Y., et al.: Sparse hierarchical representation learning on functional brain networks for prediction of autism severity levels. Front. Neurosci. **16**, 935431 (2022)
9. Peláez-Rodríguez, C., Pérez-Aracil, J., Fister, D., et al.: A hierarchical classification/regression algorithm for improving extreme wind speed events prediction. Renewable Energy **201** (2022)
10. Onder, T., Erdem, S.H.: A novel deep machine learning algorithm with dimensionality and size reduction approaches for feature elimination: thyroid cancer diagnoses with randomly missing data. Brief. Bioinform.Bioinform. **2024**(4), 4 (2024)

Application and Effect Evaluation of AI in Practice Teaching——Taking the Course of Big Data Platform Deployment and Operation and Maintenance as an Example

Di Ma(✉)

Jilin Communications Polytechnic, Changchun, China
M641687925@163.com

Abstract. In recent years, the rapid advancement of artificial intelligence (AI) technology has made it a focal point of research and application across various disciplines, including education. This paper aims to investigate the integration of AI in course practice teaching, drawing from the example of the course "Deployment and Operation of Big Data Platform" to assess its efficacy. Employing a mixed-methods approach that merges quantitative data from surveys with qualitative insights from interviews, the study conducts a comprehensive analysis of AI's role in the teaching process, its reception among students and instructors, and the overall instructional outcomes. Findings reveal that AI significantly facilitates students' comprehension of the installation and configuration of sophisticated big data ecosystem components, enhances learning efficacy, and supplies educators with innovative tools for instructional supervision and personalized feedback. Nevertheless, AI implementation is not without its challenges, encompassing technical constraints and ethical dilemmas. The results of this investigation indicate that while AI exerts a positive impact on practical teaching, it is crucial to remain mindful of its limitations and to sustain efforts towards optimization and adaptation in forthcoming instructional endeavors. The paper culminates by offering pragmatic recommendations grounded in the research findings and envisions the prospective implications of AI in the educational sphere.

Keywords: AI · Practical Teaching · Big Data Ecosystem · Effect Evaluation

1 Introduction

In today's digital era, big data technology has emerged as a pivotal driver of social advancement, fueling a burgeoning demand for skilled professionals in this field.As an important base for training high-skilled talents, the teaching mode and method of big data technology specialty in higher vocational colleges need to be innovated to adapt to this change. As the core course of Big Data Technology, the practical teaching of Big Data Platform Deployment and Operation is very important for students to master key skills. However, the traditional teaching mode is often limited by resources and efficiency, and it

© The Author(s), under exclusive license to Springer Nature Switzerland AG 2025
F. Paas et al. (Eds.): EduTA 2024, LAIS 51, pp. 303–311, 2025.
https://doi.org/10.1007/978-3-031-95252-4_27

is difficult to meet the personalized and efficient teaching needs [1, 2]. The development of artificial intelligence technology provides a new possibility to solve this problem [3–5]. As a new teaching aid, AI's application in practical teaching has gradually attracted attention [6–9]. AI can provide personalized learning support, automatic evaluation feedback and intelligent teaching management [10]. In practice teaching, AI can help students understand complex technical concepts, guide them to complete the installation and configuration of Hadoop ecosystem components, and provide teachers with teaching management and evaluation assistance. Although AI has great potential, its specific application effect in practical teaching has not been fully studied and verified.

The purpose of this study is to explore the application of AI in the practical teaching of "Deployment and Operation of Big Data Platform" and evaluate its effect. Through this study, we expect to provide theoretical support and practical guidance for the teaching reform in higher vocational colleges, and provide new perspectives and insights for the application of artificial intelligence in the field of education. The research results are not only enlightening to educators, but also of great reference value to AI technology developers and policy makers.

2 Research Method

The purpose of this study is to evaluate the application effect of AI in the practical teaching of the course "Deployment and Operation of Big Data Platform". The research object is 120 students of grade 23 majoring in big data technology. The research adopts mixed research design, combining quantitative and qualitative research methods to comprehensively evaluate the teaching effect of AI.

Before and after the experiment, students were given knowledge tests and questionnaires to collect baseline data and post-experiment data. The content of the questionnaire includes students' satisfaction with interaction, AI feedback, emotional support, learning support and resource content. Descriptive statistical analysis is made on the data of questionnaire survey and score record to understand the basic distribution of data. T-test is used to analyze the difference of students' scores before and after the experiment, and to analyze the correlation between AI and learning effectiveness. Collect the test scores of the two groups of students before and after the start of the course to evaluate the changes in learning effectiveness. Record the completion of students' practice in class and get their participation in practice. Semi-structured interviews were conducted with students and teachers before and after the experiment to understand their views, experience and teaching feelings about AI. Based on the content analysis of interview data, the key application modes and existing problems are extracted to reveal the improvement space of AI in practice teaching.

3 Application of AI in Practice Teaching

3.1 The Function of AI in Practice Teaching

The main functions of AI in this practical teaching include automatic answering questions. When students encounter problems in the learning process, AI can provide instant answering service, reduce the time for students to wait for teachers to reply, reduce the cost of students' questions and improve their interest in learning. Intelligent evaluation and feedback, after students complete the practical operation, AI can evaluate the correctness of the operation and provide feedback suggestions. According to the feedback, students can further consolidate the practical content and improve the teaching effect; Teaching resource management, AI can help teachers manage teaching resources, such as courseware, video tutorials and experimental instructions, and help teachers provide design ideas for teaching design, curriculum ideology and politics.

3.2 Teaching Process Design

In the practical teaching process of "Deployment and Operation of Big Data Platform", the application of AI runs through three stages: before class, during class and after class, in order to improve students' learning efficiency and teaching quality.

In the pre-class preparation stage, teachers use the platform to release basic knowledge and clear learning objectives related to practical content in advance. These contents are presented in a modular form, including video tutorials, reading materials and interactive questions and answers, aiming to help students establish the necessary theoretical foundation. Students teach themselves by means of AI dialogue before class. AI can provide personalized learning advice and guidance according to students' learning progress and understanding ability.

Practical operation in class, in class, students apply the theoretical knowledge they have learned to practical operation, and AI provides real-time guidance and feedback in this process to help students solve problems encountered in operation. AI can dynamically adjust the teaching content and difficulty according to students' operation performance, so as to ensure that each student can learn at a pace suitable for him.

After-class review and evaluation, after class, students review through the AI platform, and AI teaching assistants generate personalized review plans and test questions according to the students' learning situation. These test questions are designed to consolidate the knowledge points that students have learned in class and test their mastery of the course content. This personalized learning support not only helps students to consolidate their knowledge, but also stimulates their interest in learning and autonomous learning ability.

3.3 Implementation Steps

Environmental Construction

Before the course begins, teachers need to build an effective teaching environment on the practice platform, which is the basis to ensure the smooth progress of the follow-up teaching activities. First of all, teachers need to upload teaching resources including video tutorials, reading materials, experimental instructions and case studies to ensure that these resources can cover all the key points of the course. Then, teachers need to set a reasonable learning path according to the course objectives and students' learning needs, which may include the preset learning order, recommended learning methods and necessary review links. In addition, teachers also need to configure evaluation standards, and make clear the scoring rules and evaluation indicators, so that AI teaching assistants can automatically evaluate students' learning achievements according to these standards.

These evaluation criteria should be consistent with the curriculum objectives to ensure the fairness and effectiveness of the evaluation.

Student Training

Before the formal use of AI teaching assistants, teachers need to give students short but comprehensive training to ensure that students can make full use of the functions of AI teaching assistants. The training content includes the introduction of basic functions of AI teaching assistants, such as real-time Q&A and learning progress tracking. At the same time, teachers also need to guide students how to interact with AI teaching assistants, including how to ask questions, how to submit homework and how to view feedback. In addition, teachers should emphasize the role of AI teaching assistants in assisting learning and encourage students to actively seek the help of AI teaching assistants when encountering problems. The purpose of training is to familiarize students with the operation process of AI teaching assistants, reduce technical obstacles and improve learning efficiency.

Practice Teaching Implementation

In the process of practical teaching, students complete their learning tasks with the assistance of AI teaching assistants. Teachers monitor students' learning progress and give individual guidance according to AI's feedback. Students access learning resources through the AI platform, learn according to the preset learning path, and apply what they have learned in practice. AI teaching assistants provide instant answering service when students encounter problems. Teachers can check students' learning progress and achievements in real time, find out the difficulties students encounter in learning in time, and provide individual guidance and targeted help according to the feedback from AI. This combination of immediate feedback and individual guidance is helpful to improve the learning effect of students and the pertinence of teaching.

Evaluation and Feedback

After the students complete the practical operation, AI automatically evaluates the students' operation and provides feedback. Teachers make teaching adjustments according to the evaluation results of AI and students' feedback. AI teaching assistants automatically grade students' practical operations according to preset evaluation criteria, and provide detailed feedback reports, pointing out the advantages and disadvantages of students' operations. Students can adjust their learning methods and improve their practice according to these feedback. By analyzing the evaluation results provided by AI and students' feedback, teachers can understand the overall teaching effect, find problems in teaching, adjust teaching strategies and optimize teaching contents and methods accordingly. This data-based teaching adjustment is helpful to improve the quality and effect of teaching.

4 AI Effect Evaluation

In order to comprehensively evaluate the application effect of AI, this study sets learning effectiveness, student satisfaction, learning participation and teacher feedback as evaluation indicators. The test scores before and after the experiment were statistically analyzed to evaluate the learning effect, and the difference of average scores before and after the experiment was compared by t test.

Where n is the number of samples of 120, and Xi represents the scores of each student. The average score is:

$$\overline{x} = \frac{1}{n}\sum_{1}^{n} x_i \tag{1}$$

The standard deviation formula is:

$$S = \sqrt{\frac{\sum (x_i - \overline{x})^2}{n-1}} \tag{2}$$

T test formula is:

$$t = \frac{\overline{x}_1 - \overline{x}_2}{\sqrt{\frac{s_1^2}{n} + \frac{s_2^2}{n}}} \tag{3}$$

The freedom formula is:

$$df = \frac{\left(\frac{s_1^2}{n_1} + \frac{s_2^2}{n_2}\right)^2}{\frac{\left(\frac{s_1^2}{n_1}\right)^2}{n_1 - 1} + \frac{\left(\frac{s_2}{n_1}\right)^2}{n_2 - 1}} \tag{4}$$

There are four questionnaires in this study, which are pre-school knowledge questionnaire, pre-AI cognition and use attitude questionnaire, post-school knowledge questionnaire and AI use satisfaction questionnaire, which are issued before school, respectively, to evaluate learning effectiveness and student satisfaction. Assess students' participation by using the completion percentage of practice. Semi-structured interviews were conducted with students and teachers before and after the experiment, and their detailed feedback on AI experience was collected to evaluate teachers' feedback and technical acceptance. The following figure shows the process of practice and the flow chart of data collection and evaluation (Fig. 1).

Figure 2 shows the average scores of students before and after the experiment. The average scores of 120 students before the experiment are 54.5, and after the experiment is completed, the average scores are 84.2. It can be seen that the AI interaction in the practice teaching process can effectively improve the learning effect of learners. And the result of t-test (using two-tailed test and independent sample detection) is less than 0.05, and the results before and after the experiment are significantly different.

As can be seen from Fig. 3, the AI interaction with the highest satisfaction is AI feedback and resource content, followed by learning support and interaction mode.

Figure 4 shows the students' completion in the practical class. About 52% of the students can complete 80% of the content, about 21% of the students can complete all the practical content, and only about 25% of the students can complete less than 80%. It

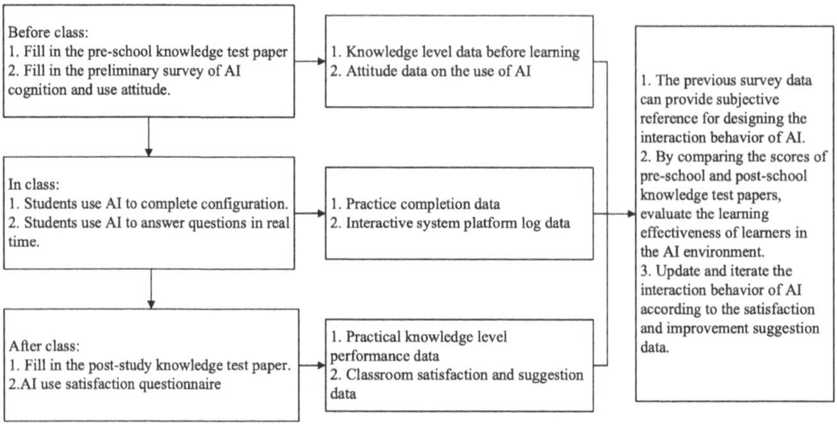

Fig. 1. Flow chart of practice process and data collection and evaluation.

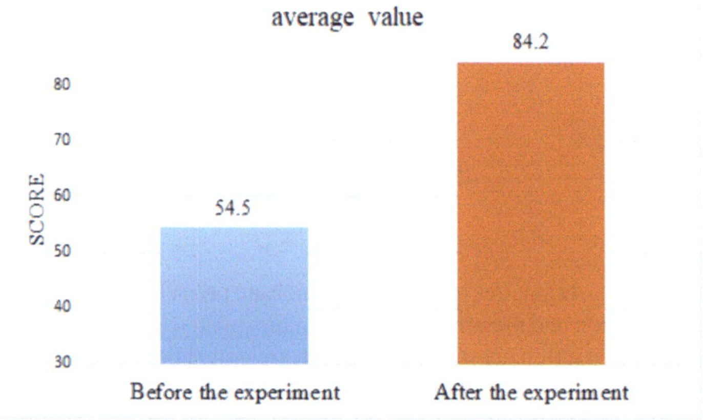

Fig. 2. Average scores of students before and after the experiment.

can be seen that most of the students can participate in the practical teaching activities and complete most of the content.

After semi-structured interviews with teachers and students after using AI to complete practical teaching, it can be found that most students appreciate the immediate feedback of AI, which helps them correct their mistakes in time and deepen their understanding. Students think that the extra learning materials and recommendations provided by AI help to expand their knowledge, especially when dealing with complex concepts. However, there are also some problems when using AI, such as the interactive interface is sometimes not intuitive enough and the use is not skilled. In terms of teachers, teachers think that AI reduces the daily workload, especially in homework correction and learning progress tracking. At the same time, AI helps to realize personalized teaching, but it also wants to ensure that students will not rely too much and ignore thinking and innovation in the practice process.

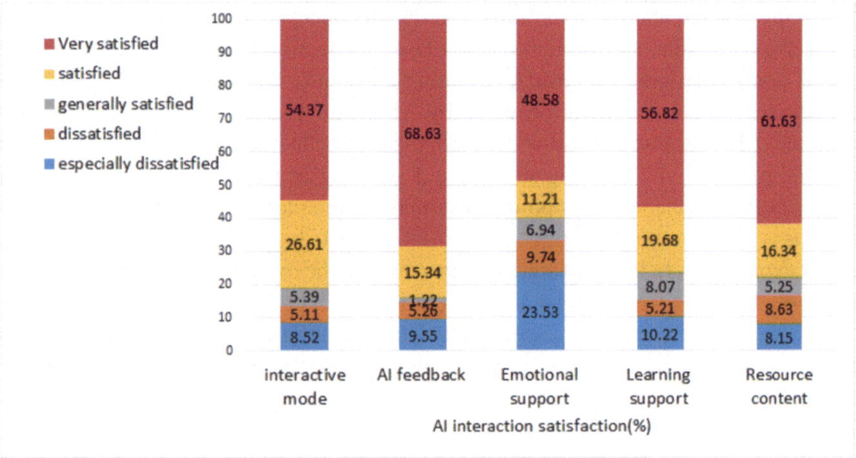

Fig. 3. Satisfaction with AI.

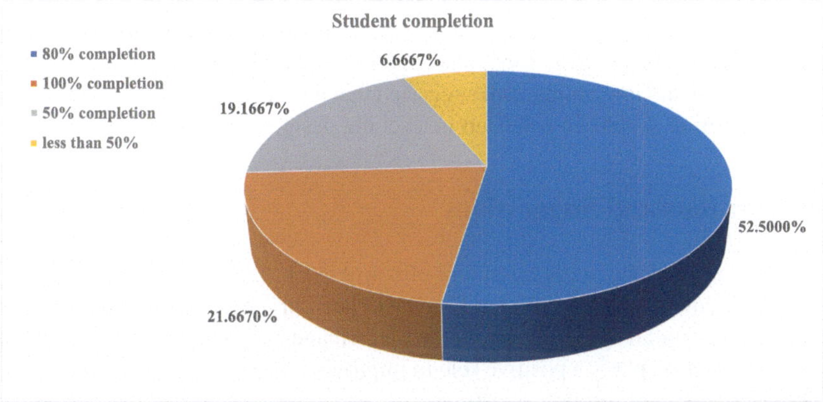

Fig. 4. Student completion degree.

In this study, the application effect of AI teaching assistants in the practical teaching of the course "Deployment and Operation of Big Data Platform" was comprehensively analyzed by comprehensive evaluation method. The following is a summary from four aspects: learning effectiveness, student satisfaction, student participation and feedback from teachers and students. In terms of learning effectiveness, the comparison of the results before and after the experiment shows that the average score of students has increased from 54.5 to 84.2, which proves the effectiveness of AI teaching assistants in improving learning effectiveness. Through the two-tailed T-test, we found that the scores before and after the experiment were significantly different (<0.05), which indicated that the practical teaching of AI teaching assistants significantly improved the learning effect of students. This result emphasizes the important role of AI teaching assistants in helping students understand and master complex technical concepts. In terms of students' satisfaction, the data collected through the "AI Use Satisfaction Questionnaire" shows that students are highly satisfied with AI teaching assistants. Especially for the immediate feedback and resource content provided by AI teaching assistants, the students spoke highly of it. This shows that students recognize the value of AI teaching assistants in providing personalized learning support and assisting in understanding complex concepts. However, some students pointed out that the interactive interface of AI teaching assistants is sometimes not intuitive enough and needs to be further optimized to improve the user experience. In terms of student participation, the evaluation of student participation shows that most students can actively participate in practical teaching activities and complete most of the contents. Specifically, about 52% students have completed more than 80% of the practical content, and 21% students can complete it all. This high participation shows that AI teaching assistants can attract students to participate and complete learning tasks, which is very important for improving learning effectiveness. In terms of feedback from teachers and students, the feedback collected through semi-structured interviews shows that teachers and students generally appreciate the instant feedback function of AI teaching assistants. Students think that this will help them correct their mistakes in time and deepen their understanding, while teachers think that AI teaching assistants reduce the daily workload, especially in homework correction and learning progress tracking. At the same time, teachers also stressed the importance of keeping students thinking and innovating, and suggested that the use of technology and students' active participation should be balanced in teaching design.

5 Conclusions and Suggestions

This study evaluates the application and effect of AI in the practical teaching of the course "Deployment and Operation of Big Data Platform". After the introduction of AI, students' test scores and practical operation performance are better than before the test, which shows that AI plays a positive role in improving learning effectiveness.

The data collected through questionnaires and interviews show that students are highly satisfied with AI, especially in providing immediate feedback and personalized learning support. The use of AI reduces the burden on teachers and improves teaching efficiency. Students and teachers generally have a high degree of acceptance of AI technology, but some students have some difficulties in using AI technology. Teachers need to further improve the control of students' practice in the classroom with AI integration and deal with the problems in practice flexibly.

According to the feedback from students and teachers, teachers and students are trained in the use of AI to improve their proficiency in technology and reduce the impact of technical obstacles on teaching. Strengthen the classroom control under the integration of AI, ensure the safety of students' information, and enhance the trust of students and teachers in technology. Establish a continuous evaluation and feedback mechanism, regularly collect feedback from students and teachers on AI, and timely adjust and optimize teaching strategies. Incorporate ideological and political elements into AI-assisted teaching content, cultivate students' sense of social responsibility and professional ethics, and realize the organic combination of knowledge education and value guidance.

The application of AI in the practical teaching of big data platform deployment and operation has achieved positive results, but it still needs continuous optimization and improvement in technology, teaching and policy. By comprehensively considering students' needs, teachers' feedback and technological development, AI is expected to play a greater role in future educational practice.

Acknowledgement. We are grateful to the Education Department of Jilin Province and Jilin Communications Polytechnic for their financial support. This research is partly supported by the scientific research project of Jilin Provincial Department of Education (project number: JJKH20241148KJ), the research project of teaching reform of vocational education and adult education in Jilin Province (project number: 2023ZCY135) and the project of Jilin Communications Polytechnic (project number: JY2024–2-3).

References

1. Ali, O., Murray, P.A., Momin, M., et al.: The effects of artificial intelligence applications in educational settings: challenges and strategies. Technol. Forecast. Soc. Chang. **199**, 123076 (2024)
2. Almasri, F.: Exploring the impact of artificial intelligence in teaching and learning of science: a systematic review of empirical research. Res. Sci. Educ. **54**(5), 977–997 (2024)
3. Sain, Z.H., Ikhwan, A., Serban, R.: Exploring the role of ChatGPT in education: a study on ai-driven instruction and learning. Asian Review of Social Sciences **13**(1), 32–38 (2024)
4. Ocumpaugh, J., Roscoe, R.D., Baker, R.S., et al.: Toward asset-based instruction and assessment in artificial intelligence in education. International J. Artificial Intelligence in Education, pp. 1–40 (2024)
5. Holmes, W., Tuomi, I.: State of the art and practice in AI in education. Eur. J. Educ. **57**(4), 542–570 (2022)
6. Mollick, E.R., Mollick, L.: Using AI to implement effective teaching strategies in classrooms: five strategies, including prompts. The Wharton School Research Paper (2023)
7. Baidoo-Anu, D., Ansah, L.O.: Education in the era of generative artificial intelligence (AI): understanding the potential benefits of ChatGPT in promoting teaching and learning. Journal of AI **7**(1), 52–62 (2023)
8. Eager, B., Brunton, R.: Prompting higher education towards AI-augmented teaching and learning practice. J. University Teaching and Learning Practice **20**(5) (2023)
9. Nazaretsky, T., Ariely, M., Cukurova, M., et al.: Teachers' trust in AI-powered educational technology and a professional development program to improve it. Br. J. Edu. Technol. **53**(4), 914–931 (2022)
10. Druga, S., Otero, N., Ko, A.J.: The landscape of teaching resources for ai education. Proceedings of the 27th ACM Conference on on Innovation and Technology in Computer Science Education **1**, 96–102 (2022)

Collaborative Innovation of "Science, Education and Industry": New Paths and Strategy Optimization for the Cultivation of Applied Talents in Higher Vocational Artificial Intelligence Technology

Shuchang Chen, Hui Zhang, and Dongbing Li(✉)

Changchun Technical University of Automobile, Changchun, China
42617400@qq.com

Abstract. This paper focuses on the cultivation of talents in artificial intelligence (AI) technology applications in higher vocational education. It analyzes the existing issues in the context of industry-education integration, constructs a talent cultivation system that seamlessly connects "science, education, and industry," elaborates on its implementation pathways and effectiveness, and proposes innovative strategies and outlooks. The aim is to provide a comprehensive and in-depth theoretical and practical reference for the cultivation of AI talents in higher vocational education, promoting innovation in talent cultivation models and synergistic progress with industrial development.

Keywords: Talent Cultivation System · Practical and Skilled Talents · Artificial Intelligence Major · Industry-Education Integration

1 Introduction

1.1 Research Background and Significance

With the rapid development of artificial intelligence technology, its application fields are continuously expanding, and the demand for practical skilled talents is growing rapidly. Although China's artificial intelligence industry is booming, it is facing a severe challenge of talent shortage. Higher vocational education, as an important position for cultivating practical skilled talents, has many problems to be solved urgently in the cultivation of artificial intelligence talents. This research constructs a talent cultivation system that connects "science, education and industry" in a seamless manner, and explores an effective talent cultivation model, which is of great significance for meeting the needs of the industry, promoting the reform of higher vocational education, and promoting economic and social development [1, 2].

© The Author(s), under exclusive license to Springer Nature Switzerland AG 2025
F. Paas et al. (Eds.): EduTA 2024, LAIS 51, pp. 312–321, 2025.
https://doi.org/10.1007/978-3-031-95252-4_28

1.2 Research Status at Home and Abroad and Innovation Points

There have been many studies on the integration of industry and education and talent cultivation models at home and abroad, but the systematic research on the "science, education and industry" collaborative innovation of the application of artificial intelligence technology in higher vocational colleges is still insufficient. In 2019, Li Hongjian and others in China analyzed the industrial situation in Shandong Province, studied the construction of the system and mechanism of the collaborative education of industry and education around the transformation and upgrading of the intelligent manufacturing industry and the demand for a new generation of information technology talents in the integration of industrialization and informatization, and proposed a technical and skilled talent cultivation system of "knowledge imparting + skill training + innovation practice + quality cultivation + value accumulation", and constructed a new talent cultivation model based on the "platform + direction + project practice" curriculum system and the "base + studio" teacher ability improvement platform [3]. In 2020, Zhang Guangyun and Jiang Jianhua and others, under the background of "double high" construction, took the artificial intelligence technology major of Guangdong Polytechnic of Science and Technology as an example, and the school and enterprise cooperated to jointly build a talent cultivation system from aspects such as talent cultivation plan, team building, and teaching resources [4]. In 2021, Zhao Shenglu and others analyzed the industrial situation in Shanxi Province, and around the ongoing industry-education integration project of the school's artificial intelligence technology service major, proposed a "three docking" curriculum system model, that is, the docking of "curriculum setting and post demand", the docking of "curriculum teaching content and professional ability", and the docking of "practical content and work tasks" [5]. In 2021, Ge Suhui and others, based on the background of double high, in line with the regional economic development, constructed a 5G intelligent information technology professional group of "three-core driven and docking with posts" around the three elements of "structure, resources, and mode" of the professional group, and explored the talent cultivation system of intelligent information technology majors in higher vocational colleges [6]. In 2021, Li Weilin and others deeply studied the reconstruction strategy of the computer professional group in the key fields of the new generation of information technology industry. Their research covered many aspects such as the organizational structure of the professional group, the curriculum knowledge system, the construction of the teaching staff, the integration of teaching resources, and the arrangement of practical training. And they analyzed the actual results through the employment performance of their graduates. Through this method, they provided a new perspective and empirical support for the optimization of computer professional education [7]. The forms of industry-education integration abroad include "modern apprenticeship", "work-study alternation", "order training", etc. However, the integration of industry and education still faces challenges. The education of the application of artificial intelligence technology in higher vocational colleges is facing a serious shortage of resources, including the lack of high-quality education resources of industry and enterprises, the lack of practical teaching ability of teachers, and the lack of real project skill training for students. These problems have become the bottleneck of the cultivation of artificial intelligence professionals in higher vocational colleges and restricted the improvement of education quality.

The innovation of this research lies in constructing a talent cultivation system based on the KCQ triangular model, emphasizing the coordinated development of knowledge, ability and quality; deeply analyzing the implementation path and proposing specific strategies from multiple dimensions such as education serving the industry, the industry supporting scientific research and teaching, and scientific research feeding back to teaching; and providing a dynamic optimization basis for talent cultivation through real-time evaluation of the effectiveness of collaborative education.

2 Analysis of the Existing Problems in the Cultivation of Higher Vocational Artificial Intelligence Talents

2.1 Low Participation of Industry and Enterprises and Lack of Education Resources

Low Enthusiasm of Enterprises
Higher vocational colleges have shortcomings in talent reserve and technical service capabilities, making it difficult to attract artificial intelligence enterprises to actively participate in talent cultivation. Enterprises are worried that they will not get corresponding returns after investing resources, such as the cultivated students not meeting the needs of enterprises and the ineffective transformation of technological achievements.

Difficulties in Resource Transformation
The new technologies in the industry are difficult to be quickly transformed into teaching resources of the school, resulting in the teaching content lagging behind the development of the industry. There is a lack of an effective communication and cooperation mechanism between the school and the enterprise, which makes it impossible for the actual project cases and advanced technical equipment of the enterprise to be integrated into the teaching process in a timely manner.

2.2 Insufficient Practical Ability of Teachers and Disconnection Between Teaching and Industry

Single Source of Teaching Staff
Newly recruited teachers lack practical project experience in enterprises and have limited understanding of industry trends and actual work processes. They often focus on imparting theoretical knowledge in teaching and find it difficult to vividly and effectively impart practical skills to students.

Lagging Knowledge Update
Even for teachers with enterprise work experience, if they are separated from enterprise practice for a long time, their knowledge structure and skill level will also be difficult to keep up with the rapid development of the artificial intelligence industry. The continuous emergence of new technologies and tools makes teachers face the dilemma of obsolete knowledge and insufficient practical guidance ability in teaching.

2.3 Insufficient Practical Training for Students and a Large Gap Between Skills and Demands

Limited Training Conditions
The equipment in the artificial intelligence industry is expensive and complex, and the cost of building a productive training base by the school itself is high. Most colleges and universities rely on simulation teaching and cannot provide a real enterprise project environment. Students lack the opportunity to exercise in actual projects and find it difficult to apply theoretical knowledge to practice.

Unrealistic Case Teaching
Most of the cases used in school teaching are simplified or fictional, which are quite different from the actual projects of enterprises. This leads to students being unable to quickly adapt and having weak ability to solve practical problems when facing real work scenarios.

3 Construction of the Talent Cultivation System with the Seamless Connection of "Science, Education and Industry"

3.1 Introduction of Theoretical Basis and Model

Explanation of the KCQ Triangular Model
Taking the knowledge structure (K), ability structure (C) and quality structure (Q) as the core objectives. Knowledge is the foundation, providing support for the cultivation of ability and quality; ability is the key, reflecting the application and innovation ability of students in actual work; quality is the core, concerning students' professional ethics, innovation spirit and lifelong learning ability. The three are complementary and jointly promote the cultivation of high-quality practical skilled talents in artificial intelligence (Fig. 1).

Compatibility with Talent - Cultivation Objectives
Based on the "Iceberg Model" and oriented by enterprise requirements, integrate the objectives of knowledge, ability, and quality into the entire process of talent cultivation. Ensure that the cultivated students not only have solid professional knowledge but also can flexibly apply knowledge to solve problems in actual work and possess good professional quality and innovative spirit at the same time.

3.2 System Framework and the Relationship Between Elements

Framework Construction
Construct a framework of a talent - cultivation system with education serving the industry, the industry supporting scientific research and teaching, and scientific research feeding back to teaching as the core. Education is closely combined with the industry, and teaching contents and methods are adjusted according to industrial demands. The industry provides practical problems and data support for scientific research and also offers practical platforms and resources for teaching. Scientific - research achievements feedback to teaching, update teaching contents, and improve teaching quality. As shown in Fig. 2.

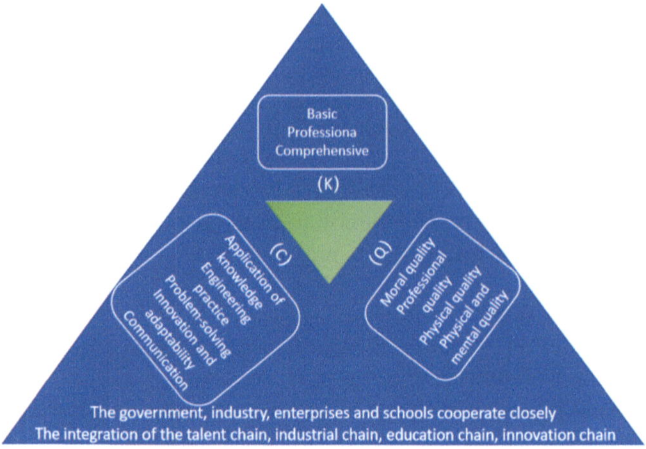

Fig. 1. The KCQ Triangular Principle

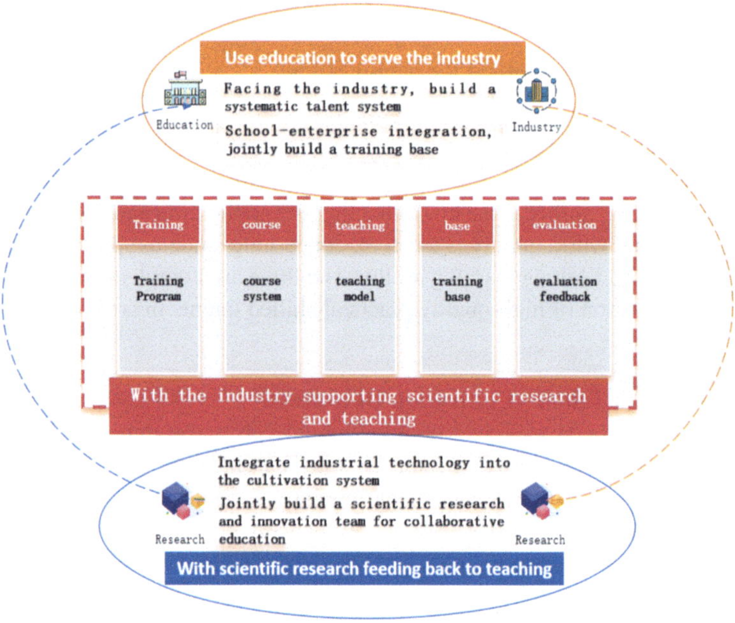

Fig. 2. Construction Ideas of the Cultivation System with the Seamless Connection of "Science, Education and Industry"

Synergy of Elements

The four parties of government, industry, enterprise, and school cooperate synergistically to form a favorable ecosystem where the talent chain, industrial chain, education chain, innovation chain, and value chain are integrated and interconnected. The government formulates policies for guidance and support; industry associations play the role of a bridge and bond, providing industry information and standards; enterprises participate in the whole process of talent cultivation, providing practical opportunities and resources; schools integrate various resources and implement teaching reforms to cultivate talents that meet the needs of the industry.

4 Implementation Path of the Talent Cultivation System

4.1 Strategies and Practices of Education Serving the Industry

Precise Docking with Industrial Demands
Conduct in-depth research on the development trend of the artificial intelligence industry and the employment demands of enterprises and construct a dynamically adjusted systematic talent system. Maintain close cooperation with enterprises and update teaching contents in a timely manner to ensure that the knowledge and skills learned by students meet market demands.

Construction and Function Expansion of Training Bases
Jointly build a "production-teaching-research-application-competition-innovation" training base with first-class enterprises to create a multi-functional practical platform. The training base is not only used for students' skill training but also conducts application development, technical services, competition activities, etc., to enhance students' comprehensive abilities and promote the in-depth integration of education and industry.

4.2 Measures and Achievements of the Industry Supporting Scientific Research and Teaching

Optimization of Talent Cultivation Plans
Formulate precise talent cultivation plans based on the demands of industrial positions and career development paths. Clearly define the knowledge and skills that students should master at different stages, focus on cultivating the adaptability across occupations and positions, so that students can quickly adapt to the industrial transformation and upgrading.

Innovation and Dynamic Adjustment of the Curriculum System
Construct a stepped dynamic curriculum system based on "post-course-competition-certificate", realizing the integration of post and course, competition and course, and certificate and course. The curriculum content is closely combined with the actual engineering projects of enterprises and vocational skill competitions, and the curriculum settings are dynamically adjusted according to employment feedback, competition results, and certification ratios to ensure the practicality and timeliness of the curriculum.

Reform and Practice of Teaching Models
Adopt a "online + offline + PDCA" hybrid hierarchical and phased teaching model, combining the advantages of online resources and offline practical projects to improve teaching effects. Continuously improve teaching quality through the PDCA cycle, and

cultivate students' basic, post, and comprehensive abilities at different levels to meet the learning needs of different students.

Deepening the Sharing and Cooperation of School - Enterprise Resources
Schools and enterprises jointly develop teaching resources, compile serialized textbooks and work manuals, and conduct horizontal research projects. Realize the organic combination of enterprise resources and school teaching, improve the quality and practicality of teaching resources, and at the same time promote enterprise technological innovation and achievement transformation.

4.3 Mechanism and Achievements of Scientific Research Feeding Back to Teaching

Integration of Technology and Innovation Practice
Integrate new technologies and methods in scientific research projects into teaching content. Encourage teachers and students to participate in scientific research activities and improve students' innovation and practical abilities through innovation and entrepreneurship projects and skills competitions.

Team Building and Ability Improvement
Build a collaborative education and scientific research innovation team and improve teachers' teaching and scientific research abilities through training and enterprise practice. Teachers introduce scientific research achievements into the classroom, enrich teaching content, improve teaching quality, and at the same time provide technical services and solutions for enterprises.

4.4 Evaluation and Feedback of the Effectiveness of Collaborative Education

Construction of the Evaluation Index System
Establish a six - in - one evaluation index system covering the resource input of enterprises and industries, production and service benefits, technical service capabilities, achievement transformation capabilities, teachers' post competency, and talent cultivation effectiveness. Comprehensively and objectively evaluate the effectiveness of collaborative education and provide data support for the optimization of the talent cultivation system.

Dynamic Feedback and Continuous Improvement
According to the evaluation results, timely discover the problems and deficiencies in the process of talent cultivation and put forward targeted improvement measures. Through the dynamic feedback mechanism, realize the continuous optimization of the talent cultivation system and ensure the continuous improvement of the cultivation quality. As shown in Fig. 3.

Taking the application of Changchun Technical University of Automobile as an example, through the talent cultivation model with the seamless connection of "science, education and industry", the school has cumulatively cultivated nearly 300 junior college

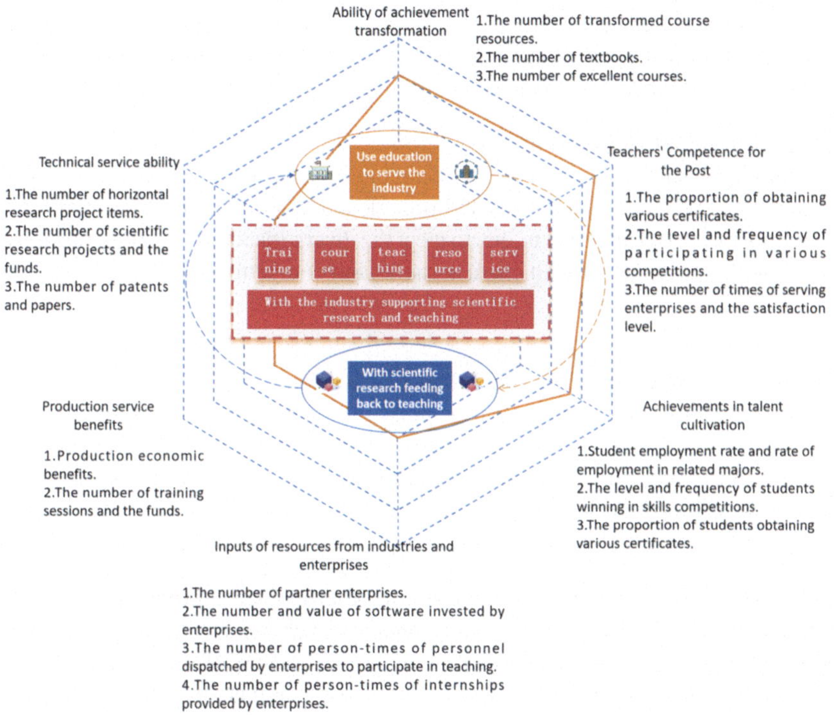

Fig. 3. Radar Chart for Real - time Evaluation of the Effectiveness of Collaborative Education

students majoring in artificial intelligence in recent years. It has cooperated with ecological enterprises such as China FAW, Huawei, and China Unicom to provide internship positions for students. A total of 5 awards (including 1 s prize and 4 third prizes) have been won in national competitions, including teachers. More than 30 awards have been won in provincial competitions. Nearly 80 people have passed the vocational skill level examination. It has undertaken more than 50 technical services and enterprise training projects, initiated 8 teaching and scientific research projects, obtained 3 authorized invention patents, 1 utility model patent, 5 software copyrights, developed 1 provincial teaching resource library course, published 5 textbooks, and cultivated 6 "double – qualified" teachers.

5 Innovation Strategies and Future Prospects

5.1 Innovation Strategies

Collaborative Governance by Multiple Subjects

Establish a collaborative governance mechanism with the joint participation of the government, industry, enterprises, and schools, and clarify the responsibilities and rights of each party. The government strengthens policy guidance and financial support; industry associations formulate industry standards and norms to promote cooperation among enterprises; enterprises deeply participate in talent cultivation and technological innovation; schools play the role of the main body of education, integrate various resources, and achieve coordinated development.

Personalized Talent Cultivation
Use technologies such as big data and artificial intelligence to analyze students' learning characteristics and career tendencies and provide personalized cultivation plans. Offer elective courses, practical projects, etc. to meet the diverse needs of students and cultivate artificial intelligence talents with characteristics and competitiveness.

5.2 Future Prospects

Technology - Driven Innovation in Talent Cultivation
With the continuous development of artificial intelligence technologies, such as the application of virtual reality and augmented reality technologies in the education field, it will bring new opportunities and challenges for talent cultivation. Future efforts will explore how to use these technologies to construct a more immersive and interactive teaching environment and improve the quality of talent cultivation.

Talent Cultivation Concept of Green and Sustainable Development
Incorporate the concept of green and sustainable development into the cultivation of artificial intelligence talents and cultivate students' awareness of environmental, social, and economic sustainability. Encourage students to develop green artificial intelligence technologies and promote the sustainable development of the industry.

Trend of Interdisciplinary Talent Cultivation
The interdisciplinary integration of artificial intelligence and other disciplines is deepening. Future efforts will further strengthen interdisciplinary talent cultivation. For example, combine with medicine, environmental science, sociology, and other disciplines to cultivate comprehensive talents who can solve complex practical problems and meet the diverse needs of society.

6 Conclusion

The talent cultivation system of the application of artificial intelligence technology in higher vocational colleges with the seamless connection of "science, education and industry" constructed in this research effectively solves the key problems of talent cultivation in the context of the integration of industry and education. Through the implementation of a series of innovative strategies and measures, the quality of talent cultivation has been improved, and remarkable achievements have been made. In the future, continuous deepening of reform should be carried out to actively respond to the challenges brought by new technologies and new demands, cultivate more high - quality practical skilled talents suitable for the development of the artificial intelligence industry, and provide strong support for economic and social development. At the same time, it is expected that more research and practice can continuously explore in this field and jointly promote the development of the cause of cultivating artificial intelligence talents in higher vocational colleges.

Acknowledgment. This paper is the research result of the 2023 Key Research Project on Vocational Education in Jilin Province, "Construction and Practice of a Talent Cultivation System for AI Technology Applications in Higher Vocational Colleges with Seamless 'Science, Education, and Industry' Integration" (Project No. 2023XHZ016), the 2022 Research Project on Vocational and Adult Education Teaching Reform in Jilin Province, "Research on the Construction of a New Generation of Information Technology Professional Groups in the Context of 'Double High'" (Project No. 2022ZCY165), the 2024 Research Project of the Chinese Electronic Education Society (Project No. DJ2024B03), and the 2023 Vocational Education Research Project in Jilin Province (Project No. 2023XHY262).

References

1. Xie, Z., Zeng, S.: The implementation value, mechanism construction and realization path of the integration of science, education and vocational education. Vocational and Technical Educ. **22**, 33–40 (2023)
2. Lin, J., Han, H.: Research on the talent cultivation model based on Huang Yanpei's thought of the integration of industry and education—taking the application of artificial intelligence technology major in higher vocational colleges as an example. J. Shanxi Finance & Taxation College **25**, 74–77 (2023)
3. Li, H., Yu, X.: Research and practice on the construction of a new generation of information technology professional group based on the development needs of the intelligent manufacturing industry. Computer Education (02), 99–102 (2019)
4. Zhang, G., Jiang, J.: Construction and practice of the "integration of industry and education, precision education" talent cultivation system for the artificial intelligence technology service major. J. Higher Educ. **24**, 159–162 (2020)
5. Zhao, S., Yu, Y.: Construction of the curriculum system for the artificial intelligence technology service major under the background of the integration of industry and education. Computer Knowledge and Technology (17), 232–234 (2021)
6. Ge, S., Gao, Q., Wan, Q., Wei, Q.: Innovative research on the construction of the 5G intelligent information technology professional group under the background of "double high." J. Anhui Vocational College of Electronics & Information Technol. **20**, 92–96 (2021)
7. Li, W., Wu, Q.: Reconstruction of the computer professional group based on the demands of the new generation of information technology industry—taking guangdong polytechnic of engineering as an example. Education Teaching Forum (31), 41–44 (2021)
8. Luo, M., Chen, Y.: Research on the talent cultivation plan for the application of artificial intelligence technology major in higher vocational colleges—text analysis of 119 higher vocational colleges based on ROST-CM6 technology. Vocational Education **22**(15), 22–26 (2023)
9. Li, J., Gao, W., Zhan, S.: Research on the training model of on-site engineers for the mould major under the background of digital and intelligent transformation. Education and Vocation **04**, 105–111 (2024)
10. Xiao, Z.: The Training Requirements And Paths Of The Core Competencies Of Higher Vocational Students From The Perspective Of Career Development. Education and Vocation **16**, 73–78 (2021)
11. Li, D., Zou, S., Zhang, Y.: Research on the improvement of the social service capacity of higher vocational colleges from the perspective of the integration of science, education and society. Education and Vocation **04**, 21–26 (2024)
12. Qiu, C., Xu, Z., Li, C.: Exploration and practice of solving the difficulties in the cultivation of integrated circuit talents in higher vocational colleges through the integration of "industry, science and education". Chinese Vocational and Technical Education (25), 72–78 (2023)

Probe into the Path of Teachers' Professional Development and Ability Improvement for Digital Education

Pengrun Chen(✉)

School of Management, Guilin University of Aerospace Technology, Guangxi, China
pengrunchen68@gmail.com

Abstract. With the rapid development of information technology, digital education has become the key force of educational modernization, which puts forward new requirements for teachers' roles and abilities. Firstly, this paper analyzes the current situation of teachers' professional development in the digital education environment and points out the challenges faced by teachers in technology application and interdisciplinary integration. Subsequently, the paper discusses the key paths to improve teachers' digital ability, including technology application ability, teaching innovation ability, interdisciplinary integration ability and data-driven decision-making ability. Through the analysis of practical cases, this paper evaluates the effect of the "Professional Development Project of Digital Education Teachers" implemented by H Middle School and finds that the project has significantly improved teachers' digital literacy and teaching innovation ability, and at the same time improved students' academic performance and comprehensive quality. The research of this paper provides practical guidance and theoretical reference for educators, policy makers and researchers, and helps teachers to achieve self-transcendence in the digital age and promote the digital transformation of education.

Keywords: Digital Education · Teachers' Professional Development · Ability Improvement

1 Introduction

In today's era, the rapid development of information technology is promoting profound changes in all fields of society at an unprecedented speed, and the field of education is no exception. Digital education, as an important part of this wave of change, is gradually becoming a key force to improve education quality, promote education equity and realize education modernization. It not only transforms conventional teaching approaches and learning styles, but also introduces new demands for teachers' roles, professional competencies, and educational philosophies.

Facing this historic change, teachers' professional development and ability improvement have become the core issues to be solved urgently in the field of education. Under the background of digital education, teachers are no longer just imparting knowledge,

© The Author(s), under exclusive license to Springer Nature Switzerland AG 2025
F. Paas et al. (Eds.): EduTA 2024, LAIS 51, pp. 322–331, 2025.
https://doi.org/10.1007/978-3-031-95252-4_29

but also integrating learning resources, designing learning activities and cultivating students' autonomous learning ability [1]. Thus, enhancing teachers' digital literacy, ability to innovate in teaching, capacity for interdisciplinary integration, and data-driven decision-making skills has become crucial for advancing the digital transformation of education and achieving high-quality educational outcomes.

The digital competence of educators refers to their capacity to effectively integrate digital technologies into teaching and learning processes. This includes the use of information and communication technology (ICT), the integration and management of digital resources, and the design and execution of online instruction [2]. Research indicates that teachers' digital competence has a direct impact on the quality of instruction and the academic outcomes of students [3]. As a result, improving teachers' digital skills is crucial for advancing educational modernization. On a global scale, there are various established frameworks and standards for teachers' digital competence, such as UNESCO's "Information and Communication Technology Competency Framework for Teachers" (ICT CFT), the EU's "Digital Competence Framework for Educators" (DigCompEdu), and the US "Standards for Digital Literacy for Educators" [4, 5]. These frameworks offer specific objectives and strategies for fostering digital competence among educators. As education continues its digital transformation, the professional development of teachers is also adapting and evolving [6]. Researchers put forward a variety of teachers' professional development models, such as the seven practical forms of "internet plus" teachers' professional development summarized in literature [7] and emphasized the digital transformation of teachers' large-scale learning paradigm and its specific practices [8]. The improvement of teachers' digital literacy needs various supports and guarantees. Relevant seminars pointed out that the education administrative department should build a guaranteed mechanism for improving teachers' digital literacy, the education and scientific research department should promote evidence-based teaching research and evaluation based on big data, and schools and teachers should focus on the application scenarios of "learning, teaching, management and evaluation" [9]. In addition, to improve the digital literacy of university teachers, we should also build a quality assurance system from the dimensions of people, machines, materials, methods, evaluation and environmental factors [10].

This study deeply explores the effective path of teachers' professional development and ability improvement for digital education, aiming at providing practical guidance and theoretical reference for educators, educational policy makers and educational researchers. By systematically analyzing the present situation and challenges of teachers' professional development under the current digital education environment and combining with advanced experience and practical cases at home and abroad, this study will put forward a series of targeted and operable strategies and suggestions, with a view to helping teachers achieve self-transcendence in the digital age and contributing to the digital transformation of education in China and even the world.

2 Analysis on the Current Situation of Teachers' Professional Development Facing Digital Education

2.1 Analysis of Existing Circumstance

The release of the education industry standard "Teachers' Digital Literacy" has provided a clear definition of digital literacy for educators, emphasizing their awareness, skills, and responsibilities in enhancing, innovating, and transforming teaching and learning through the use of digital technologies. Teachers are actively developing their abilities to acquire, process, apply, manage, and assess digital information by participating in various training programs and practical experiences. This enables them to more effectively identify, analyze, and address challenges in education and teaching. Within the digital learning environment, educators are increasingly adopting advanced technologies such as smart interactive whiteboards, 3D simulation modules, and automated grading and analysis systems to continuously refine their teaching methods, ultimately enhancing both teaching effectiveness and quality. Through practice-oriented and classroom-oriented training, teachers have improved their teaching ability in real teaching scenes. Teachers' scientific research ability has also been developed under the promotion of digitalization. Teachers can share high-quality resources and grow in clusters through cross-school, cross-regional and cross-academic research and training. At the same time, teachers are also actively exploring the application of scientific research results to teaching practice to realize the mutual promotion of teaching and scientific research.

Although digitalization brings opportunities for teachers' professional development, there are also challenges. Some teachers may hesitate to adopt new technology and prefer traditional teaching methods. In addition, the incomplete technical infrastructure, limited access to new technologies and lack of high-quality digital literacy professional development opportunities are all problems that teachers need to face in the process of digital transformation. In order to meet these challenges, the state and local governments have issued a series of policies and measures, such as the implementation of digital empowerment of teachers to promote teachers to actively respond to new technological changes. Schools all over the country are also exploring and practicing. For instance, the new campus of Haidian Teachers' Training School in Beijing serves as an advanced teacher training hub, offering both physical and virtual spaces that facilitate interactive teaching and research [11]. In addition to advancing educational digitalization, efforts are being made to bridge the disparity in information technology proficiency between urban and rural teachers. By creating teaching and research communities, such as the "1 + 1" dual school model, rural teachers' teaching skills have been enhanced, helping to reduce the digital literacy gap between urban and rural educators.

2.2 Exist Problem

The rapid advancement of digital education has brought both new opportunities and challenges to teachers' professional growth. Currently, teachers face two main issues in the digital transformation process: limited technical proficiency and a lack of strong interdisciplinary integration skills. First of all, although many teachers have realized the importance of digital technology in teaching, there are still problems of unskilled and unsystematic application of technology in actual operation, which limits the deep integration and innovative application of digital technology in teaching. Secondly, with the increasing integration and complexity of educational content, teachers need to have stronger interdisciplinary integration ability to meet the needs of talent training in the new era. However, at present, teachers' preparation and training in this field are obviously insufficient, and it is difficult to effectively realize the integration and application of knowledge in different disciplines. It is crucial to address these issues and challenges promptly to facilitate the digital transformation of teachers' professional growth and enhance the quality and effectiveness of education and teaching.

3 Path Exploration

3.1 Technical Application Energy

The study of teachers' professional growth and skill enhancement in digital education seeks to assist educators in meeting the evolving educational demands of the modern era, thereby enhancing the overall quality of teaching and learning. Among them, technical application ability is a key link. Teachers should be familiar with the use of modern educational technology tools, including but not limited to multimedia teaching equipment and online teaching platforms. By using multimedia teaching methods flexibly, teachers can enrich teaching forms and stimulate students' interest in learning, thus improving teaching effect. At the same time, teachers should have the ability to develop and manage digital resources, effectively integrate and optimize online education resources, provide strong support for teaching, and promote teachers' professional growth and the continuous improvement of education and teaching level.

3.2 Teaching Innovation Ability

The digital age has endowed education with brand-new forms and opportunities, and at the same time it has put forward brand-new challenges to teachers' roles and abilities. Teaching innovation means not only adopting novel teaching methods and tools, but also stimulating students' curiosity and thirst for knowledge, guiding them to think independently, dare to question and solve problems through practical operation. This necessitates that educators possess the skill to merge knowledge across various disciplines, creating engaging and effective teaching activities. This approach enables students to discover the enjoyment and significance of innovation during their learning journey.

Teachers should actively explore and practice diversified teaching strategies, such as project-based learning, cooperative inquiry, flipping the classroom, etc., encourage students to actively participate, solve practical problems through teamwork, and cultivate their team spirit and communication skills. At the same time, the use of digital platforms and tools, such as virtual laboratories and online resource libraries, enriches teaching methods, realizes personalized learning and instant feedback, and improves teaching efficiency and quality. Facing the uncertainty of future education, teachers need to keep a keen learning attitude, constantly update their knowledge system and master cutting-edge educational concepts and technologies. By participating in professional training, academic seminars and other forms, exchange experiences with other educators and jointly explore the direction and ways of educational reform. In addition, teachers should also pay attention to social needs and industry trends, combine educational goals with social reality, and cultivate future citizens with social responsibility and innovative spirit.

Under the wave of digital education, teachers, as leaders and reformers, shoulder the important mission of cultivating students' comprehensive quality and innovative ability. Through continuous professional development and ability improvement, teachers can better adapt to the educational needs of the new era and contribute to the realization of high-quality talent training and educational equity.

3.3 Interdisciplinary Integration Ability

The cultivation of interdisciplinary integration ability is a key link. This ability requires teachers to cross the boundaries of traditional disciplines, organically combine knowledge from different fields and design innovative and comprehensive teaching activities (Fig. 1). This not only helps to broaden students' knowledge horizons, but also effectively cultivates students' comprehensive literacy and interdisciplinary thinking ability.

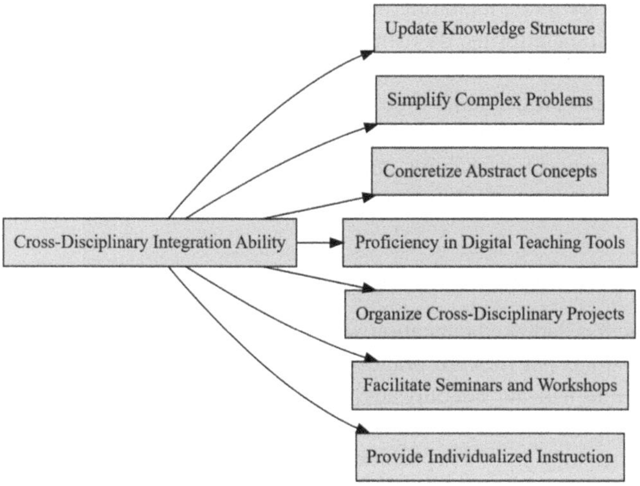

Fig. 1. Composition of teachers' interdisciplinary integration ability

In order to achieve this goal, teachers need to constantly update their knowledge structure and master the basic knowledge and frontier trends of multi-disciplines. At the same time, teachers should also have the ability to simplify complex problems and visualize abstract concepts in order to better guide students to understand and master interdisciplinary knowledge. In addition, teachers need to master digital teaching tools and methods, such as online collaboration platform and virtual laboratory, to support interdisciplinary teaching activities. On the practical level, teachers can organize interdisciplinary projects, seminars, workshops and other forms to promote exchanges and cooperation between teachers and students and stimulate students' innovative thinking and problem-solving ability. At the same time, teachers should also pay attention to the individual differences of students, teach students in accordance with their aptitude, and provide each student with learning opportunities and resources suitable for their development.

3.4 Data-Driven Decision-Making Ability

Under the background of digital education, data-driven decision-making ability is one of the key skills that teachers must master. This requires teachers to skillfully use data analysis and data mining technology to deeply understand students' learning status, progress and needs. By collecting and analyzing all kinds of data in the teaching process, such as learning behavior data and performance data, teachers can identify students' learning difficulties and interests more accurately, thus formulating more personalized and effective teaching strategies. In addition, data-driven decision-making can also help teachers optimize teaching content and methods, improve teaching efficiency and quality, and finally realize the accuracy and intelligence of education.

3.5 Collaboration and Communication Skills

With the progress of technology and the change of educational model, the role of teachers has changed from the traditional knowledge transmitter to the designer, guide and collaborator of learning environment. In order to adapt to this change, especially the needs of digital education, teachers need to develop a number of new professional abilities (Fig. 2). These include mastering digital tools and platforms to enrich teaching content and designing courses that can stimulate students' interest and promote active learning. In addition, teachers need to cultivate effective communication and collaboration skills with colleagues, students and parents, share resources through open channels, and optimize teaching quality.

Teachers should also have the ability of technology integration, evaluation and feedback, lifelong learning and leading innovation, so as to improve teaching efficiency and students' learning effect. At the same time, teachers need to pay attention to ethical and security issues to ensure that they and students can protect personal privacy and network security in the digital environment. Lifelong learning attitude and continuous professional growth are very important for teachers, which will help them to remain competitive in the rapidly developing field of educational technology and lead students to meet future challenges.

4 Practical Cases and Effect Evaluation

H middle school is located in a developed city in the east of China. It is a public school with a long history and a good reputation. In recent years, the school has actively responded to the call of the digital transformation of national education and launched the "Professional Development Project for Digital Education Teachers", aiming at improving teachers' digital literacy and teaching innovation ability through a series of innovative measures, thus promoting the leap of the overall education quality of the school.

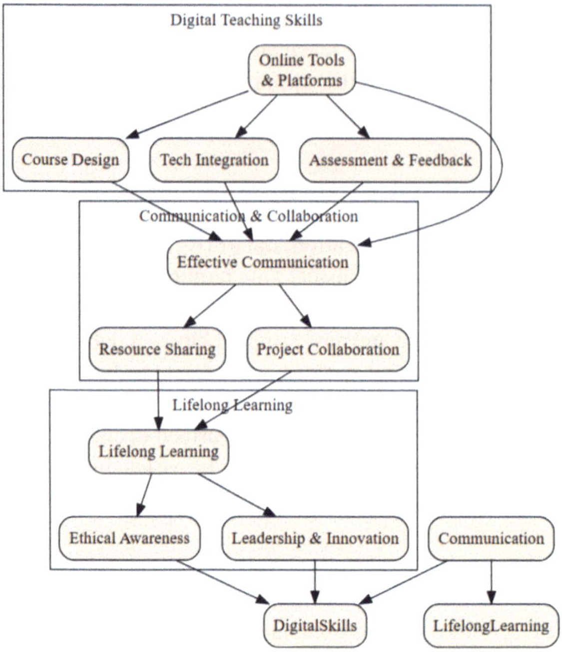

Fig. 2. Composition of teachers' ability to develop many new specialties

Through cooperation with educational technology enterprises, the school provides technical training and certification for teachers and encourages them to master digital tools; Establish an interdisciplinary team and use VR/AR technology to create an immersive learning experience; Establish a learning data platform, improve teaching strategies through data analysis, and realize personalized teaching; Build an online teaching and research community to promote experience sharing and cooperation among teachers; Introduce diversified evaluation system, including project-based learning and peer evaluation, to improve the comprehensiveness and accuracy of evaluation and promote the improvement of teachers' evaluation ability.

By designing questionnaires containing quantitative and qualitative questions, the feedback on the effect of digital teaching, teachers' ability and students' experience was collected from 500 teachers and 1000 students respectively, in which the quantitative questions were scored by Likert scale and the qualitative questions were collected with open opinions; Then, statistical software is used to analyze the data and evaluate the impact of digital education and the effectiveness of teachers' professional development.

After one year's implementation, the professional development project of digital education teachers in H Middle School has achieved remarkable results. Teachers' digital literacy has been significantly improved, and the participation rate of teachers' training is over 90%, of which 80% teachers have successfully passed at least one certification related to digital teaching; The activity of online teaching and research community has increased by 30% every month, and the number of digital teaching resources shared among teachers has doubled. See Table 1.

Table 1. Teacher training participation and certification pass rate

training program	Number of participating teachers	Participation rate (%)	Number of certified people	Certification pass rate (%)
technical training	450	90.0	400	88.9
Application training of VR/AR technology	300	60.0	280	93.3
Data analysis training	350	70.0	320	91.4
Online teaching and research community participation	480	96.0	460	95.8
Project-based learning evaluation training	200	40.0	180	90.0

Figure 3 below provides an insight into the distribution of interdisciplinary course in H Middle School and students' evaluation. In the picture on the left, interdisciplinary course (20) has the most science, followed by mathematics (15) and history (12). There are few art and physical education courses, which shows that the curriculum focuses on the integration of science and mathematics. The picture on the right shows that most students (45%) think their experience is "average", while a considerable number (100 students) think it is "good" and only a small number of students think it is "poor", which shows that they are generally satisfied with the digital education plan. The number of students who rated the experience as "excellent" (30) also showed a positive response to the plan.

In Fig. 4, we can see that the overall trend of students' average scores and scores in mathematics, physics and chemistry subjects has increased in the past year. From the beginning of the year to the end of the year, the average score of students has gradually increased from 70 to 82, showing remarkable progress. Specific to the results of various subjects, whether it is mathematics, science or English, all show a similar upward trend, indicating that students' performance in these subjects has improved. This shows that after the implementation of the professional development project of digital education teachers in H Middle School, the overall learning effect of students has been obviously improved, and the scores of all subjects have been improved in a balanced way.

After one year's implementation, the professional development project of digital education teachers in H middle school has achieved remarkable results. Teachers' digital literacy and teaching innovation ability have been significantly improved, students can participate in the learning process more actively, and their academic performance and

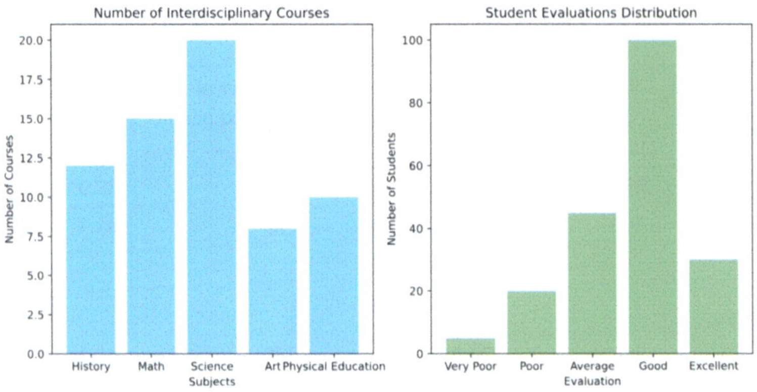

Fig. 3. The number of interdisciplinary course and the distribution of students' evaluation

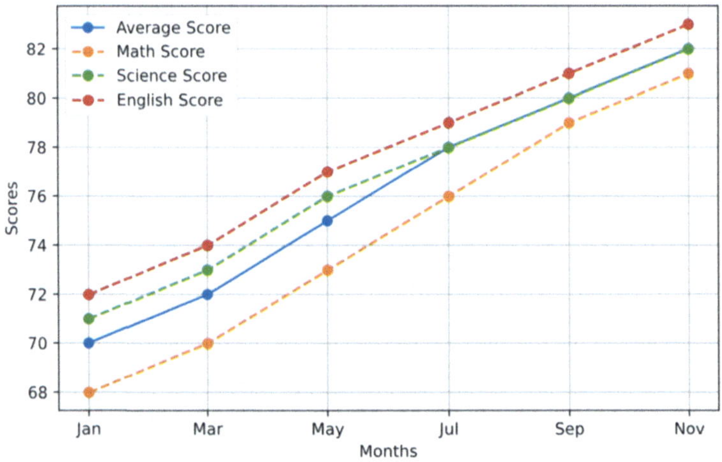

Fig. 4. The trend of students' average grades and academic achievements in the past year

comprehensive quality have been significantly improved. Additionally, the project facilitated the sharing of teaching resources and enhanced the cultural environment within the school, establishing a strong foundation for its long-term development and the continued enhancement of educational quality.

5 Conclusion

Under the background of digital education, teachers need to master modern educational technology tools to improve teaching effect, and have the ability of teaching innovation, interdisciplinary integration, data-driven decision-making and effective cooperation and communication, so as to stimulate students' curiosity and independent thinking, design comprehensive teaching activities and optimize teaching quality; The practice of H middle school proves that systematic professional development projects can significantly improve teachers' digital literacy and teaching innovation ability, improve students' academic performance and comprehensive quality, and effectively help the modernization and digital transformation of education.

References

1. Hao, J., Guo, J.: Empowering teacher professional development with emerging technologies: demands, challenges, and approaches. Open Education Res. **29**(1), 46–52 (2023)
2. Liu, G., Qin, Y.: Seeking 'symbiosis' through 'resonance': the development paths and pursuits of teachers in the context of digital transformation. Educational Research and Teaching **44**(9), 114–121 (2023)
3. Ren, R., Xu, H.: Path selection and reflections on building a high-quality vocational education teacher team in Singapore. Vocational Education Forum **40**(2), 121–128 (2024)
4. Tian, X., Ji, Y., Zhou, Y.: Reconstructing the competence structure of teachers: a key support for the digital transformation of education. J. East China Normal University: Educational Science Edition **41**(3), 91–100 (2023)
5. Wang, X., Xie, X.: Digital transformation of teacher professional development: an analysis based on technology-supported teacher development. Contemporary Education and Culture **15**(5), 40–47 (2023)
6. Huang, Q., Li, X.: Micro-certification of teacher digital literacy: connotations, system construction, and implementation paths. Educational Theory and Practice **44**(7), 39–44 (2024)
7. Zheng, X., Ma, Y., Yue, T.: Continuously promoting teacher professional development in the digital age—an examination based on the norwegian teacher professional digital competence framework. Foreign Primary and Secondary Education 139–150 (2021)
8. Li, M., Chen, S.: Exploring the professional development of higher vocational college teachers empowered by digital technology. Vocational and Technical Education **44**(7), 33–38 (2023)
9. Luo, R.: Realistic dilemmas and alleviation paths of digital transformation in teacher teaching. Educational Research and Teaching **44**(12), 102–107 (2023)
10. Xu, Q., Wu, X.: Digital competence of vocational college teachers under the background of digital transformation: development logic, connotation elements, and enhancement strategies. Vocational and Technical Education **44**(23), 13–20 (2023)
11. Yan, H., Yu, S.: Enhancing teacher digital literacy: a digital practice path with professional development training as the foundation. Educational Research and Teaching **44**(8), 115–121 (2023)

Author Index

© The Editor(s) (if applicable) and The Author(s), under exclusive license
to Springer Nature Switzerland AG 2025
F. Paas et al. (Eds.): EduTA 2024, LAIS 51, p. 333, 2025.
https://doi.org/10.1007/978-3-031-95252-4

The manufacturer's authorised representative in the EU is Springer Nature Customer Service Centre GmbH, Europaplatz 3, 69115 Heidelberg, Germany. If you have any concerns regarding our products, please contact ProductSafety@springernature.com

Printed and bound by CPI Group (UK) Ltd, Croydon, CR0 4YY
30/09/2026
02225588-0011